Haynes *for* HOME DIY

WASHING MACHINES

WASHER DRIERS
TUMBLE DRIERS

The complete guide to the installation, repair and maintenance of front loading appliances

Graham Dixon

Credits

Written by:	Graham Dixon
Edited:	Haynes Publishing
Design:	David Hermelin
	Rhian Walters
Page Layout:	Haynes Publishing with help from Steve Groves
Indexed:	Louise Pritchard & Rachel Rogers

The author and publishers would like to thank the following companies and individuals for providing assistance, information and photographs.

Oracstar Ltd
Lever Bros
Crabtree Electrical Industries Ltd
Dave Marsh Hardware
Nigel Tate
Andrew Morland & Jason Venus for photographic work.

First Published 1988
Reprinted 1988, 1989, 1990, 1991
Revised 2nd Edition published 1992
Reprinted 1993 (twice), 1994
Revised 3rd Edition 1995

© Graham Dixon

All rights reserved. No part of this publication may be reproduced, stored in a retrieval system or transmitted, in any form or by any means, electronic, mechanical, photocopying, recording or otherwise, without prior permission in writing from the publisher.

Published by:
Haynes Publishing
Sparkford, Nr Yeovil, Somerset BA22 7JJ

British Library Cataloguing-in-Publication Data:

A catalogue record for this book is available from the British Library.

ISBN 1 85960 102 2

Printed in Great Britain by J H Haynes & Co Ltd.

While every effort is taken to ensure the accuracy of the information given in this book, no liability can be accepted by the author or the publishers for any loss, damage or injury caused by errors in, or omissions from, the information given.

WASHING MACHINES
WASHER DRIERS
TUMBLE DRIERS

Contents

Chapter 4 44

Fault finding

Chapter 5 148

Drying

Chapter 6 172

Further information

Introduction

Today's modern automatic washing machines, washerdriers and tumbledriers are the sophisticated offspring of their predecessors. Many refinements have been made to aid production and to cut costs for the manufacturer, while at the same time improvements to washing and spin speeds have also been incorporated. However, the automatic washing machine has changed little in its basic operation over the years, apart from faster spin speeds, increased wash variations and obvious cosmetic changes. In recent years the washing machine and tumbledrier have been combined in the washerdrier. These changes have not significantly increased the overall cost of automatic machines and, in real terms, they are actually cheaper now than 25 years ago. With the use of electronics to control motor speeds and the high demand for machines, mass production techniques and a competitive market have kept costs relatively low.

The main drawback has been in the cost of repairs and servicing once the guarantee has expired. Many people opt for the five year cover offered by manufacturers only to find that, in some cases, only the defective part is covered and not wear and tear. Worse still, the labour charge for fitting the replacement part may not be covered.

Repairing and servicing modern domestic appliances may seem a daunting task. This is a myth which has been perpetuated for a number of different reasons.

- Manufacturers have created a mystique that their machines are more complicated than they actually are.
- Many amateurs have an in-built fear of electrical wiring. This is not wholly a bad thing as electricity should be respected whatever the voltage and whenever you are working on a machine, it must be totally isolated from the mains supply.
- There is a shortage of detailed manuals.
- Many people believe that, even if they have overcome the first three difficulties, the parts are likely to be difficult to obtain. In fact, a comprehensive range of 'blister packed' spares are now appearing in most good D-I-Y chain stores and on market stalls. As these are manufactured for the general public, they are usually clearly marked with the type of machine and application.

These reasons are often excuses to leave the repair of your washing machine, washerdrier or tumbledrier to someone who possesses the skills, manuals and parts.

Call-out and labour charges are often expensive, particularly when the repair proves to be a simple, five-minute job. However, until now the D-I-Y enthusiast has been unable to repair any major faults to such machines because manufacturers and repair companies were not prepared to disclose the relevant information. This has now been remedied: studying the following chapters and using a little common sense will not only reduce your repair costs, but give you the satisfaction obtained only from successfully completing a repair yourself.

This manual has been thoughtfully designed to help you understand the function and operation of the internal components of your automatic washing machine, washerdrier or tumbledrier. Flowcharts, diagrams and step-by-step illustrations provide a logical pattern to fault finding. It will assist you in finding the fault and will give you the knowledge to repair it. It covers the jobs of checking and maintaining your machine. This, too, is covered in the book. It is hoped that this book will also make you more aware of safety around the home through a better understanding of electrical appliances and their limitations. Regular checks for faults that can be rectified before failure or accident greatly increase the safety of your appliances. You will gain more efficient use of your machines through a proper understanding of their correct operation.

This book has been written for those who possess little practical knowledge of washing machines or their faults. Seasoned D-I-Y car mechanics will also find the book useful, as many of the problems found with the washing machine are similar to those with cars, such as worn or noisy bearings, faulty hoses and problems with suspension.

Approached in a logical step-by-step manner, repairing most faults is within the capabilities of the D-I-Y amateur. It is best to read right through the book first to familiarise yourself with procedures and the best ways to locate and rectify the faults that may occur. You can then use it as a quick reference guide before and during repairs. It is impossible to deal specifically with any particular machine as models vary considerably, each manufacturer having his own style of pumps, hoses, bearing sizes and so on, but the concept of an automatic washer, washerdrier or tumbledrier differs little from one manufacturer to another.

The machines shown in the step-by-step illustrations are a cross-section of some of the most popular. Both old and new machines have been used to highlight actual fault areas and faults to watch out for. All names and model numbers are used for customer reference purposes only.

Before beginning work on a washing machine (or any electrical appliance) always make sure that it is isolated from the electrical supply – switch off the socket and remove the plug – and from the water supply – turn off the feed taps. This ensures your safety and that of the machine and its surroundings.

Washerdriers

These, as the name implies, are a combination of automatic washing machine and tumbledrier in one machine. As with any combination, compromises have to be made in order for the machine to run smoothly, and the load size and drying time will not suit every situation.

There are two main types of automatic washerdrier. The first allows the warm, moisture laden air to escape through a vent usually positioned at the rear of the machine. This type of machine requires a vent hose to the outside, either through an open window or through a wall vent system *(see page 30)*. It can be used without venting, but moisture laden air discharged by the machine on a drying cycle will condense on cold surfaces in the same room, such as windows, cupboards, walls and tiles. For economic use and in order to avoid problems with damp cupboards and walls, it is advisable to vent the machine in some way.

Condenser washerdriers use the same system of rotating the wash load in a stream of warm air following the final spin. However, the resulting moisture laden air is not vented out of the machine but allowed to pass over an internal condenser unit. The unit is kept cool by a small amount of cold water passing through it via an extra cold water inlet valve. When the warm air meets this cold surface the moisture is condensed. The resulting liquid collects at the bottom of the outer tub ready to be discharged by the outlet pump through the drain hose of the machine. The warm, now dry air is then recycled via the drier heater making this sort of machine more economical than the straight vent to atmosphere machines. The condenser washerdrier requires no fittings in addition to those needed for a normal automatic washing machine.

The repair and maintenance of automatic washerdriers is similar to ordinary automatic washing machines and many parts, such as valves, pumps and pressure switches, may be common to other machines within the manufacturer's range. The main differences are the addition of fan and heater to supply warm air, larger vents on the outer tub, door seals with a hot air intake aperture and timers with a dry cycle position. Some machines also have separate timers for the drying cycle. One other important difference is that the space within the machine in which to carry out repairs is restricted because of the added bulk of the fan and heater assembly and, on condenser machines, the external condenser unit.

Both vented and condenser machines have many parts in common. Although the size and shape of the parts may change, depending on manufacturer, the basic principles are the same. Combining the auto washer with the tumbledrier is beneficial for the manufacturer. The automatic washing machine already has a drum for the load, a means of rotating it and an outlet pump to discharge the water produced by condenser machines. In simple terms, all that needs to be added is a fan to circulate air, a heater to warm it, a means of ducting it through the drum and a timer for the dry cycle. Condenser machines require the addition of an extra cold water valve and a condenser unit which might be a simple trickle bar within the wash tub or externally mounted on the outer of the wash tub or inner of the cabinet of the machine.

Maintenance and repair should be considered as a whole, as many faults interrelate between the washing sequence and the drying action. A simple division between the two operations is not always possible. It is essential to understand the correct operation of all the parts, the way in which they function, their interrelationship, associated faults and potential areas for trouble. Therefore, throughout this manual, each item is examined individually and details are given of how it relates to other working parts. Where possible, specific faults are described with help regarding checks and action to be carried out for rectifying faults.

Tumbledriers

The basic function of the tumbledrier is to circulate warm air through damp clothes for a period of time selected by the user, followed by a cool tumble before switching off. This is a relatively simple operation requiring a drum to hold the load, a heater to warm the air and a fan to circulate it through the drum. The drum rotates to allow the load to gain maximum benefit from the warm airflow. In most instances, the motor that revolves the drum also drives the fan used for air circulation.

Being a dedicated machine – drying only – the tumbledrier works more efficiently than the combined washerdrier. There are several reasons for this. The drum capacity can be much larger; as it is not necessary to have a watertight outer tub, the drum can take up nearly all the available space within the cabinet. Larger machines, that is, those with cabinets the same size as a washing

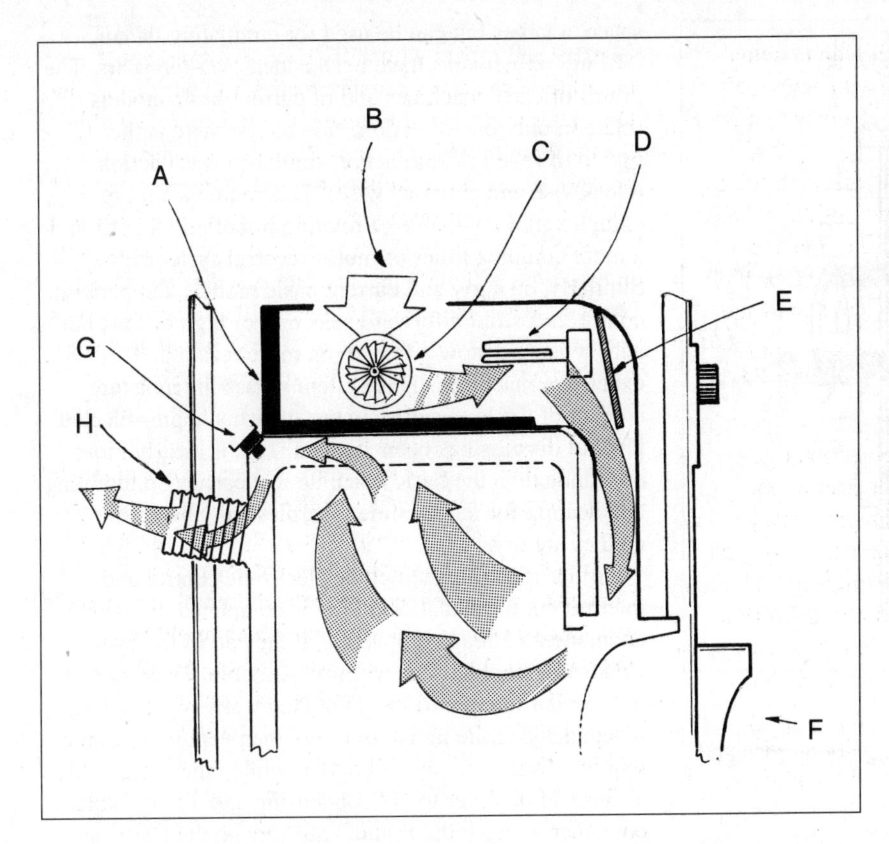

Simple vented combined washerdrier
The fan unit, duct and heaters make up part of the top tub weight (A). Cold air is drawn in from the interior shell space of the machine (B) by fan (C). The air drawn in is blown over the heating elements (D) and in doing so, lifts the flap (E). Access to the inner drum is by an extra flange/inlet to the door gasket/seal (F). The now warm air picks up moisture from the wash load on its way through the contra rotating drum. The now moisture laden air escapes through the contra rotating drum. A thermostat (G) monitors the air temperature within the outer tub. The now moisture laden air escapes through the vent positioned on the rear of the outer tub unit. A convoluted hose (H) guides the warm moist air to the exterior of the machine. A filter is normally situated on the outlet hose to the rear of the machine.

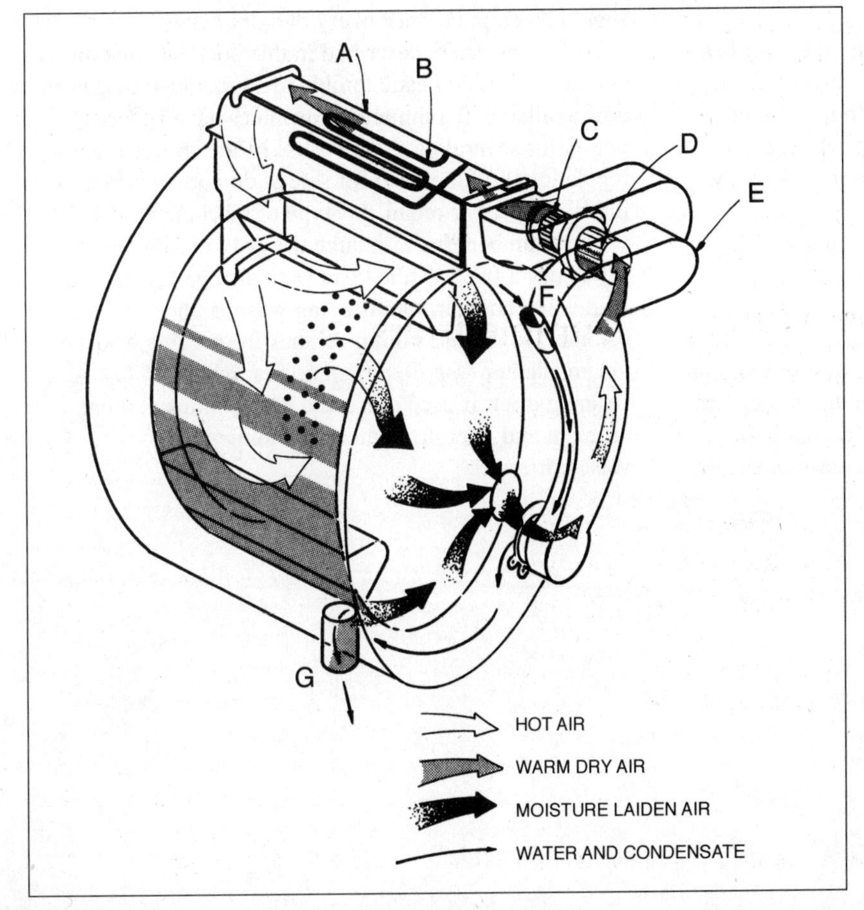

Condenser combined washerdrier

- **A** Heater duct
- **B** Heating element (maybe two)
- **C** Circulation fan motor
- **D** Air circulation fan
- **E** Condenser unit externally mounted, in this instance on the back of the outer tub
- **F** Cold water inlet to spray bar or trickle plate
- **G** Normal sump hose to pump

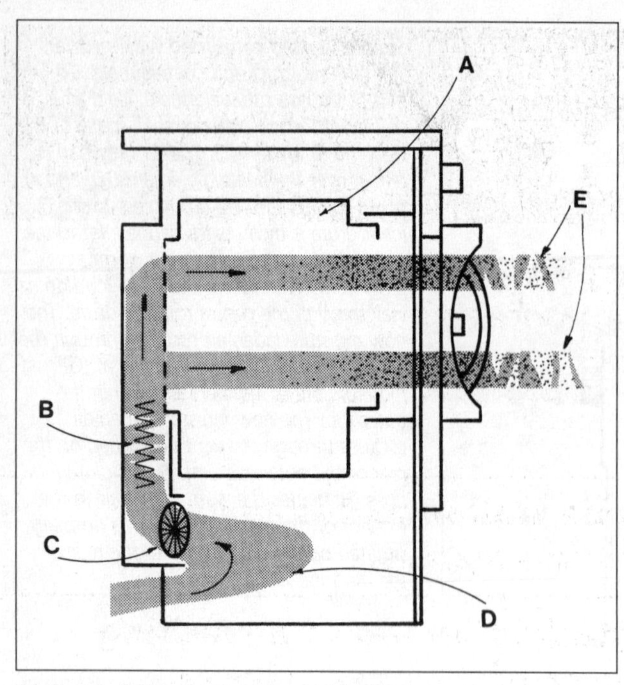

Airflow of front-vented tumbledrier
A Drum
B Heater
C Fan
D Cold air inlet
E Warm moist air vented through door grille and filter

machine, can take the same size load. In other words a 5kg (11lb) wash load can be transferred directly from your washing machine into the tumbledrier. Smaller models with a 2.7kg (6lb) load capacity, which are convenient if space is restricted, operate in exactly the same way as the larger models.

The larger drum of a dry only machine allows for better movement of the clothing through the airflow, ensuring more even and quicker drying even though clothing, towels, sheets and so on increase in bulk during drying. This increase in bulk is why the washerdrier can dry only half its wash load. A larger wattage heater can be used, if necessary, with a greater surface area for warming the air. Again, owing to the increase in usable space, a larger fan can be used for circulating the air.

Many refinements have been added over the years. The drums of early machines and of current basic models rotate in only one direction. Nowadays, there is the option of reverse drum action similar to wash action (clockwise and anticlockwise). This requires a more complex motor capable of rotating in both directions and a more complex timer or motor reversal system. Similarly, on early and current basic models, temperature control is a straightforward selection of high or low, but auto sensing is now a feature on more recent and expensive machines. This system senses the moisture content of the load and switches off when a pre-selected level of dryness has been reached. This is clearly more economic than the basic machine that carries on tumbling and heating for the set time regardless of whether the clothes are dry.

Further refinements include electronic control and intermittent tumble after drying is completed to prevent an unattended load compacting at the end of the cycle. Condenser tumbledriers are also available. These operate in a similar manner to the combined washerdrier system by condensing the moisture rather than venting it. Some require a water supply and outlet, while others use cold air instead of water to aid condensing and a removable container to catch the liquid. On average, the container should be emptied once every two dry cycles.

The tumbledriers described in this book heat the air by electricity, but domestic tumbledriers that heat by gas are now available. (Commercial machines have had this option for several years.) Because of the stringent gas regulations and the need for safety, do not attempt to install, inspect or repair this type of machine yourself.

The main benefits of a large dry only machine are that it will dry a full wash load and can also be drying one load while another load is being washed, thus saving a lot of time. Do not use both machines from a single socket via an adaptor *(see Electrical basics, pages 10-13)*. A separate drier, if used correctly, is also quicker, more efficient and, therefore, more economic than washerdriers.

SITING THE MACHINE WATCHPOINTS

1. **Finding room for another appliance the size of a washing machine can be a problem** in houses with restricted space.
2. **Some smaller machines can stand on work surfaces** if necessary.
3. **Some smaller machines can be wall mounted** using a suitable bracket obtainable from the manufacturer.
4. **Some large models can be stacked on top of a washing machine.** These usually need to be of the same make and require a stacking frame for safety.

Safety Guide

Most people have a healthy respect for electricity and understand that, although it can be put to many uses, it can be dangerous – sometimes lethal – if taken for granted or misused. At all voltages electricity must be respected. People who do not observe the basic rules are not only a danger to themselves but to those around them.

Electrical accidents should be thought of as avoidable. Most result from plain carelessness and the failure to follow basic rules. There are some 16 million homes in Britain supplied with electricity and each has about 25 electrical appliances. With so many items it is, perhaps, surprising that fatalities from electrical accidents are fewer than 80 each year. Although this is a small percentage of the total population and constitutes only one per cent of the 8,000 deaths each year resulting from accidents in the home, it is still far too high.

Avoiding accidents

The commonest causes of shocks and fires from electrical appliances fall into three categories.

- Faulty wiring of the appliance, such as frayed or damaged cable, an incorrect fuse, poor socket, damaged plug and an incorrectly wired plug.
- Misuse of the appliance, such as the incorrect installation of a washing machine in the bathroom. Rules about electrical items in bathrooms are strict for a good reason: the combination of water and electricity greatly increases the possibility of injury.
- Continuing to use an electrical appliance knowing it to be unsafe because of cracked casing, a faulty plug, a damaged cable or a faulty on/off switch, for example.

Several of these faults can be avoided completely if you are fully aware of safety. Others can be eliminated by regular inspection and immediate correction of faults, failure or wear. Misuse may be the result of a purely foolhardy approach or genuine ignorance. Overcome this by understanding and, above all, acting upon the guidelines in this book. If you feel you are unable to do a particular job yourself, then it is best not to try. You can still diagnose the problem to check that any work carried out by a repair company is correct. This can sometimes save a lot of time and expense.

DOs

- Do read all the information in this book before putting it into practice.
- Do isolate any appliance before starting to repair or inspect it.
- Do make sure that you use the correct fuse.
- Do fit the mains plug properly, ensuring the connections are in the correct position, tight, and that the cord clamp is fitted on the outer of the cable.
- Do check that the socket is in good condition and has a sound earth path.
- Do take time to consider the problem and allow enough time to complete the job without rushing.
- Do approach the stripdown of the item methodically and make notes. This helps greatly with subsequent re-assembly.
- Do double-check everything.
- Do ask or seek help if in doubt.
- Do ensure that a residual current device (RCD) is in circuit when making a functional test.

DON'Ts

- Do not work on any machine that is still plugged in even if the socket switch is off. Always isolate fully; switch off, plug out.
- Do not install or allow the use of washing machines or tumbledriers in a bathroom or shower room. It may seem harmless to run an extension lead from a convenient socket on the landing, but it is extremely dangerous and must not be done under any circumstances.
- Do not repair damaged wiring or cables with insulation tape.
- Do not sacrifice safety by effecting a temporary repair.

Caution – safety first

Use this slogan as a memory aid for the basic rules of safety when using, inspecting and repairing electrical appliances.

C **C**onsider your own safety and that of other people.

A **A**ct in a way that prevents incidents from becoming accidents.

U **U**se your common sense and think before acting.

T **T**idy workplaces make safer workplaces.

I **I**dentify hazards.

O **O**bserve the rule of safety first.

N **N**ever underestimate the dangers.

S **S**witch off!

A **A**lways pull out the plug and disconnect from the mains. Appliances vary – make sure you have a suitable replacement part.

F **F**or screws use a screwdriver, for nuts use a spanner, for knurled nuts use pliers.

E **E**xamine and clean all connections before fitting new parts.

T **T**ighten all nuts, bolts and screws firmly

Y **Y**our safety depends on these simple rules.

F **F**uses: up to 250 watts 1 amp; 750 watts 3 amp; 750 to 3,000 watts 13 amp.

I **I**nsulation is for your protection. Do not interfere.

R **R**enew worn or damaged flex.

S **S**ecure flex clamps and all protective covers.

T **T**est physically and electrically when you have finished the job.

Plug wiring

Plug wiring must be connected according to the following code to ensure safety. The colours are as follows:

Live – Brown (or Red), symbol **'L'**
Neutral – Blue (or Black), symbol **'N'**
Earth – Green/Yellow (or Green), symbol **'E'**.

The colours in brackets are those that used to be used until the present international standards were introduced and may still be found on some older equipment. Plug terminals are identified either by colour (old or new) or by the letter symbols shown.

WARNING

NEVER LEAVE BARE WIRES OUTSIDE TERMINALS

ELECTRICAL SAFETY WATCHPOINTS

1. **Faulty wiring of appliances,** such as frayed or damaged flex or cable, incorrect fuse, poor socket, damaged plug, or incorrectly wired plug.
2. **Misuse of appliances,** such as using a washing machine or tumbledrier in the bathroom.
3. **Continuing to use an electrical appliance** in spite of knowing it to be unsafe.

Emergency procedures

When symptoms such as leaking, flooding, unusual noises and a 'blown' fuse occur it is best to carry out the following procedure. Do not allow the machine to continue its programme until the fault has been located and rectified.

- Do not panic.
- Isolate the machine from the mains supply: turn the machine off, switch off at the wall socket and remove the plug from the socket. Turn off the hot and cold taps to which the fill hoses of the machine are connected. If a valve is at fault, it may be jammed in the open position so the machine will still fill, even with the power turned off.
- Although the power and water are disconnected, the machine could still be leaking if there is still water in it. This can be extracted from the machine by siphoning. Lift the outlet hose from its usual position and lower it to below the level of water in the machine. This will allow the water to drain (unless there is a blockage in the outlet hose). It is easiest if the outlet hose will reach to an outside door, where all that is needed is a little movement and the water should drain. Alternatively, the water can be caught in a bucket using the same technique. To stop the water draining, lift the hose above the height of the machine. Repeat this process until the machine is empty.
- Do not open the door to remove the clothes until all of the previous steps have been carried out and a further few minutes have elapsed to allow the clothes in the machine to cool. If the machine was on a very hot wash, wait about half an hour.

When all these steps have been carried out and the clothes have been removed from the drum you can then sit down and calmly start to work out what the problem may be and form a logical plan for dealing with it. Refer to Care and repair (pages 34-43) and Fault finding (Chapter 4) of this manual.

Tools and equipment

Maintaining and repairing modern automatic washing machines does not require very specialised tools. Many routine repairs, such as blocked pumps, renewal of door seals and replacement of hoses, can usually be completed with a selection of the following tools: cross-blade and flat-blade screwdrivers, combination pliers, simple multimeter and pliers. Most people who are interested in D-I-Y own one or more of these items already. Useful additions to this basic kit are a 'Mole' wrench, a socket and/or box spanner set, soft-face hammer and circlip pliers. These can help with larger jobs, such as motor and bearing removals.

Some jobs such as bearing removal and renewal, may also require specific tools, such as bearing pullers. These are expensive so it is better to hire them from a tool hire specialist for the short period that you require them. Local garages may also be willing to let you hire them for a small deposit.

It is fairly easy to build up a selection of tools capable of tackling the faults that you are likely to find on your machine. Most large D-I-Y stores stock the tools that you require – often at an economic price. When buying tools, check the quality. A cheap spanner or socket set is a waste of money if it bends or snaps after only a short period of use. Nevertheless, there are many tools available that are of reasonable quality and inexpensive. Try to buy the best that your budget will allow.

Remember, the tools you buy are a long-term investment and should give years of useful service. It is sensible, therefore, to look after them properly. Having spent time choosing, and money buying tools, keep them clean and in a serviceable condition. Make sure that they are clean and dry before putting them away.

Chapter 1

Understanding electricity

Electrical basics

A basic understanding of electricity is essential for the sake of safety around the home. Even if you do not intend to carry out any repairs or servicing of your appliances yourself, a sound understanding of the household electrical supply will prove invaluable in the long run. Ignorance is no protection against your own or someone else's errors, whether on repairs, servicing or installation.

The household supply

The following is a simplified description of a typical household supply. Power is supplied to the substation at very high voltage (400,000 volts) in three-phase form. There it is converted, via a transformer, to 230 volt single-phase. It is then distributed to our homes. Under normal circumstances, current flows from the live supply of the substation's transformer, through the electrical appliances being used in the house and back, via the neutral conductor (cable), to the substation transformer's neutral pole (a closed loop). The neutral terminal of the transformer is, in turn, connected to the ground (earth – meaning in this case, the general mass of the Earth).

It is usual to use the armoured sheath of the electricity supply authority's cable to provide a low impedance continuous link back to the supply transformer's start point. Various types of earthing may be encountered: connection to the armoured sheath of the authority's supply cable, own earth rod, transformer earth rod via the general mass of the Earth or the increasingly popular use of the neutral conductor of the authority's supply cable. This last is often called protective multiple earthing (PME) or TN-C-S system.

Whatever the earth system, if a fault occurs between the live conductor and the earth conductor, another loop forms. This is called the 'earth loop'. The earth loop path is designed to encourage current to flow in the event of an earth fault, so that the protective device (fuse or circuit breaker) in the consumer unit can operate to isolate the supply to the circuit. If the protective device does not operate, the appliance remains 'live' and anyone touching it will receive an electric shock. The resistance of the earth loop path must, therefore, be low enough to allow sufficient fault current to flow to operate the protective fuse or circuit breaker.

Earth loop impedance

It is not just important to have an earth; you should have one that has a low resistance/impedance, that is, perfect earth. The term used for testing for this is earth loop impedance. This means checking to see if the current flow is impeded and by how much. The test requires a specialised meter giving a resistance figure in ohms. The maximum recommended reading is 1.1 ohms.

A correct test cannot be carried out with a low voltage test meter. In this case, an existing fault, such as one tiny strand of wire poorly connected, would allow a low voltage of, say 9 volts, to pass easily. However, it would break down and go high resistance or open circuit if a true fault voltage of 230v at 13 amps tried to be passed. Low voltage testing gives an indication of earth path, but it cannot indicate quality. Only an earth loop impedance meter gives a proper indication of earth quality under realistic conditions.

What is an earth fault?

An earth fault is the condition where electricity flows to earth, which, under normal circumstances, it should not do. There are two recognized ways in which this may happen: direct and indirect.

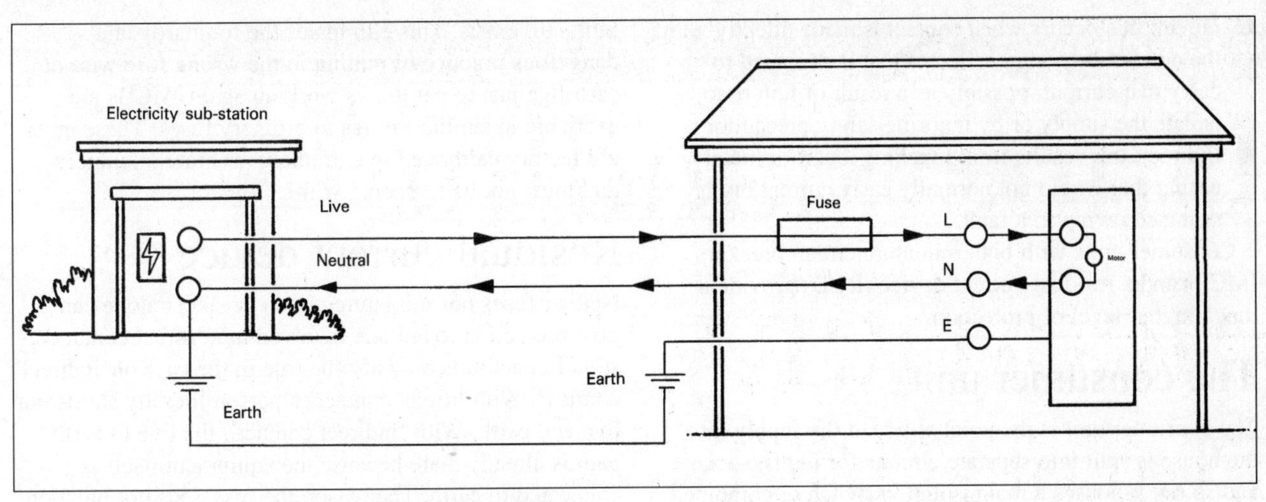

Fig. 1 Typical household supply in normal operation (simplified).

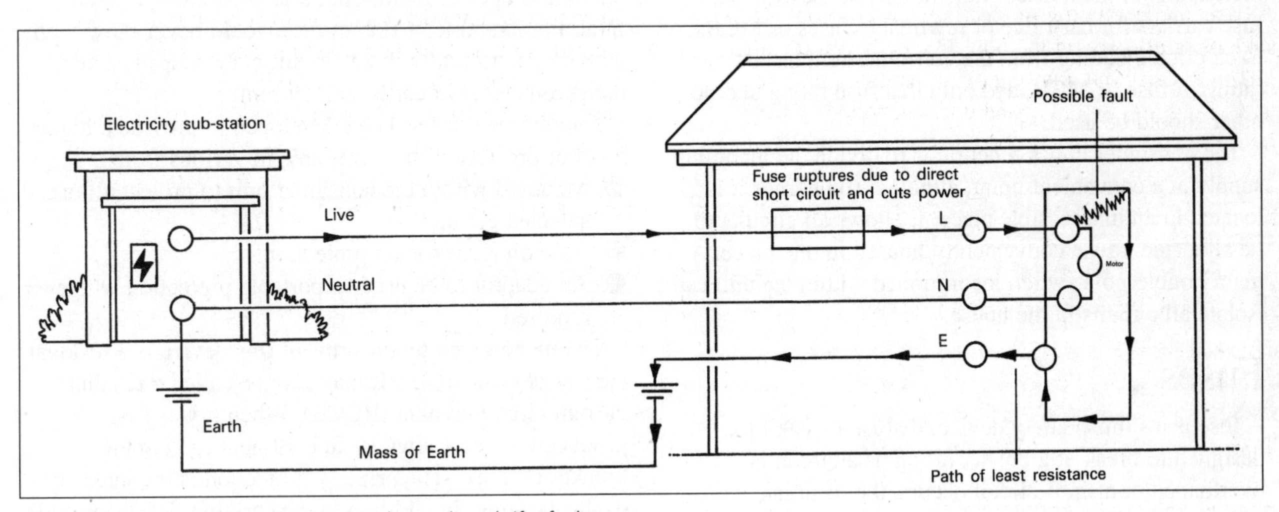

Fig. 2 Typical household supply showing the earth path if a fault occurs.

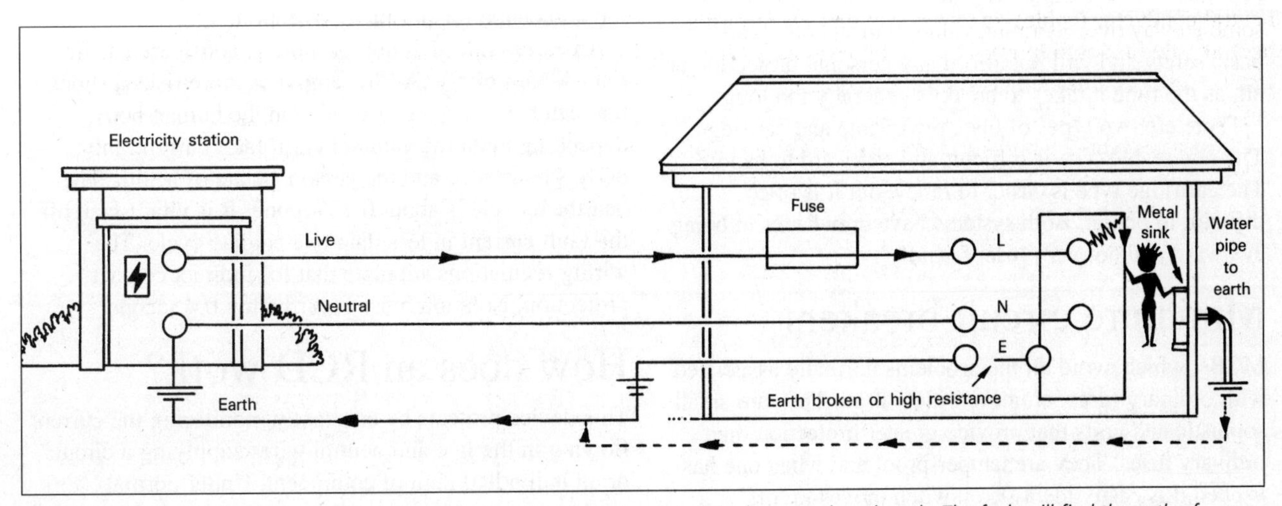

Fig. 3 Typical household supply, showing the result of high resistance or break in the normal earth path. The fault will find the path of least resistance.

- Direct: this occurs when contact is made directly with the current carrying conductor that is designed to carry that current, possibly as a result of failure to isolate the supply or by ignoring safety precautions.
- Indirect: this results from touching a part or metal casing that would not normally carry current but is doing so owing to a fault.

Consumer units with both miniature circuit breakers (MCB) and a residual current device (RCD) provide a much higher level of protection.

The consumer unit

The consumer unit is the point at which the supply into the house is split into separate circuits for lights, sockets and so on. It houses a main isolation switch or combined RCD which is used to isolate (remove power from) all the circuits in the house. The unit also houses various fuse carriers for cartridge or rewireable fuses or MCBs. Each circuit leading from the consumer unit has its own rating of fuse or MCB, and only that fuse rating and no other should be used.

The consumer unit's function is to divide the incoming supply at a convenient point, although sometimes it is located in an inaccessible place. It allows all circuits to be allocated a fuse conveniently housed in one place. A main double-pole switch incorporated within the unit can isolate all circuits in the house.

Fuses

A fuse is a simple safety device, that is, a weak link designed to break at a pre-set rating. If a circuit is overloaded or a short circuit occurs, the resulting overload causes the fuse to melt and cut the supply. Unless a direct short circuit occurs, the overload on the fuse may not be enough to cause it to 'blow' as there is some leeway over its rating value. It offers only very basic safety and will not afford any personal protection at all, as the time it takes to break is generally too long.

There are two types of fuse: rewireable and cartridge. The rewireable type is difficult and awkward to handle. The cartridge type is easier to renew but it is often difficult to obtain. Both systems have drawbacks in being awkward and not very 'user friendly'.

Miniature circuit breakers

MCBs, which avoid all the problems normally associated with ordinary fuses, are now widely used. They are small sophisticated units that provide greater protection than ordinary fuses. They are tamper-proof and when one has tripped it is easily identified (switch moves to 'off' position). Most importantly, they cannot be reset if the fault still exists. This eliminates the foolhardy and dangerous practice of putting in the wrong fuse wire or cartridge just to get things working again. MCBs are available in similar ratings to ordinary fuses. These units are factory calibrated to extremely accurate tolerances and must not be tampered with.

Residual current device

Neither fuses nor miniature circuit breakers alone can give protection to anyone involved in a 'direct contact' situation and this may also be true in the case of 'indirect contact'. With 'direct contact' a person literally shorts out live and earth. With 'indirect contact', the live to earth path is already there because the equipment itself is connected to earth. The reason the fuse does not blow or the circuit breaker trip is because the fault is not large enough to operate them – but it is large enough to be fatal. For example, a 10amp fuse would never blow with an 8amp earth fault on the circuit, but 8 amps is a very dangerous level of earth fault current.

Another device has been developed to provide a higher level of protection. It is available in various forms.

- Mounted within the consumer unit to protect all or selected circuits.
- As individual socket protection.
- An adaptor to be used as portable protection wherever required.

The name given to all forms of this device is a residual current device (RCD). It may also be called a residual current circuit breaker (RCCB). When it was first produced it was known as an earth leakage circuit breaker (ELCB). The primary protection is the integrity of the earthing. In addition to the earthing, RCDs provide a much higher level of protection, depending upon their sensitivity. A sensitivity of 30mA ($^{30}/_{1000}$ amps) is recommended for personal protection.

An earth fault of 1 amp or more is considered a fire risk. A fault of 50mA ($^{50}/_{1000}$ amps) or more risks a shock that can have varying effects upon the human body, depending upon the value of earth fault current, the body's resistance and the person's state of health. The heartbeat cycle is about 0.75 second. It is vital to cut off the fault current in less than one cardiac cycle. The wiring regulations stipulate that for indirect contact protection, isolation must occur within 0.4 second.

How does an RCD work?

This device protects by constantly monitoring the current flowing in the live and neutral wires supplying a circuit or an individual item of equipment. Under normal circumstances, the current flowing in the two wires is equal. However, when an earth leakage occurs as a result of a fault or an accident, an imbalance occurs. This is detected by the RCD which automatically cuts off the power within 200 milliseconds (the rated sensitivity).

To be effective the RCD must operate very quickly and at a low earth leakage current. Those most frequently recommended are designed to detect earth leakage faults in excess of 30mA ($\frac{30}{1000}$ amp) and to disconnect the power supply within 30ms ($\frac{30}{1000}$ second); these limits are well inside the safety margin.

RCDs are designed to sever mains current should your electrical appliance develop an electrical fault. They are not designed to let you increase the risk to yourself by being cavalier about safety. They are simply fail safe devices. Used correctly they are an invaluable asset to your household.

An RCD must be used in addition to and not instead of normal overload protection – fuses or MCBs. All residual current devices have a test button facility; check regularly that the device operates. If you are using an RCD with an adaptor or socket, test before each operation. If failure occurs, that is, it does not trip or tripping appears sluggish, have the unit tested immediately. This requires an RCD test meter and is best left to a qualified electrician.

EMERGENCY PROCEDURES WATCHPOINTS

1. **There is still a live supply to the consumer unit** even when it is switched off.
2. **Do not remove the covers** of the consumer unit.
3. **Do not attempt any inspection or repair to the consumer unit** without seeking further information. Leave faults, other than fuse renewal, to skilled electrical engineers.

Plugs and sockets

Problems with washing machines and tumbledriers are not always the result of a failure of the appliance itself but with the electrical supply to it via the socket. A three-pin socket must have a live supply, a neutral return and a sound earth path. When a plug is inserted into the socket, contact must be firm at all three points. If the live or neutral pins of the plug or connection point within the socket fail to make adequate contact or are free to move, localised heating occurs within the socket.

Avoiding problems

Problems with plugs and sockets are varied and may be caused by one or a combination of any of the reasons listed below.

- Frequent use of the socket opening up the contact points within it (general wear and tear).
- Poor quality socket, plug or both.
- Loose pins on plug.
- The use of a double adaptor can cause a poor connection simply through the weight of cables and plugs preventing a tight fit in the socket.
- The use of a double adaptor to allow two high current draw appliances, such as a washing machine and a tumbledrier, to be run through one socket overloads a single 13 amp socket. Whenever possible, avoid the use of adaptors by providing an adequate number of sockets. Never exceed 3kW load on a single socket.
- Use of multi-point extension lead when the total load on the trailing socket can easily exceed the safe 3kW load of the single socket.

Do not use a faulty socket until the problem has been rectified. If the socket shows any of the faults described here, it must be renewed completely. If it is a single socket, it may be wise to replace it with a double socket. Make sure you buy a good quality replacement socket, as there are many of dubious quality to be found. Price is a good indicator of quality.

You should also renew any plugs that have been used in the faulty socket as damage may have been done to them. Continuing to use the old plugs may result in the premature failure of your new unit.

There are many styles and qualities of plug, too. Some poorer quality plugs may be reliable on low current consumption items, such as lamps, televisions and radios, but they may not be so good for washing machines and tumbledriers. Although British Standards apply to these items, quality varies considerably. When buying plugs and sockets, go to outlets that can give advice and that carry a good selection. This will allow you to compare quality and build of the products. Look for the ASTA mark which proves that the design and manufacture has been approved by the Association of Short Circuit Testing Authorities. Replacement fuses for plugs should also carry this ASTA mark.

Socket highlighting overheating. Both plug and socket require replacing.

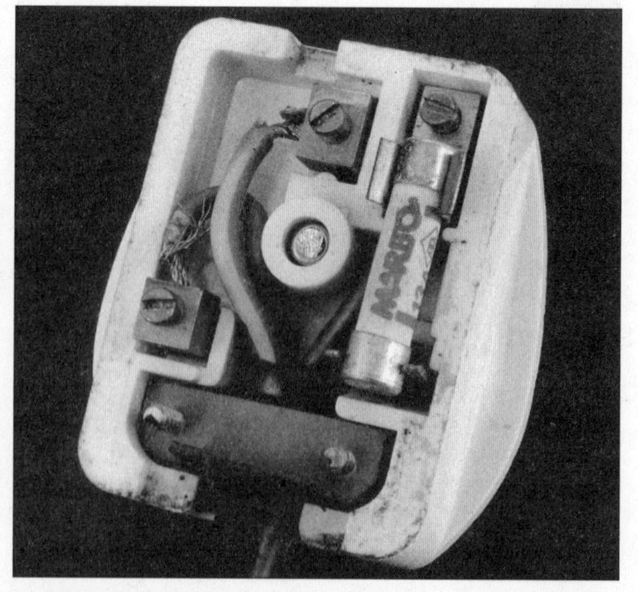

Internal view of severe burn-out caused by poor connection to terminal. A new plug is required and the cable should be cut back to sound wire or renewed.

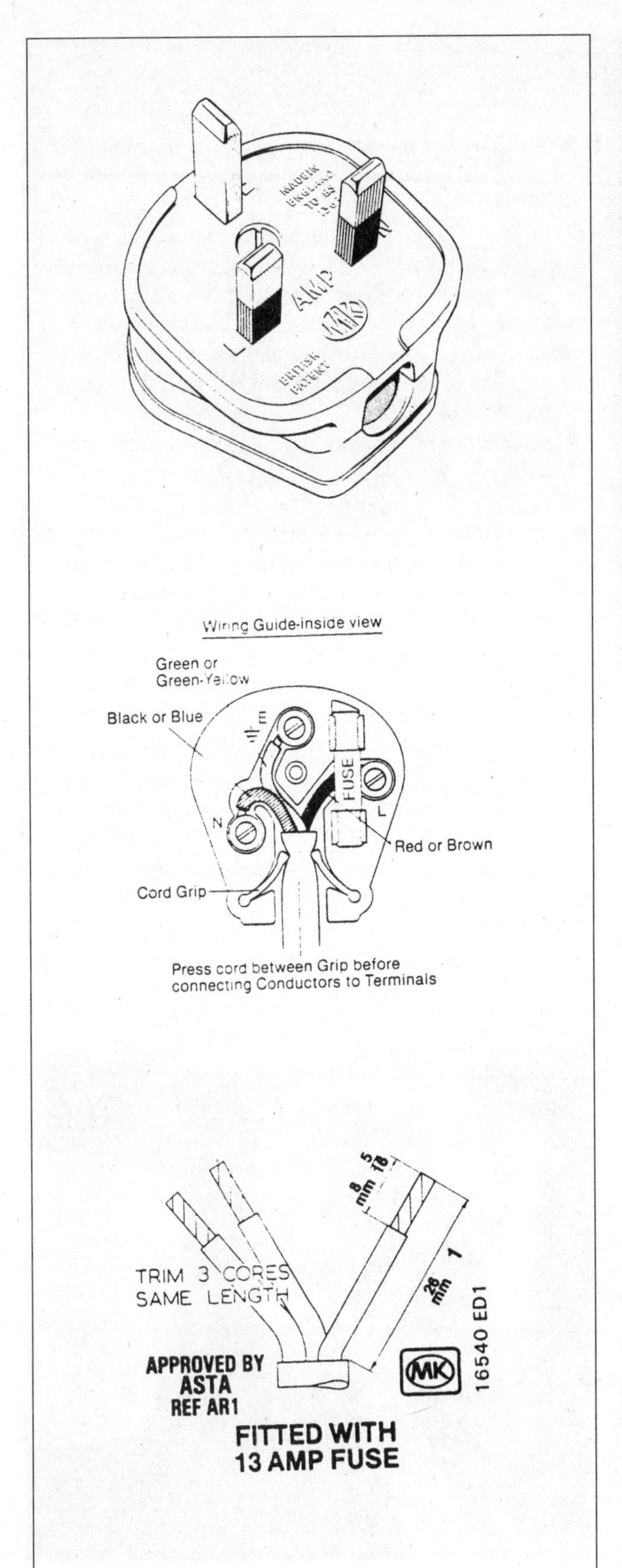

Specific requirements for MK plugs.

Typical plug-in socket tester.

The earth

All the previously mentioned faults relate to the live supply and neutral return on the socket, plug or both. There is, of course, a third pin, although it takes no active part in the operation of the appliance. Nevertheless, it is the most important connection of all. Products that have three core cable must have the yellow and green earth wire securely connected to the earth pin of the plug or pin marked E.

Testing the earth path

The earth path of any appliance can easily be checked with a simple test meter *(see Electrical circuit testing, pages 22-25)*. A path of low resistance is required from all items in the product that are linked into the earth path via the yellow and green cable. A maximum reading of 1ohm from any point to the earth pin is recommended.

Checking the socket requires a correctly operated earth loop test meter. These meters are expensive and there may be problems with distribution boards fitted with an RCD *(see Electrical basics, pages 12-13)*, so it is advisable for sockets to be tested by a qualified electrical contractor. All household wiring must conform to the Institute of Electrical Engineers' (IEE) recommendations. The requirement for the earth path is that it has as low a resistance as possible. The present regulations allow a maximum of 1.1 ohms. This can be checked fully only with an earth loop test meter.

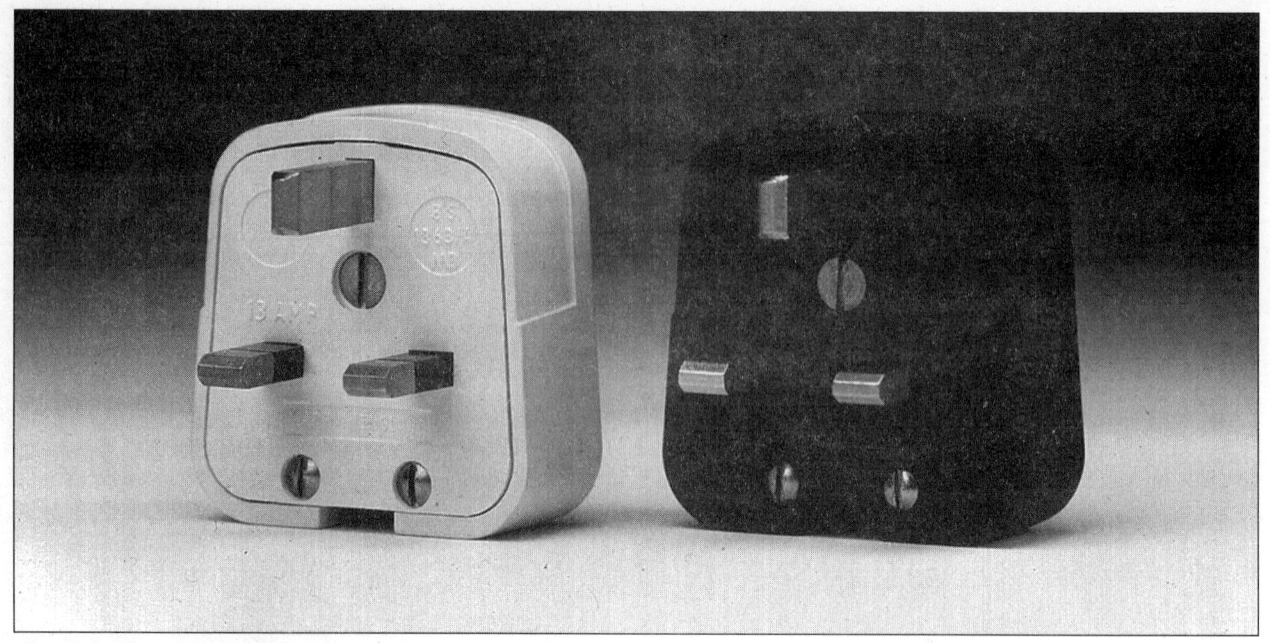

Typical resilient plugs which can stand up to rugged use without cracking.

Simple plug-in testers are available from most good electrical shops and D-I-Y outlets. They are most useful for checking a socket for reverse polarity, indicating whether it has been incorrectly wired. A wrongly wired socket can still work and show no outward sign of a problem. This type of fault is dangerous and quite common.

The plug-in tester also indicates if an earth path is present, but does not show its quality. That is to say it may have a very high resistance but would still allow the neon of the tester to light. If the earth resistance is high, this may result in a failure to blow the fuse which may, in turn, cause overheating at the high resistance point or allow a flow of electricity through anything or anyone else that can give a better route to earth.

Wiring a plug

Many people think fitting a plug is a completely straightforward task that needs little or no explanation. On the contrary, this is an area where many problems may occur and that may prove dangerous if the fitting is not done correctly. Do not neglect this most important – even vital – item. Taking time to do it properly also prolongs the life of your machines, avoiding unnecessary faults.

The text and step-by-step pictures on pages 18 and 19 deal specifically with modern, 13amp flat pin plugs. If your home has round pin plugs and sockets, the wiring is probably old and you should have it thoroughly checked.

When wiring a plug, leave the earth wire (yellow/green) longer than the live and neutral wires. Form the extra length into a slight loop within the plug. This allows the earth wire to be the last to pull out if the cable is pulled or caught. If the cord grip is unable to hold the cable in position, the inner wires tear free from their fixing points.

However, the post and nut plug shown does not allow for this extra length and the manufacturer recommends that all wires be cut to the same length.

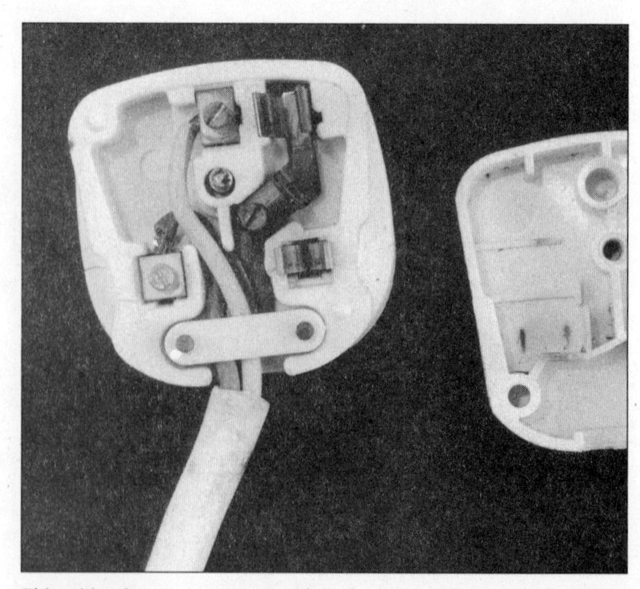

This wiring is not cut to correct length. As a result, the cord grip is fixed across inner wires not outer sheath.

FAULTY PLUGS AND SOCKETS WATCHPOINTS

1	**Burn marks around one or both entry points on the socket.**
2	**Plug hot to the touch** after using the machine in that socket.
3	**Pungent smell from socket** when the machine is in use.
4	**Pitting and burn marks** on and around the plug pins.
5	**Radio interference** to nearby equipment caused by internal arcing.
6	**Intermittent or slow operation** of the appliance.
7	**Failure of the fuse** in the plug caused by heat being transferred through the live pin and into the fuse.
8	**All these problems are pronounced** with higher current-draw appliances, such as washing machines and tumbledriers.

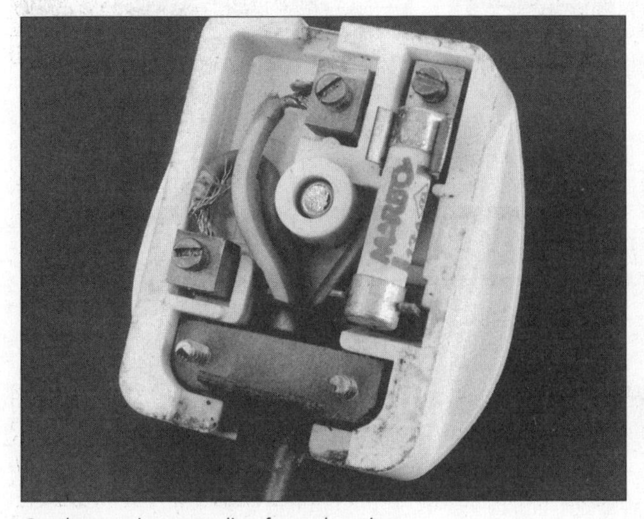

Conductor wire protruding from plug pins.

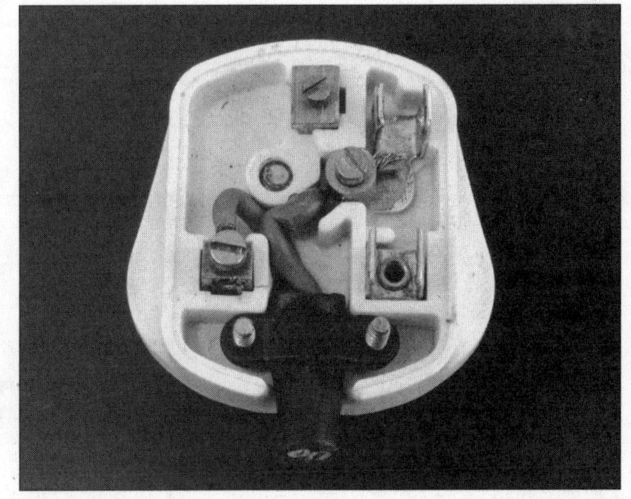

Wiring incorrectly bunched into plug to allow cord grip to hold outer sheath.

Wiring a pillar type plug

TOOLS AND MATERIALS

- ☐ Screwdriver
- ☐ Wire cutters
- ☐ Wire strippers
- ☐ Plug
- ☐ Correct rating fuse

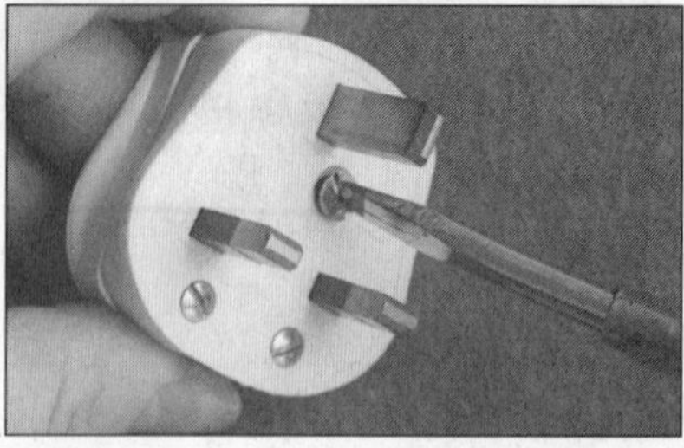

1 Remove the screw holding the plug top/cover in position, taking care not to lose it.

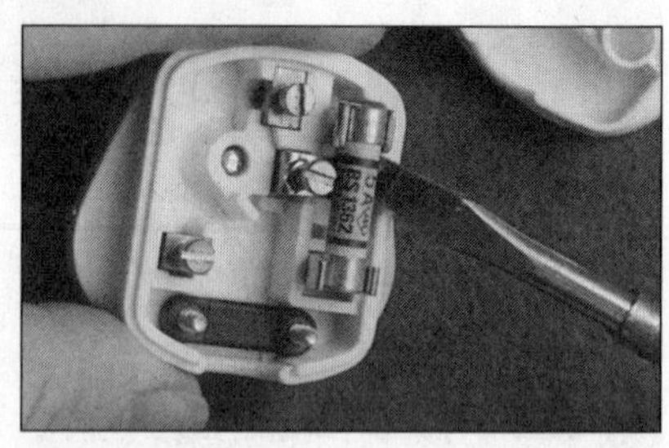

2 Ease out the fuse. If you use a screwdriver, take care not to damage it.

3 Check that the fuse supplied with the plug is the correct rating for the appliance.

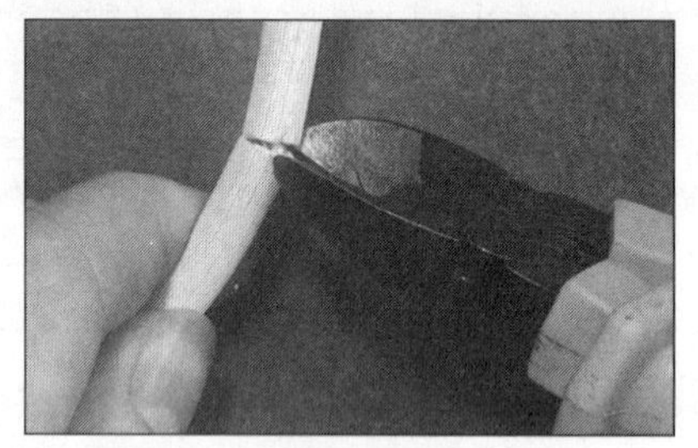

4 Carefully remove the outer cable sheath to expose the inner wires. If you accidentally damage the inner wires in the process, cut back the cable and start again.

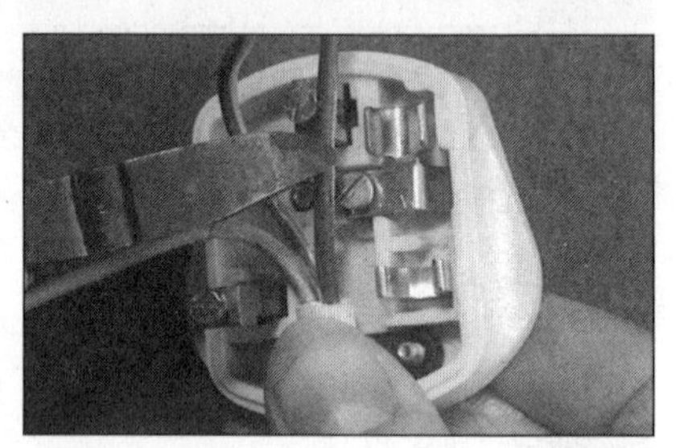

5 Offer the wiring to the plug base with the outer sheath in its correct position resting in the cord clamp area. Then cut the inner cables to suit, allowing 13mm (½in) past the fixing point. Allow a little extra on the earth cable to form a slight loop.

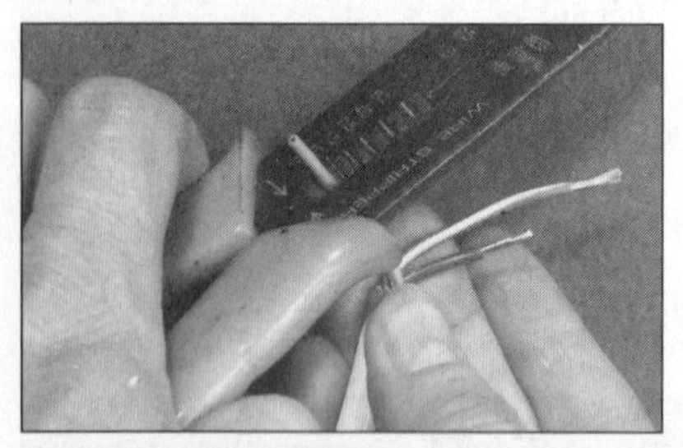

6 Carefully remove 6mm (¼in) of insulation from the end of each wire without damaging or cutting any strands of the conductor.

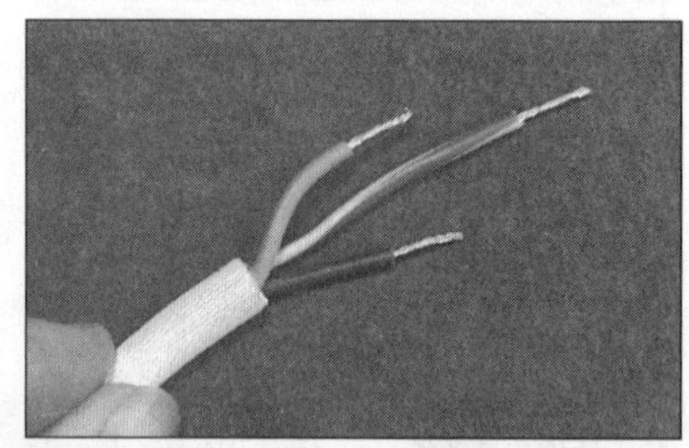

7 Twist the strands of each wire securely together, making sure there are no loose strands.

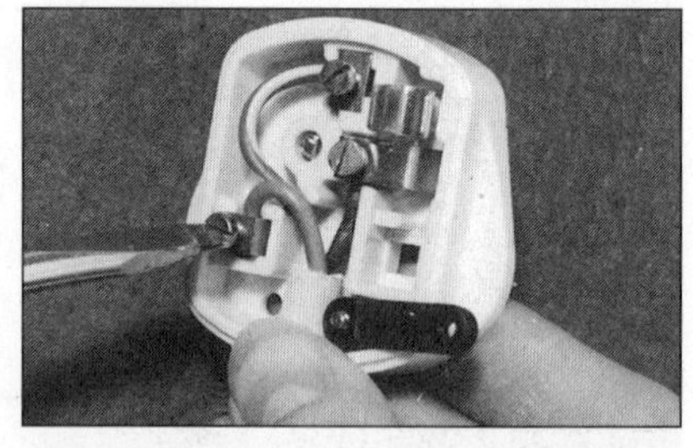

8 Fit each wire into its correct pillar and tighten each screw, ensuring that it grips the conductor firmly. Make sure the wire fits up to the insulation shoulder and no wires or strands protrude from the pillar.

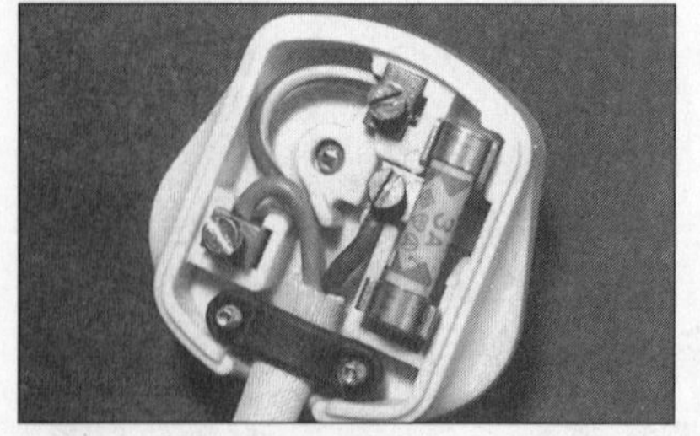

9 Fit the cord clamp over the outer sheath and screw it firmly into position, taking care not to strip the threads of the plastic grip.

10 Before refitting the top/cover, double-check all fixings. Ensure the wiring is seated and routed neatly and is not under stress or bunched.

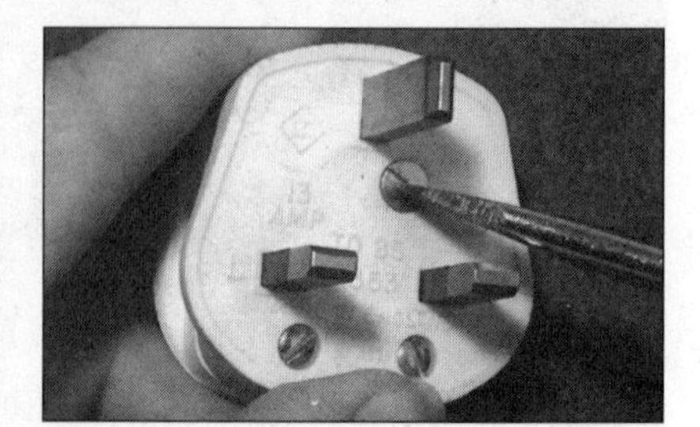

11 Refit the top/cover and tighten the securing screw.

Wiring a post and nut type plug

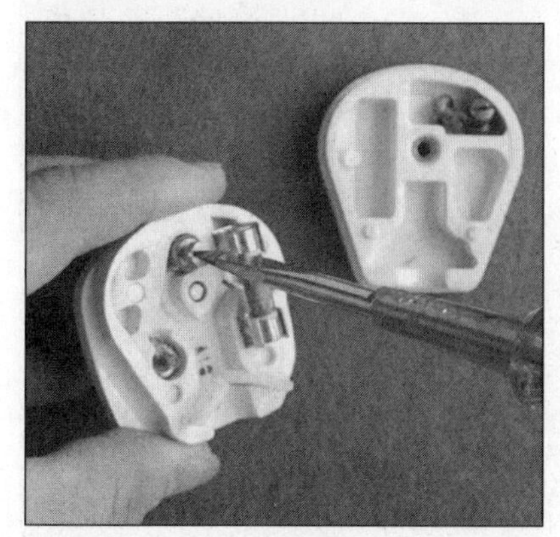

1 Remove the top cover as on previous page, remove the knurled/slotted nuts and place them safely in the top to avoid losing them.

2 Ease out the fuse. If you use a screwdriver, take care not to damage the fuse.

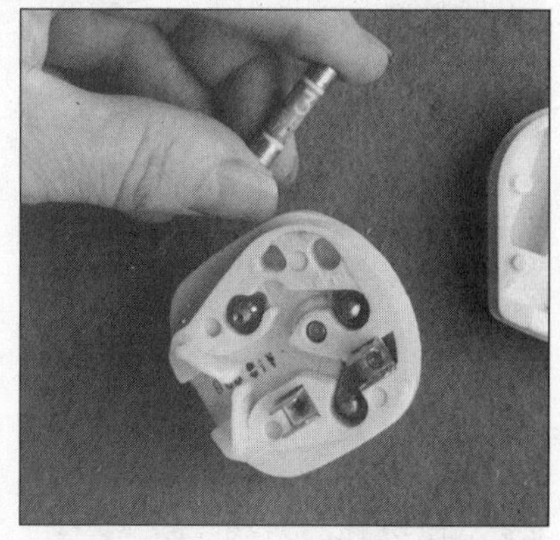

3 Check that the fuse supplied with the plug is the correct rating for the appliance.

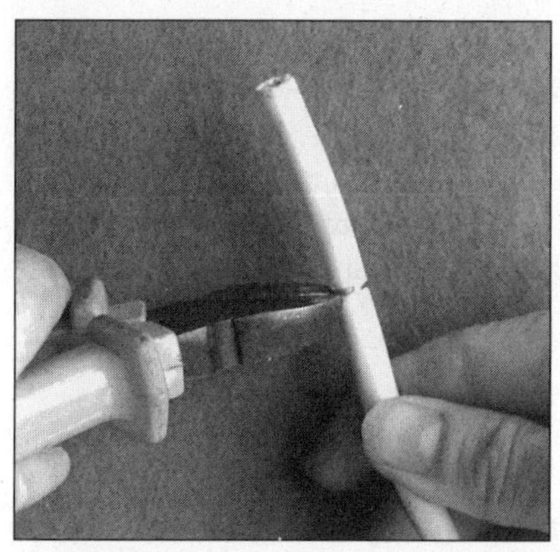

4 Carefully remove 32mm (1¼ins) of the cable sheath. If you damage the inner wires, cut back the cable and start again.

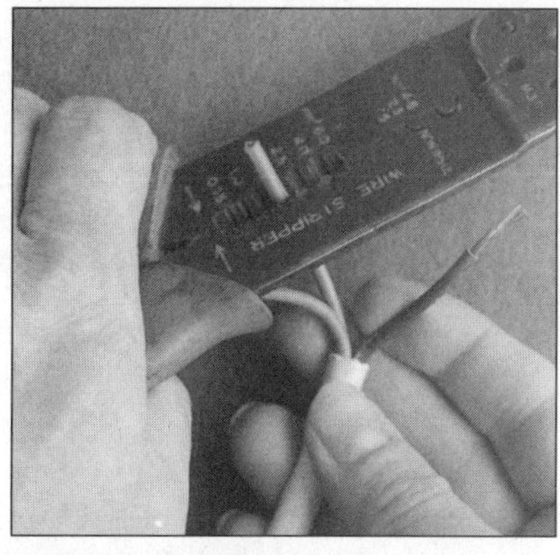

5 Remove 15mm (9/16in) of insulation from the end of each wire without damaging or cutting any strands of the conductor.

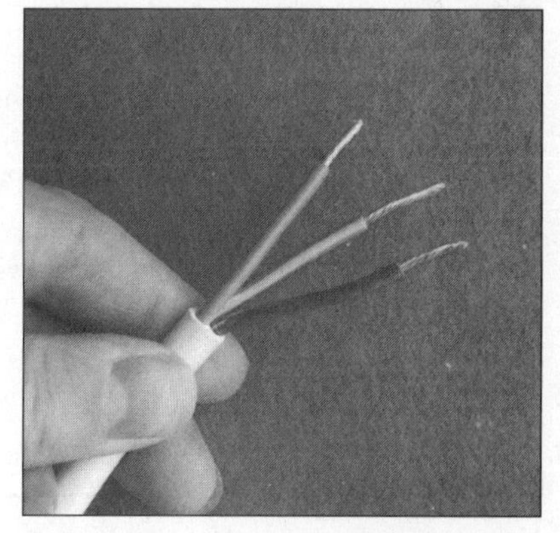

6 Twist the strands of each wire securely together, making sure there are no loose strands.

7 Insert the prepared cable into the cord grip, making sure that only the outer sheath is gripped.

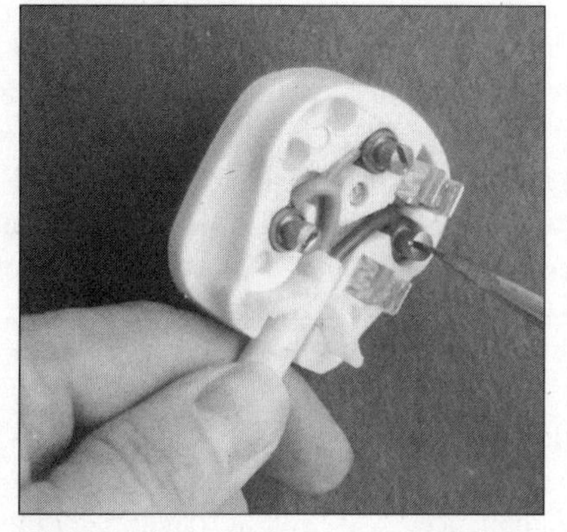

8 Fit each wire to its correct terminal in a clockwise direction; otherwise it will be pushed out as the nut is tightened. Ensure only the conductor is gripped and not the outer insulation.

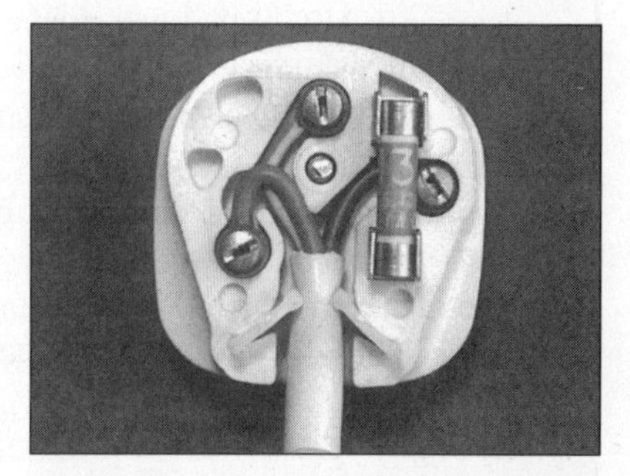

9 Securely tighten all three nuts. Check that the wire fits up to the insulation shoulder and that no wires or strands protrude from the terminal. Before refitting the top/cover, double-check all fixings. Ensure the wiring is seated and routed neatly and is not under stress or bunched. Fit the correct rated fuse, making sure that it is firmly and securely positioned.

Typical moulded plug.

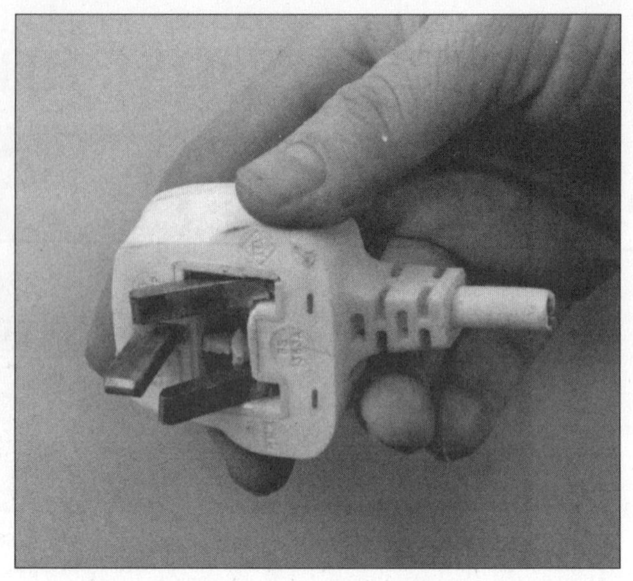

Make sure a moulded plug removed from an appliance cannot be plugged in. Remove the fuse and bend the pins.

Moulded plugs

Some appliances are supplied fitted with moulded 13 amp plugs. If this plug has to be removed to allow the cable to be slotted through a hole in a work surface or because of damage, it cannot be removed in the normal way. It must be cut off with suitable wire cutters and a new plug fitted.

Dispose of a removed moulded plug immediately. Take out the fuse and bend the pins of the plug with pliers as soon as you have removed it to render it completely unusable. This makes sure that it cannot be plugged into a socket. Do not leave it lying about, dispose of it where children or the unwary cannot find it and plug it in.

DOs

- Do ensure the cable insulation is removed carefully. Use of the correct wire strippers is strongly recommended.
- Do make sure that connections are the right way round.
- Do trim the wires to suit the plug fixing points so that no bunching is present.
- Do make sure that all connections are tight and no strands of wire protrude from the terminals. To prevent this, twist the strands together before fitting.
- Do fit the cord grip correctly around the outer insulation only.
- Do use the correct fuse to suit the appliance.
- Do check that the plug top or cover fits tightly and securely with no cracks or other damage present.

DON'Ts

- Do not damage the inner core of wires when removing the outer or inner insulation. If you do, cut back and start again.
- Do not fit tinned ends of cables into plugs (some manufacturers tin the ends of the exposed inner conductors, i.e., dip them in solder), as they will work loose and cause problems. Also the excessive length of exposed inner wire which the manufacturer usually provides can prevent the cord clamp working correctly.
- Do not allow strands of wire to protrude from any fixing points.
- Do not fit incorrect fuses. Always follow the manufacturer's instructions.
- Do not reuse overheated or damaged plugs.
- Do not by-pass the internal fuse.

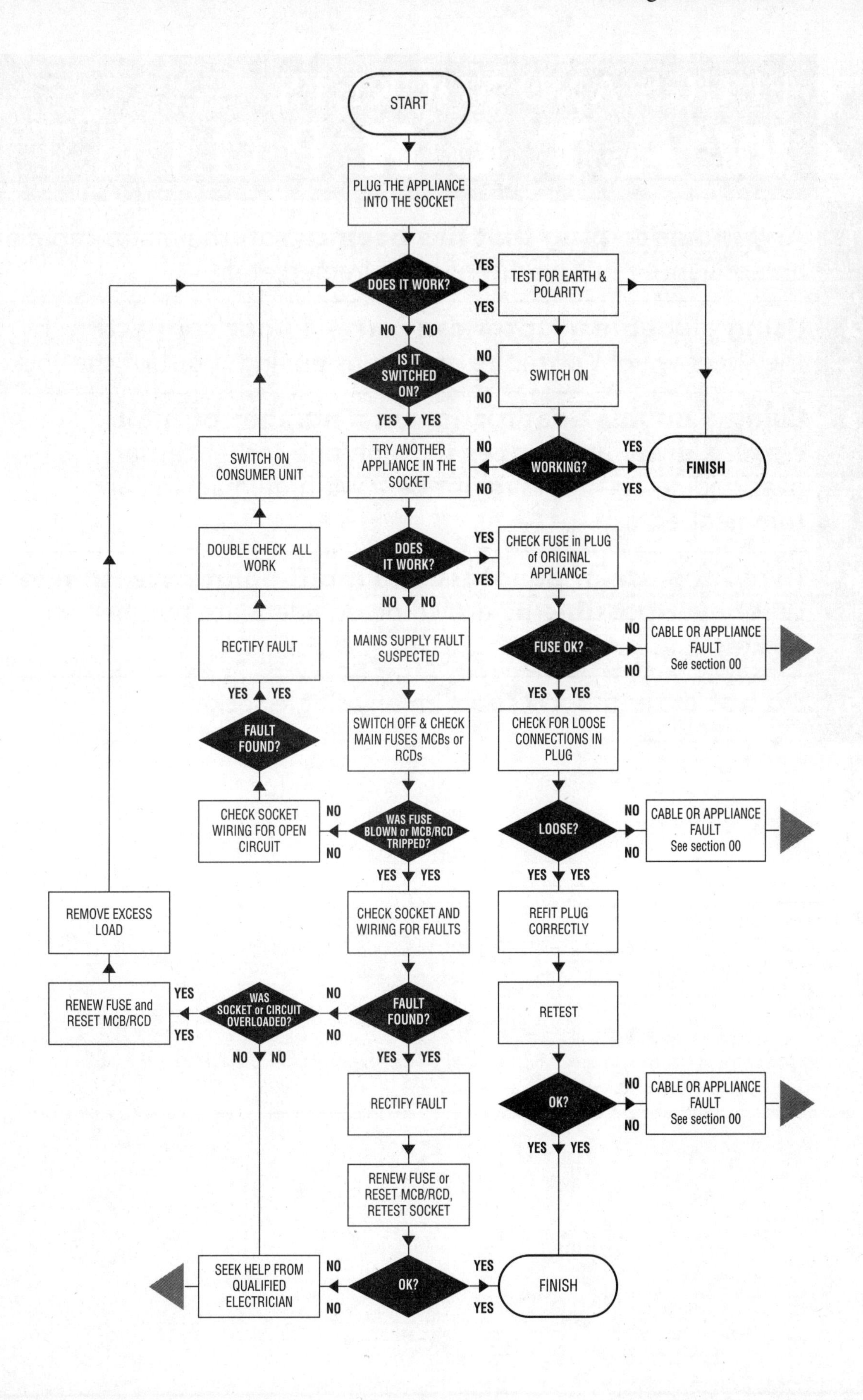
START
PLUG THE APPLIANCE INTO THE SOCKET
DOES IT WORK?
YES
TEST FOR EARTH & POLARITY
NO
IS IT SWITCHED ON?
SWITCH ON
WORKING?
FINISH
TRY ANOTHER APPLIANCE IN THE SOCKET
SWITCH ON CONSUMER UNIT
DOUBLE CHECK ALL WORK
DOES IT WORK?
CHECK FUSE in PLUG of ORIGINAL APPLIANCE
FUSE OK?
CABLE OR APPLIANCE FAULT See section 00
RECTIFY FAULT
MAINS SUPPLY FAULT SUSPECTED
FAULT FOUND?
SWITCH OFF & CHECK MAIN FUSES MCBs or RCDs
CHECK FOR LOOSE CONNECTIONS IN PLUG
CHECK SOCKET WIRING FOR OPEN CIRCUIT
WAS FUSE BLOWN or MCB/RCD TRIPPED?
LOOSE?
REMOVE EXCESS LOAD
CHECK SOCKET AND WIRING FOR FAULTS
REFIT PLUG CORRECTLY
RENEW FUSE and RESET MCB/RCD
WAS SOCKET or CIRCUIT OVERLOADED?
RETEST
OK?
RENEW FUSE or RESET MCB/RCD, RETEST SOCKET
SEEK HELP FROM QUALIFIED ELECTRICIAN

PLUGS AND SOCKETS
WATCHPOINTS

1	**Any moulded plug that has been cut** off the mains cable of an appliance must be disposed of immediately.
2	**Using a double adaptor can cause a poor connection** by the sheer weight of cables and plugs pulling it out of the socket.
3	**Using a double adaptor to run a number of high current-draw appliances** through one socket causes overloading. Examples might be a washing machine and tumbledrier.
4	**Avoid the use of adaptors and multi-point extension leads whenever possible** by providing an adequate number of sockets.
5	**Do not exceed 3 kW load** on any single socket.

Electrical circuit testing

Throughout this book references are made to meters and their use in continuity testing of individual parts of appliances and their connecting wires. All such testing and checking for 'open' (not allowing current to flow) or 'closed' circuit (allowing current to flow), must be carried out using a battery-powered multimeter or test meter. Testing should never be carried out on 'live' machines, that is, appliances connected to the mains supply. Isolate the appliances from the mains supply before starting any repair work or testing.

Although some meters and testers can check mains voltages, they are not recommended for use in repairs to domestic appliances. Faults can be easily traced with simple low-voltage (battery power) continuity testing. The simplest meters – even a home-made one like that described below – are perfectly adequate for some faults. The home-made continuity tester helps trace faults in the wiring of the appliance only. A multimeter, like the ones shown here, is required for component testing.

Safety is paramount and should never be compromised in any way. Always double-check that the appliance is unplugged; a good tip is to keep the plug obviously in view so that no-one else can inadvertently plug it in.

Choosing a meter

If you decide to buy a test meter, you may well be faced with a bewildering variety of choices. Do not be tempted to buy an over-complicated one as it may only confuse and mislead you. It will probably also prove unnecessarily expensive.

Before using your new meter, read the manufacturer's instructions thoroughly and make sure that you understand them fully. The meter shown here is simple to use for continuity testing and has a scale that reads 'open' circuit or 'closed' circuit. It was purchased from a local D-I-Y store at an economic price. The meter can also help locate faults with car electrics, but should never be used on live mains circuits.

Some multimeters can show the resistance value of the item being tested as well as indicating continuity. This can be extremely useful if the correct value of the item being tested is known, that is, correct resistance of motor winding, armature or element, for example. However, this facility is by no means essential.

Details about using the multimeter for this function will be found in the manufacturer's instructions.

Continuity testing

This simple continuity device can be used to trace wiring faults in most appliances and is very easy to make. It uses the lack of continuity to its full advantage. To make it, you need a standard battery, a small torch bulb and three wires: one 13cm (5in) and two 25cm(10in). Connect the short wire to the positive terminal of the battery and the other end of that wire to the centre terminal of the bulb. Attach one of the longer wires to the negative terminal of

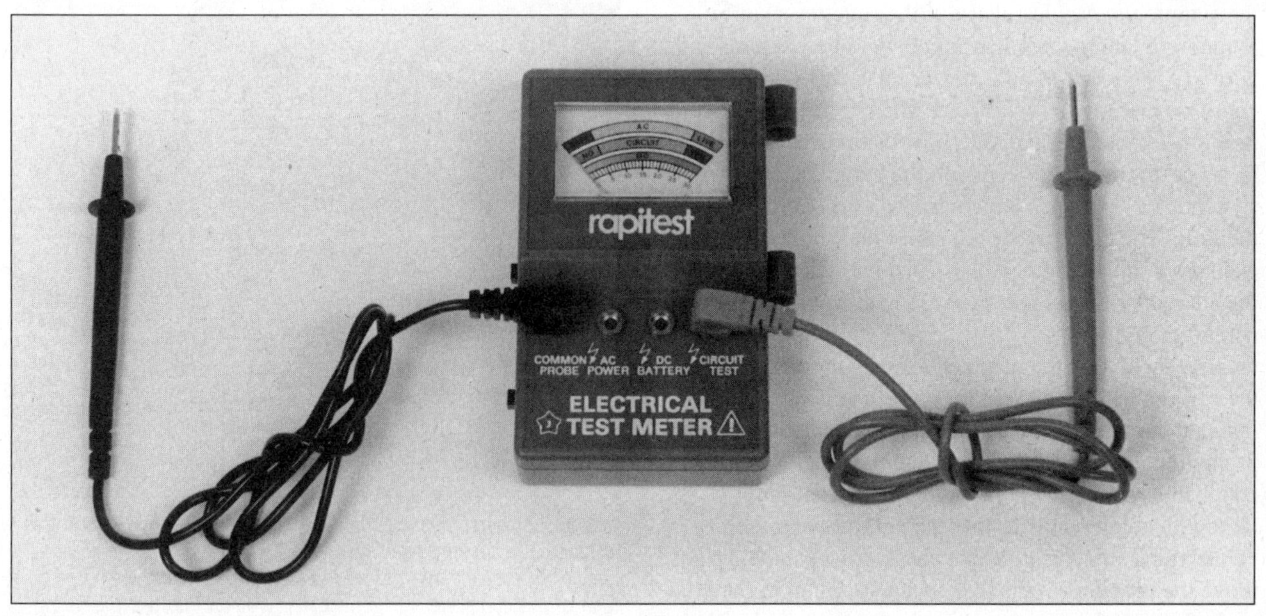

A typical test meter of the type available in most D-I-Y stores. Try to obtain a meter with a good informative booklet. The meter shown was inexpensive and proved useful for many other jobs around the house and in the car.

the battery and leave the other end free. Attach the other longer wire to the body of the bulb and leave the end free.

The two free ends now act as the test wires. Press them together, and the bulb will light. If it does not, check that the battery, bulb and all connections are alright. To indicate an 'open circuit' the light will stay off; and to indicate a 'closed circuit' the light will go on.

Low-voltage bulb type testers of 1.5 volts or 3 volts are unsuitable for testing the continuity of components within the machines. A test meter like the ones shown here is required to test high resistance items, such as pumps, timer coil and valves. Ensure that the machine is isolated from the mains supply before attempting to use a meter.

To test for an open circuit, note and remove the original wiring to the component to be tested. Otherwise false readings may be given from other items that may be in circuit. Attach the ends of the two wires of the test meter to the suspect component. For example, to test a heater for continuity, place the metal probes on the tags at the end of the heater and watch the meter. The needle should move. If the heater is open circuit, that is, no movement, the heater should be suspected and tested further. If it is closed circuit, the heater continuity is all right.

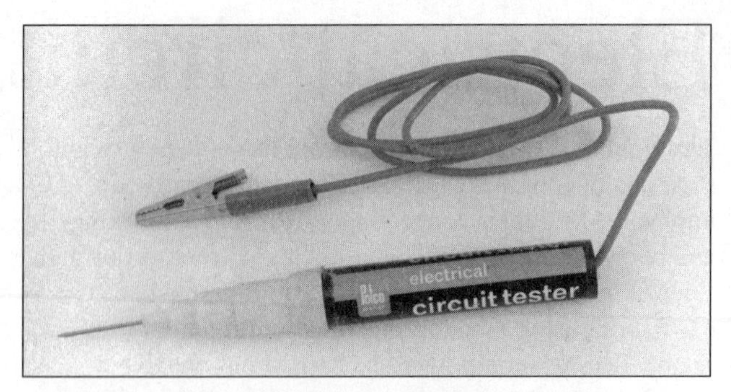

This simple continuity tester was purchased from a local automart. It is a manufactured version of the home-made type described.

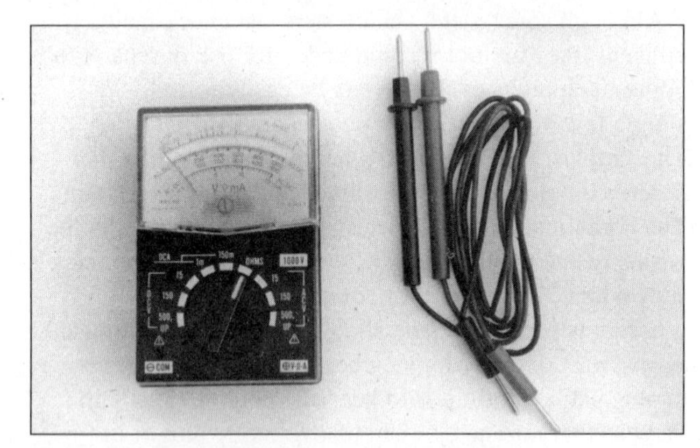

This type of multimeter can be found in most D-I-Y stores. Try to obtain a meter with a good information booklet.

Leap-frog testing

Often the most effective way to trace a fault is to use a very simple, but logical approach. One such approach, called the leap-frog method, can be used to find the failed/open circuit part or parts. Let us assume that the appliance does not work at all when functionally tested, so you cannot deduce where the problem lies purely from the symptoms. A quick check of the supply socket by plugging in another appliance known to be working will verify (or not) that there is power up to that point. This confirms that the fault lies somewhere in the appliance, its supply cable or plug. We know that power normally flows in through the live pin on the plug, through the switched on appliance and returns via the neutral pin on the plug. The fact that the appliance does not work at all even when plugged in and switched on indicates that an open circuit exists somewhere along this normal live to neutral circuit.

First, check that the meter is working correctly; when the test probes are touched together the meter should indicate continuity. Isolate the machine and connect one probe to the live pin of its plug and the other to the live conductor connecting point in the plug. Continuity should be found which confirms that the pin, fuse and their connections are alright. If this check proves to be alright, move the probe from the live conductor point in the plug to the live conductor connection in the terminal block

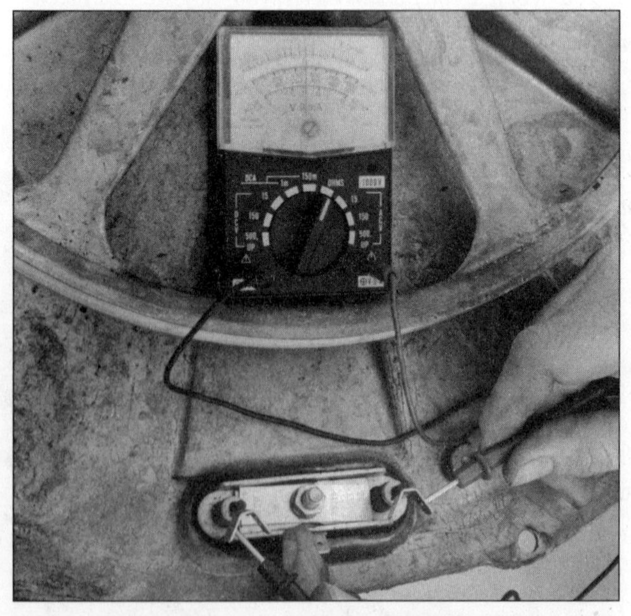

Testing the heater element to check for a circuit 'through it'. In this instance, the heater does have a circuit as shown by the meter needle. This means that the no heat fault on the machine is not a fault of this component. The next step would be to test the wiring and connections to and from the heater in the same way. Also check the timer and/or the thermostat if in the heater circuit.

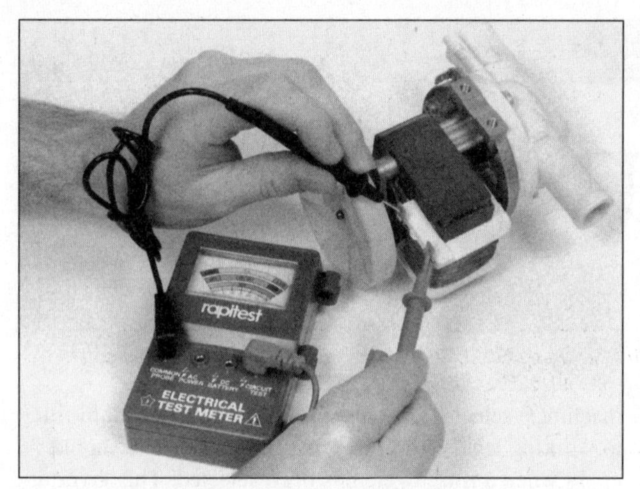

The test on this outlet pump stator coil is alright, that is, closed circuit/continuity. This means that the supply or neutral to and from the pump requires checking to discover the reason why the pump fails to work at any point in the programme.

within the appliance. Again, continuity should be found; if not, a fault between plug and terminal block is indicated. On cable continuity testing, it is best to move the cable along its length during the test to ascertain if an intermittent fault may exist.

If this test is alright, move the probe to the next convenient point along the live conductor – in this instance, the supply side of the on/off switch, which is part of the main programme switch on some machines (usually the front terminals). Again, continuity is required. An open circuit indicates a fault between terminal block and switch connection. The next step is to move the probe to the opposite terminal of the switch. Operate the switch to verify correct action: on gives continuity, off gives open circuit. If this is alright, proceed to the next point along the wire – in this instance, the interlock connection. Again continuity is required. If this is alright, move the probe to the terminal on the return side of the heater within the interlock *(see: Door locks, pages 75-78)*. This again should indicate continuity through the heater of the interlock. At this point, we will assume that an open circuit has been indicated, so go back to the last test point and verify continuity up to that point. If this is alright, then a fault has been traced that lies within the interlock which requires renewal.

This simple, methodical approach is all that is required to find such problems. It is best to break down more complex circuits into individual sections: motor, heater, switch, etc., and test continuity of each section from live through the timer and the individual parts and back to neutral. This may involve moving the live probe that would normally remain on the plug live pin to a more convenient supply point within the appliance to avoid misleading continuity readings from other items within the appliance circuit. With practice and common sense, even faults in complex wiring can be found in this way.

The action of switching within the interlock of power back to the timer cannot be verified, but continuity of the wiring can be checked in a similar leap-frog manner. In this instance, as the heater is open circuit, the interlock would fail to operate and the action of power being returned (switched) to the timer for distribution to other parts could not take place. A fault with the main switching action of the interlock would have been indicated during the functional test *(see: Functional testing, pages 32-33)*. In other words, when the machine is switched on, the door locks but nothing else operates. This is because most, but not all, machines have the interlock as the first item in circuit when switched on, so incorrect latching of the door or failure of the interlock (other than short circuit of the internal switch) will render the appliance inoperable.

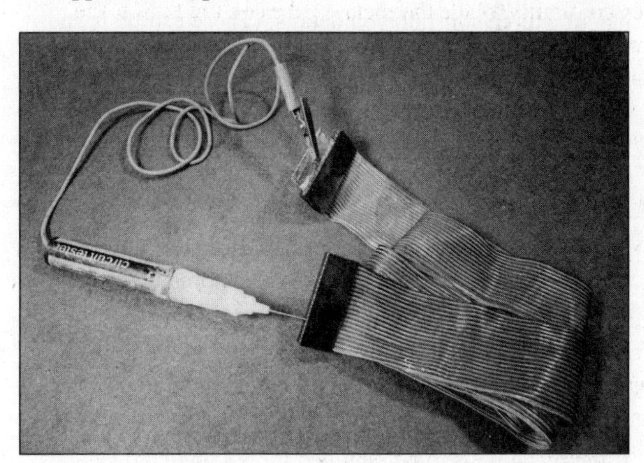

A simple test is all that is required to trace harness faults (see pages 142-143).

Do not trace faults by looking for mains voltages. There is no need to consider or use such dangerous techniques. All testing can and should be carried out with the appliance completely isolated, using only a battery powered meter or tester to indicate continuity or open circuit.

The ohms reading will differ from item to item. Test for open or closed circuits only. Any reference to an ohm (Ω) reading is a guide only as resistances differ from machine to machine. The objective is simply to test for continuity or lack of continuity of the item being tested.

Chapter 2

Installation

Plumbing basics

Although your machine may appear to work properly, if the installation was not correct in the first place, faults may occur many months later. Because of this time lag people do not always associate the faults with bad plumbing and tend to look for other causes, which is very time consuming and annoying. It is worth spending a few minutes examining pipework and checking the manufacturer's installation details. (These can be found in the manufacturer's booklet that came with the machine.) Even if the installation of your machine was left to an 'expert', it is still advisable to read this section, as it quite likely that the expert will not have read the installation details either!

For those who cannot find the manufacturer's installation instructions, a brief description of plumbing requirements that apply to nearly all automatic washing machines follows, together with the reasons why they should be observed.

Water supply

If the machine is to be plumbed in 'hot and cold', isolation taps must be fitted. This enables the water supply to be cut off (isolated) between the normal house supply and that of the washing machine.

The rubber hoses connected to the isolation taps should be positioned so that they do not get trapped when the machine is pushed back or rub against any rough surfaces during its operation. These conditions can cause the pipe to wear because of the slight movement of the machine in use. Also ensure that no loops have been formed in the hot inlet hose. This would not cause any trouble initially, but as the pipe gets older and the hot water takes effect, it will soften and a kink will form, causing a restriction or even a complete stoppage of water to the machine. This can also happen to the cold inlet hose, although it is very much rarer because of the increased pressure in the cold system.

Next, you should check that there is adequate water pressure to operate the hot and cold valves *(see Functional testing, pages 32-33)*. On hot and cold machines, select a hot only fill. The machine should fill to working level within four minutes. The same should apply when a rinse cycle has been selected. This gives a reasonable indication that the water pressure is adequate to open and close the valves. The valves are pressure operated; a minimum of 0.28kg/cm^2 (4psi) is required for their correct operation *(see Water inlet valves, pages 58-63)*. The cold pressure is usually governed by the outside mains pressure, but the hot water pressure is governed by the height of the hot water tank or the header tank. Problems can arise when the tanks are less than 2.5m (8ft) higher than the water valve they are supplying. This is often found in bungalows and some flats.

If a slow fill is suspected, check the small filters inside the hot and cold valves by unscrewing the inlet hose. These can be removed and cleaned by simply pulling them out gently with pliers. Take care not to damage the filter or allow any small particles to get past when you remove it. Clean water is normally supplied to the valves, but old pipework or the lime scale deposits from a boiler can collect at these points. Kinks and loops can also affect the outlet pipe and cause problems.

Outlet hose

The outlet hose must fit into a pipe larger than itself to provide an 'air brake' that eliminates siphoning. The height of the outlet hose is also important if siphoning is to be avoided. Siphoning can occur when the end of the outlet hose is below the level of water in the machine. This results in the machine's emptying at the same time as it is filling. Furthermore, if the machine were to be turned off, siphoning would continue to empty the water from the machine down to its siphon level.

D-I-Y plumbing in

When a machine is to be installed near an existing sink unit, you can take advantage of the new 'self-bore' taps and outlet systems now available. These simple and effective D-I-Y fittings save both time and money. In most cases, fitting the taps requires only a screwdriver and no soldering is necessary. You do not even need to drain or turn off the main water system. Follow the step-by-step diagrams (shown below) for this easy-to-manage straightforward procedure.

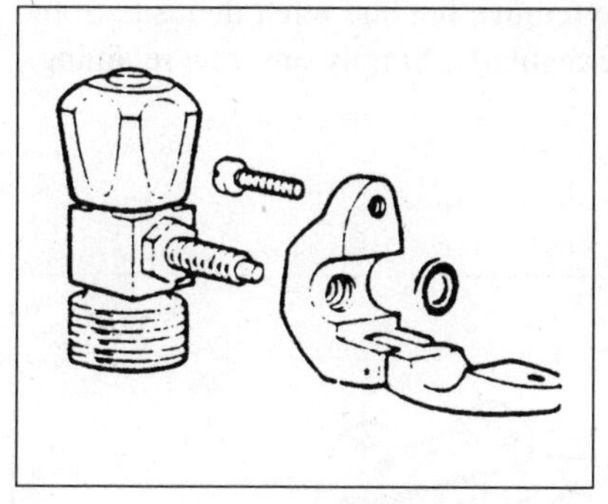

1 Unscrew tap and open clamp.

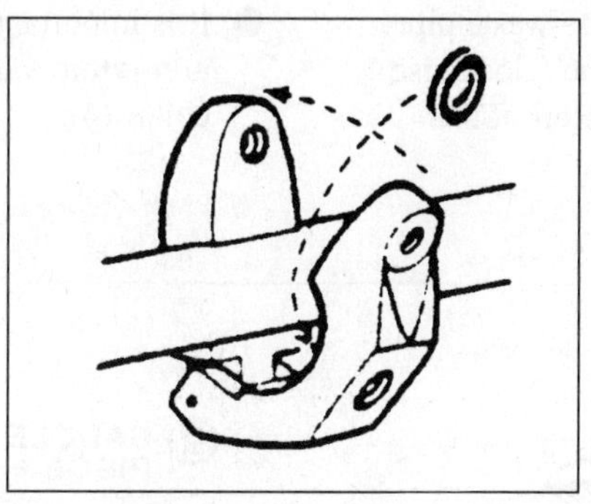

2 Fit clamp around copper pipe in required position. Make sure washer is in position.

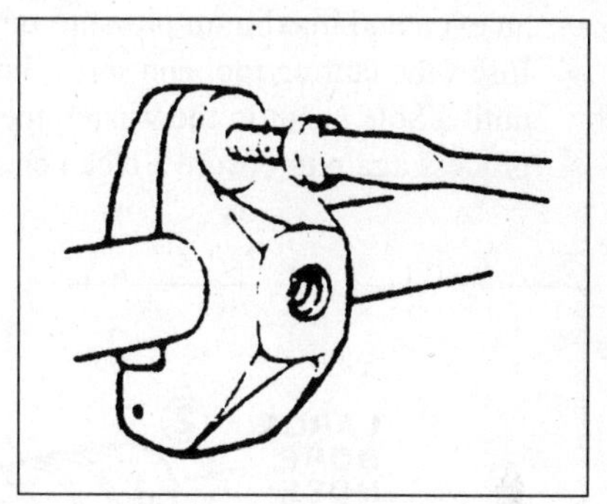

3 Engage screw and tighten until clamp is secure. Do not over tighten.

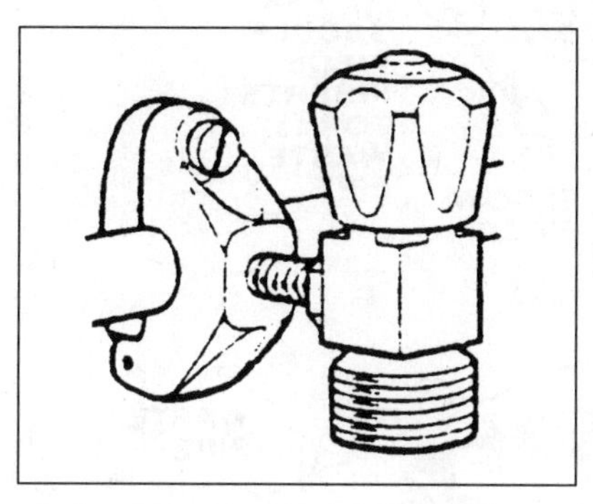

4 Insert tap assembly into clamp. Ensure that the tap is in off position.

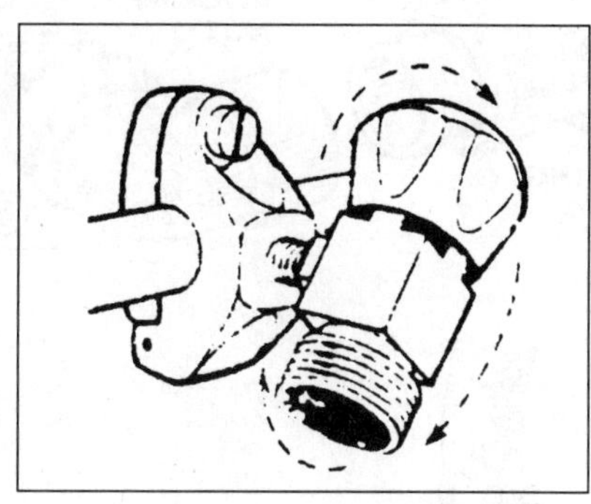

5 Turn clockwise until the pipe is penetrated. Set tap to position required.

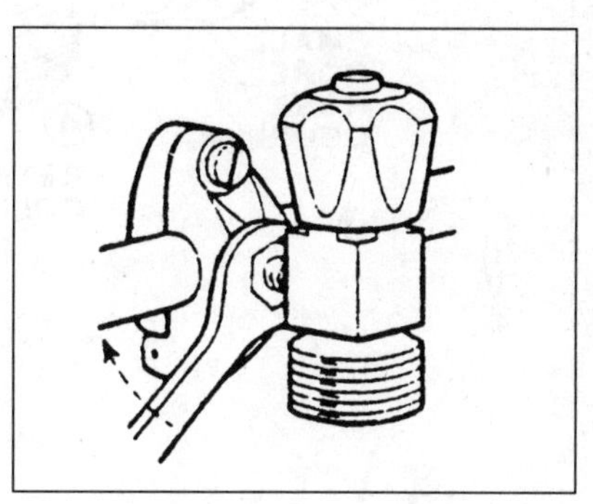

6 Tighten hexagonal nut towards the pipe. This secures the tap in position.

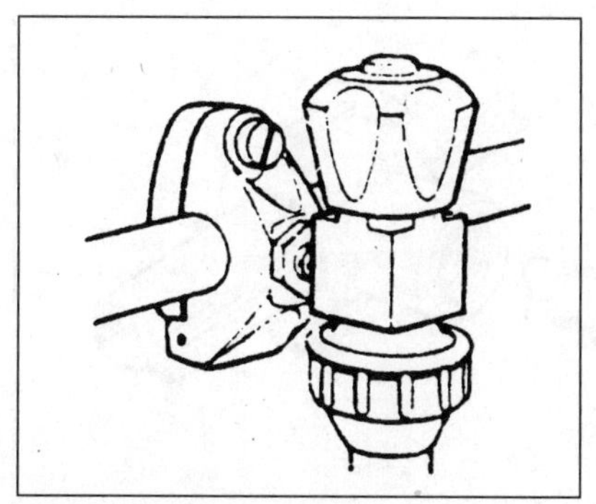

7 The tap is now ready for use. Connect hose to ¾ BSP thread on tap and turn on.

D-I-Y plumbing out

- Select the most convenient place in the waste pipe. 31mm (1¼ins) or 38mm (1½ins) diameter.
- Disconnect components, as shown. Place saddle half around the waste pipe, removing saddle inserts if pipe is 38mm (1½ins) diameter. Ensure that 'O' ring is seated in recess. Tighten the screws by stages to give an even and maximum pressure on the waste pipe. Insert the cutting tool and screw home (clockwise) until a hole is cut in the waste pipe. Repeat this process again to ensure a clean entry.
- Remove cutter and screw in elbow. Use a locking nut (5) to determine final position of elbow and tighten or screw non-return valve (3) directly into saddle piece.
- To complete installation choose correct size hose coupling to suit drain hose and secure hose with hose clip (not included in the illustration).
- It is important to remove lint and other deposits from non-return valve regularly. Simply unscrew retaining collar (4).

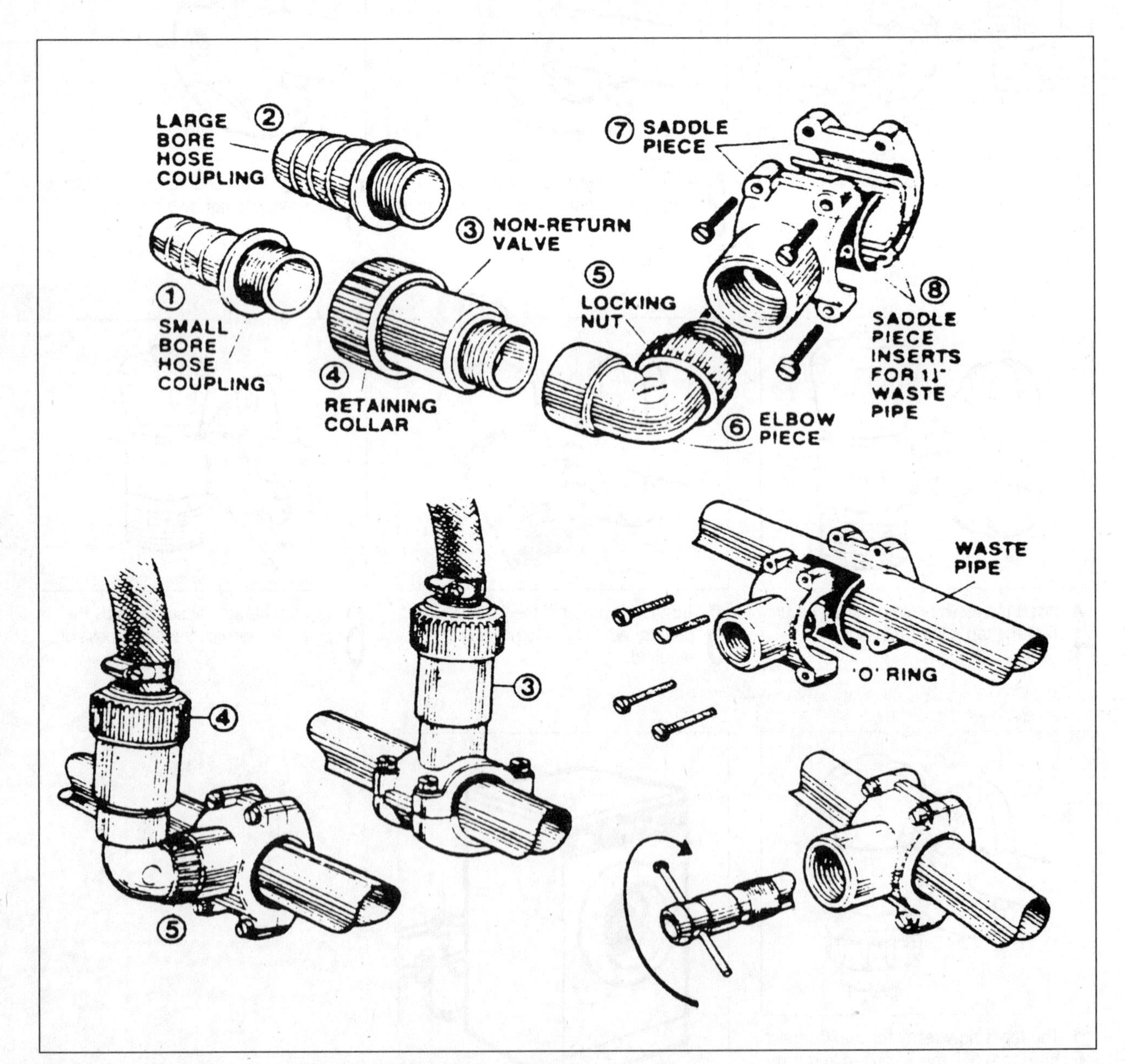

Plumbing out. All components can be unscrewed by turning anticlockwise.

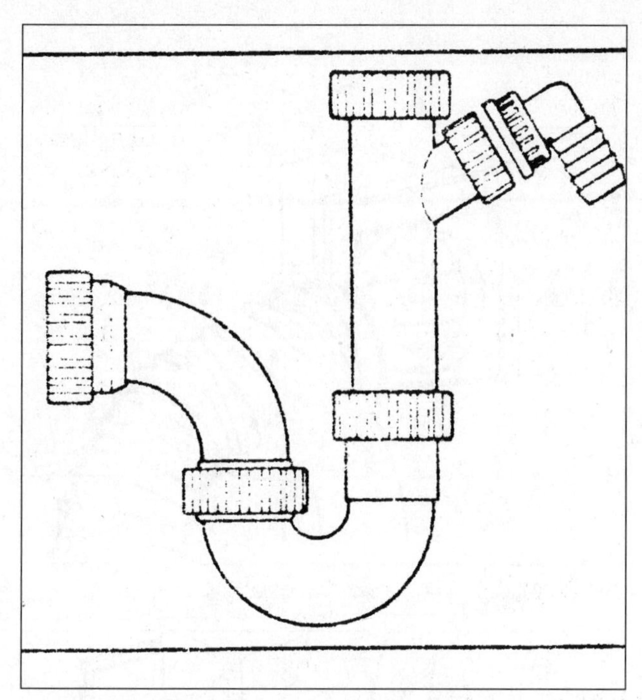

Discharge into a combined sink and washing machine trap. This trap allows water from the sink to drain away as normal, but has an extra branch for attaching the washing machine hose.

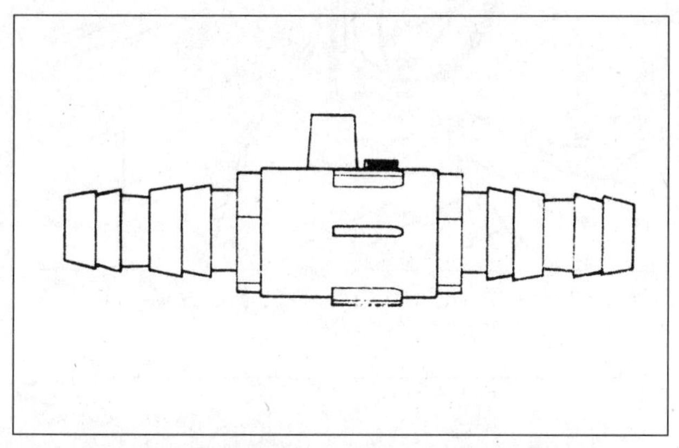

Siroflex anti-siphon unit. This unit provides an in-line air break to prevent siphoning occurring via the drain hose. Full fitting instructions are supplied with every unit.

Venting

Many washerdriers and tumbledriers vent the warm moist air produced during the drying cycle into the room where they are situated. This can amount to considerable quantity of water vapour which will readily condense on the nearest cold surface, such as cupboards and windows. Most appliances are situated in the kitchen where moisture from cooking is already produced, so extra condensation is unwanted, especially as it may result in damage to the surrounding areas and encourage the growth of mould. Condenser washerdriers and tumbledriers alleviate this problem, but they are more expensive than ordinary vented tumbledriers.

External venting is the answer to problems caused by ordinary machines. This may simply be a case of attaching one end of a flexible hose to the vent of the machine and hanging the other end out of an open window (both front and rear venting machines can be vented in this way). Although this solves the problem of moisture condensing in the room, it has its drawbacks. There has to be a nearby window that can be left open with the hose hanging through for as long as the machine is in use. As tumbledriers are used mainly in the autumn and winter this may not be convenient. The machine may also be left unattended or used at night on Economy 7 – either way, it is not wise to leave a window open. (The setting of domestic appliances to take advantage of off-peak electricity is commonplace. However, it is sensible to install a smoke detector in the vicinity of any machines that are used in this way in case a serious fault develops while the machine is unattended during operation.)

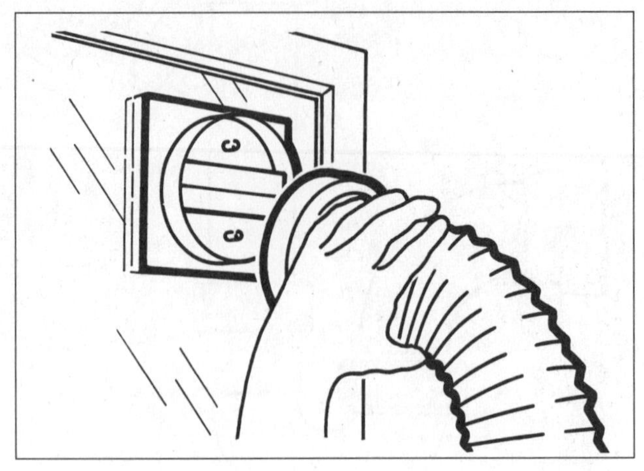

A window-mounted permanent vent system.

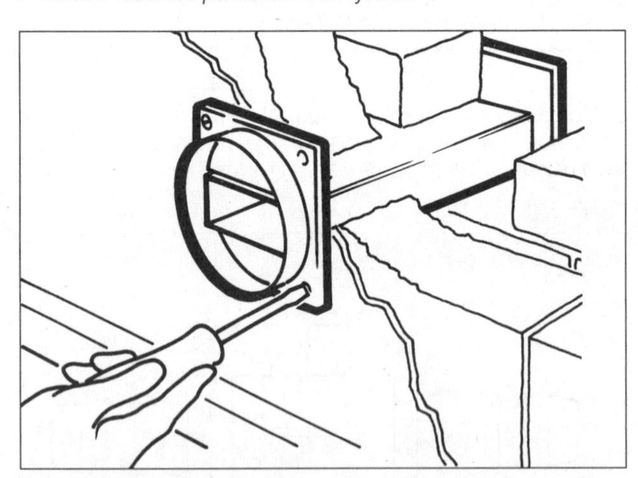

A through-the-wall vent system showing the new style square ducting, which is much easier to fit.

If an anti-draught cover is fitted, make sure it remains clear by checking it regularly for blockages and fluff build-up.

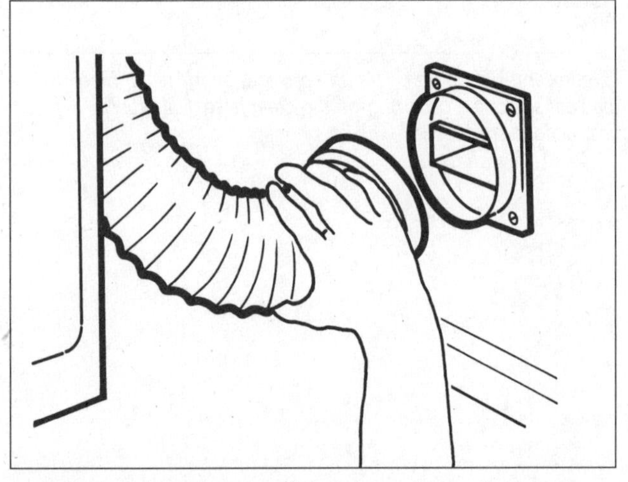

Both through-the-wall and window fittings take the circular vent hose of the machine, and adaptors are available to suit most sizes of hose and machine combinations.

The most convenient way around all these problems is to vent the machine permanently. This is not difficult and with the square, through-the-wall vent system, as opposed to the large round trunking, fitting is quite simple. As shown by the illustrations, several variations are available, with ends to fit most popular makes. If a convenient outside wall is not available, the kit can be used for a permanent vent through glass, avoiding the need to leave windows open. Each kit comes with fitting instructions. Remember to check for hidden pipes and cables before making the hole in the masonry.

VENTING WATCHPOINTS

1. **Clean the removable filter** within the exhaust tube that many driers use to vent to the atmosphere.
2. **Do not forget** to remove the filter of a permanently venting drier that vents to the atmosphere or it will quickly clog.
3. **Ensure that the flexible** section of hose is not flattened between the machine and the wall when you slide a machine with a rear fixed vent hose back into position.
4. **Do not fit a hose that is too long.** It may kink and condensation will collect, eventually causing a blockage.
5. **Always buy a suitable vent kit and install it properly.** Do not push the hose through a hole in the wall because it may kink. Condensation will then collect and cause a blockage.

Functional testing

There are references throughout this book to functional testing to ensure that the action of the machine and its installation are correct. Use this sequence as a guide to ensure correct operation after installation, repair or servicing. The purpose is to test all functions of the machine and plumbing in the most efficient manner. The test will suit most types of machines that use mechanical timers/programmers. Some slight modification may be required to suit model variations. Electronically controlled machines – those with non-mechanical timers – will have self-test programmes similar to this sequence *(see Timers, pages 98-109)*.

A typical installation and functional test

First test the hot fill valve.

- With the machine in its correct operating position and levelled correctly, and with all panels fitted, check that both hot and cold taps and power supply socket are turned on.
- Check the door is correctly closed and latched.
- It is not necessary to put a wash load into the machine or to pour detergent in the dispenser for this type of test.
- Select a hot wash cycle (90-95°C). On most machines this will energise only the hot fill valve and therefore test the flow rate/pressure supplied to it.

The machine should fill to low level within four minutes if the water pressure is adequate. If it takes longer, check that the hot supply tap is fully on. In most instances, little can be done for slow hot filling (see Plumbing basics, pages 26-29). In general, it causes only minor problems, such as failing to dispense detergent powders effectively from the drawer. In severe cases either connect to the cold supply with a 'Y'-joint or sprinkle the powder over the wash load in the drum (only if the wash is to be started immediately). The latter still allows the machine to take the hot fill at its reduced rate but without any problems.

Some machines can be modified internally. Instructions for this are given in the manufacturer's handbook/installation. Follow them carefully and do not attempt alteration without detailed instructions. As with any maintenance, repair, or inspection, isolate the machine thoroughly before removing any panels – switch off, plug out.

Next check for correct drum action.

- When the machine has filled to low level, check the drum action.
- Some machines require the timer to be advanced slightly to avoid a paused heat-only cycle before drum rotation takes place *(see Timers, pages 98-109)*.
- Check for clockwise and anticlockwise drum action.
- Check that the door interlock operates and that the door will not open during this cycle.
- Switch the machine off and move the timer to the special treatments (fabric conditioner) position. This selects a higher level fill via the cold valve.

Some machines possess a choice of the positions for this: one before the short or delicates spin and one before the normal or fast spin. Remember on delicates cycles the machine will stop full of water after the special treatments cycle and will move on to the spin only when instructed. Each machine has its own way of impulsing to the spin from this 'hold' position so make yourself aware from your handbook. *(See also Not emptying, pages 48-49)*.

Now check the cold inlet, the emptying procedure, the spin and the interlock mechanism.

- If you choose the position before the normal spin, the machine will fill to its high level using the cold valve. This checks the cold inlet.
- Wait for a maximum of four minutes for the high level to be reached. As the cold inlet pressure is usually governed by the street mains to the house, the pressure is normally well above the minimum required.
- When the high level is reached, check for rotation, again clockwise and anticlockwise.
- As the programme of special treatments is short, allow the machine to impulse normally to the pump out stage. (If on delicates cycle, use the normal advance mechanism.)
- Check that the machine empties within one minute (verifying that the pump rate and outlet are correct).
- Allow the machine to impulse on to the spin and through to the off position. Time and spin speeds will differ depending on which cycle or spin speed was selected.
- Check immediately after the off position has been reached to see if entry to the machine is correctly inhibited by the interlock.
- Time the delay of the interlock; minimum one minute up to two minutes.
- Check that after the proper time has elapsed, the door can be opened correctly. Do not be impatient and force the latch mechanism or handle.

If the machine is a washerdrier, go on to check the drying function.

- Select a drying programme. Refer to the manufacturer's instruction booklet, as machines differ in the way they are set.
- During the cycle, check for drum rotation clockwise and anticlockwise, blower motor operation (background noise), warming up of the door glass (do not touch the door glass directly as some makes get extremely hot).
- On condenser type machines, check for intermittent pump action.
- With vented washerdriers, check the rear vent (or if a vent hose system is fitted, check hose) for free flow of warm air.
- Do not allow the dry sequence to continue for too long without a load.
- Move the drier timer to the cool tumble position (five minutes before the end of drying programme) and allow the sequence to finish. This allows both drum and heater unit to cool correctly while confirming correct operation of drier timer.

This simple, logical sequence is likely to take about 10-15 minutes and is time spent usefully. If at any point a fault should occur, the correct action can then be implemented. These steps effectively confirm (or otherwise) the operation of both water valves and pressure supplied to them, the pressure system for water level control, hoses and seals, outlet pump and waste, programmer (although not in depth), spin speed, drier blower motor (if fitted), heater and drier timer.

What this simple test does not show is whether the wash heater works and if the wash thermostat operates correctly. Check these on the next full wash cycle after the basic functional test. Note the time taken to heat up and the correct impulse from the thermostat for the particular programme setting It is advisable to check both heater and correct temperature advancing during the course of a normal wash cycle.

Chapter 3

Care and repair

Regular maintenance

For many repairs it is best to lay the machine on its front face or side, preferably the side opposite the timer. (The timer is located directly behind the main programme knob.) This is to avoid the tub and drum assembly coming in contact with the timer.

Always ensure that the outer shell is protected with a suitable cover when you lay the machine down. It should be lowered slowly to avoid excessive movement of the suspension. It is a good idea to place a strip of wood under the top edge of the machine as you lower it, to provide room for the fingers when lifting it back after the repair.

When laying the machine over, care should be taken to protect oneself from injury:-

Firstly, ensure that the machine is completely disconnected from the main supply, and that the inlet and outlet hoses are removed

Secondly, before attempting to lay the machine over, decide if you need any help. Washing machines are very heavy and a little help may prevent a slipped disc

Thirdly, before attempting to move the machine, ensure that the floor is dry. A wet floor has no grip, especially if any of the water is soapy

CARE AND SAFETY WATCHPOINTS

1	**Protect yourself from injury** when laying down the machine.
2	**Ensure that the machine is disconnected from the mains supply** and that you have removed the inlet and outlet hoses.
3	**Seek help with moving the machine.** These machines are very heavy and a little help may prevent a slipped disc.
4	**Make sure the floor is dry** before attempting to move the machine. A wet floor has no grip and a soapy floor is very slippery.

Regular inspection points

Regular internal inspection of your machine may enable you to identify a part that may not be working correctly or find a perished hose before a leak occurs. It is recommended that the following points be checked regularly.

INSPECT	WHEN	SPECIAL NOTES
Pump filter (if fitted)	Weekly	According to manufacturer's instructions. Often depends on usage.
Vent filter (if fitted)	After every dry cycle	Must be kept clean for efficient drying.
Valve filters (hot & cold)	6 months	If dirty, pull out carefully with pliers and wash carefully.
Door seal	6 months	If seal is tacky to the touch it may need to be renewed soon.
Door glass		Remove sticky fluff or scale deposits from door glass inner surface with non-abrasive pad. Check inlet duct for cracking.
All hoses	6 months	Ensure that all corrugations in all hoses are checked thoroughly.
Pump and sump hose catch pot	6 months	Check for items that may have collected in or at these points. Remove as necessary.
Condenser unit (if fitted)	6 months or as required	Check for fluff build-up in fan chamber and ducting. Some makes are more prone than others to problems. Also dependent on amount of usage.
Suspension	6 months	Check suspension mounts on tub and body of machine. If slide type, see Suspension (pages 93-97).
Motor brushes (if fitted)	6 months or yearly	Check for wear and/or sticking in slides. Renew if less than half normal length.
Belt tension	6 months or yearly	Check and adjust if necessary. See Belts (pages 123-126).
Level machine	Yearly/after every repair	Check that the machine is standing firmly on the floor and does not rock. Adjust by unscrewing adjustable feet or packing under the wheels.
Plug and connectors	Before and after every repair	After repair look for poor connections in the plug and socket. Also look for any cracks or other damage. Renew as necessary.
Taps and washers	Before and after every repair	Check taps for free movement, corrosion and/or leaks.

Note: For inspection points on tumbledriers, see Dry only machines (pages 154-167).

Diagnosing faults

Whenever possible the symptoms of a fault should be confirmed by operating the machine up to the point of the suspected fault using the appropriate test sequence. Then the machine should be stopped, disconnected from the mains supply and the relevant flowchart or fault sequence followed.

However, this is not practical for such faults as major leaks and blown fuses; more damage may result by repeated operation of the machine. In these cases, the fault is known and further confirmation would be of little benefit. Continuing to use a machine with a known fault may, in fact, result in further damage to the machine or its surroundings.

Locating a fault

Assessing and locating a fault may, at first, seem a difficult thing to do, but carrying out a few simple procedures before starting the work will help cut down on the time spent on the machine. Hopping at random from one part of the machine to another, hoping that you will come across the fault and subsequently repair it, is hardly the best way to tackle any job. The best method of fault finding is to make full use of your own experience of your particular machine and of all the available information. A methodical approach saves time and effort by eliminating unnecessary replacements based on guesswork.

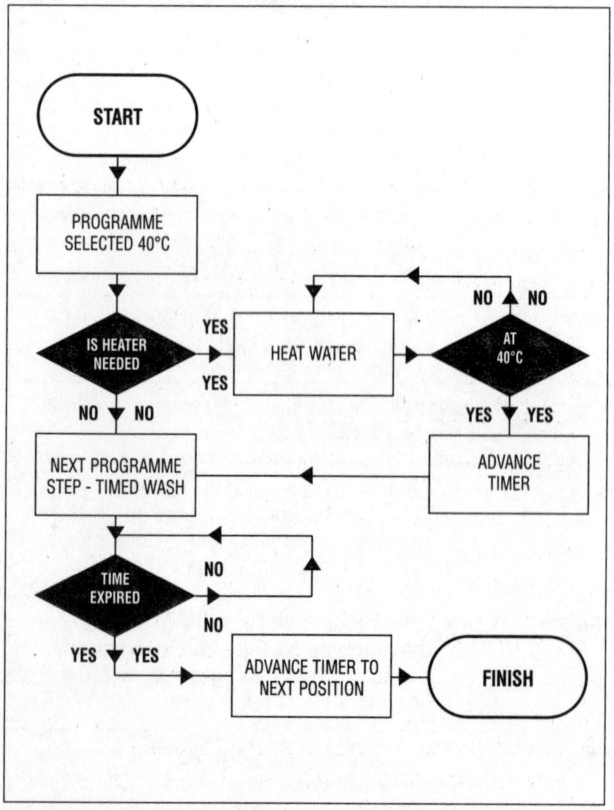

However there are a few things that can be done before such testing to check whether the machine itself is at fault or if an external fault or even misuse is the source. A large percentage of repair calls are, in fact, not for machine faults at all. Before jumping to conclusions, pause for a moment and double-check that the problem is not a simple oversight. You will not only save your time and effort, but, possibly, quite a sum of money as well.

If, after checking for oversights, the fault still remains, the next step is to determine its true nature and subsequent repair.

Remember these points when starting a repair.

- Allow yourself enough time to complete the task.
- Do not cut corners at the expense of safety.
- Try to ensure adequate working space wherever possible.
- Make notes about the position of the part/s to be removed, the colours and position of wires, bolts, etc.

If this practice becomes a matter of routine, it will help you in all repairs that you carry out, not only with your washing machine or tumbledrier.

A simplified flowchart of the operation described.

Determining the fault

Throughout this manual, flowcharts are used to aid the fault finding process. The location of faults will become much simpler as you become more conversant with your machine. Selecting the correct flowchart or fault guide for the job will be made easier if you remember that faults fall into three main categories. They may be categorised as mechanical, electrical or chemical.

Mechanical faults

These normally become apparent through a change in the usual operational noise level of the machine. For instance, faulty suspension may cause a banging or bumping noise, a broken or slipping belt (incorrect tension) may give rise to excessive spin noise or little or no drum rotation. This may also indicate a drum or motor bearing fault.

Electrical faults

These fall into two major categories: impulse path and component faults.

A fault is classed as an impulse fault when an internal,

DIAGNOSING FAULTS WATCHPOINTS

1. **Check that the machine is turned on** at the socket.
2. **Check that the fuse in the plug is intact** and working by replacing the suspected fuse with one out of a working item of the same rating.
3. **Check that the taps are in the on position.**
4. **Check that the door is closed correctly,** that a wash cycle is selected and the knob or switch has been pulled/pushed to the on position.
5. **Check that the machine is not on a 'rinse hold'** position. This will cause most machines to stand idle until instructed to do otherwise.

predetermined instruction has failed. For example, the thermostat does not close or open at the required temperature or the timer fails to move on after a given time sequence.

A component fault occurs when a complete unit fails. If, for instance, the pump, heater or motor should fail, this is said to be a component fault. Such faults may not always be immediately apparent. For instance, a failure in the coil of the water valve supplying the condenser unit of a washerdrier would manifest itself as a failure of the machine to dry the clothes. The wash cycles remain unaffected and all obvious functions of the dry cycle, such as heating up, circulation of air (noise of motor) etc. appear to be alright. However, with no water to cool the condenser, drying will not take place. This highlights the need not only to understand the basic function of each item, but also to appreciate its relationship with others. Read all sections of this manual thoroughly to help you understand the way in which components function together. This will assist you greatly in determining the cause of the trouble from symptoms shown by the machine.

Chemical faults

These are normally associated with the detergent powder that is being used and include such problems as poor washing, scaling problems and blocking. A comprehensive guide to washing problems and some other useful hints will be found later in this manual *(see Poor washing results, pages 42-43)*.

Noise faults

Noise can be one of the first and most useful signs that something is going wrong with your washing machine or washerdrier. Noise faults, however, are easily ignored and, over a length of time, become accepted as the norm. It is, therefore, important that noise faults are examined immediately they are first noticed.

As with other faults, noise faults become easier to locate the more conversant you are with your own particular machine.

- A loud grating or rumbling noise indicates a main drum bearing fault *(see Bearings, pages 129-143)*.
- A loud, high-pitched, screeching noise indicates either a main motor bearing or pump bearing fault *(see*

Motors, pages 110-122 and Pumps, pages 70-75).

- A noise just before and after spin, indicates wear or water penetration of the suspension mounts *(see Suspension, pages 93-97).*
- A squeaking noise mainly during the wash cycle indicates a poorly adjusted drive belt *(see Belts, pages 123-126).*
- Noise on the dry cycle may indicate a blockage in the fan housing, perhaps a build-up of fluff or distortion of moulding *(see Drying components, pages 144-153).*
- A metallic or grating noise on the dry cycle may denote a failure of one or both bearings of the fan motor. Such faults may often be accompanied by intermittent operation of the motor as a result of jamming, poor and/or lengthy drying or too high a temperature during the drying cycle caused by a reduction in the airflow as the fan rotates slowly *(see Motors, pages 110-122).*

A rather odd, although not, in fact, uncommon fault may be found on machines with cast aluminium pulleys. If a crack or break develops in one leg or spoke of the pulley, a noise very similar to that of main bearing failure can be heard. Check such pulleys closely for this type of defect and also for a tight fit on the shaft of the drum *(see Bearings, pages 129-143).* This type of fault may become apparent on any cycle.

Coin damage

Coins and other small metal items are easily trapped in the machine and can cause a great deal of damage. They can become caught between the inner drum and outer tub, and should be removed before damage to the drum occurs.

Coin damage can be identified by small bumps on the inner drum or rattling during spinning. On early enamel drums, this may be accompanied by small flakes of enamel in the wash load. It is not uncommon for stainless steel drums to be torn open by coin damage. Take care when checking the drum interior for this kind of damage as the torn metal edges can be extremely sharp. In such instances, the inner drum has to be renewed as a repair is not possible. Small pips of plastic found in the wash load may indicate the presence of a coin or similar item trapped between the inner and outer tub.

All of these items should have been removed before the washing was put in the machine.

Where there is no damage to the drum, it may be possible to avoid a full stripdown by removing the coin from the drum/tub gap by removing the heater *(see Heaters, pages 79-84).* The item can either be removed via this opening or manoeuvred into the sump hose, where it can be extracted easily.

Any large items such as bra underwires or keys can be removed via the heater opening, avoiding the sharp 90 degree angle into the sump hose. Almost all such problems could of course be avoided by first removing all items from pockets before washing. This simple action can save a lot of time and money in the long run.

Typical coin damage to a drum. This kind of fault can be easily avoided by careful checking of pockets, etc., before loading the washer. If this type of damage has been caused, the only cure is to carry out a complete drum renewal.

Typical coin damage to an outer tub. Although this does not seem so bad as the damage to the drum, the chips on the enamel will cause corrosion. Treat the affected areas with an anti-corrosion compound. This is readily available from motorists' shops and is usually used for minor bodywork repairs on cars. When applying the compound, follow the manufacturer's instructions carefully and do not allow the compound to come into contact with rubber hoses or seals. Work in a well-ventilated area. Stainless steel and plastic tubs are not generally affected by this type of damage, although regular inspection is important. Do not apply anti-corrosion compound to plastic tubs.

FAULT FINDER (WASH CYCLE)

The lists below each of the main faults indicate the sequence in which they should be examined.

Symptom: Machine will not work at all

See: Electrical basics ....pages 10-13
Locating a fault..................page 36
Door locks ....pages 75-78

Symptom: Machine leaks

See: Emergency procedures........page 9
Locating a fault..................page 36
Leaks ....pages 44-47
Water level control.............pages 63-68
Pumps ....pages 69-74
Motors...............................pages 110-122
Water inlet valves...............pages 58-62
Suspension ....pages 93-97

Symptom: Machine will not empty

See: Not emptying ....see pages 48-49
Emergency procedures........page 9
Locating a fault..................page 36
Pumps ....pages 69-74
Plumbing basics ....pages 26-29
The wiring harness.............pages 142-143

Symptom: Machine washes but no spin

See: Locating a fault..................page 36
Door locks ....pages 75-78
Pumps ....pages 69-74
Motors...............................pages 110-122
(main motor speed control)

Symptom: Machine will not turn drum

See: Belts ....pages 123-126
Motors...............................pages 110-122
(main motor speed control)
Door locks ....pages 75-78

Symptom: Machine spins in all positions

See: Locating a fault..................page 36
Motors...............................pages 110-122
(module control)
Timers ....pages 98-109

Symptom: Machine will not fill or take powder

See: Electrical basics ....pages 10-13
Locating a fault..................page 36
Water inlet valves...............pages 58-62

Symptom: Machine does not wash clean

See: Locating a fault..................page 37
Belts ....pages 123-126
Poor washing results..........pages 42-43
Plumbing basics ....pages 26-29
Pumps ....pages 69-74

Symptom: Machine is noisy

See: Noise faults........................page 38
Bearings.............................pages 129-141
Motors...............................pages 110-122
Suspension ....pages 93-97

Symptom: Machine will not move through programme

See: Plumbing basics ....pages 26-29
Temperature control...........pages 85-92
Heaters ....pages 79-84
Water inlet valves...............pages 58-62
Timers ....pages 98-109

Symptom: Machine sticks through programme

See: Locating a fault..................page 36
Plumbing basics ....pages 26-29
Pumps ....pages 69-74
Water inlet valves...............pages 58-62
Timers ....pages 98-109

Symptom: Machine blows fuses

See: Emergency procedures........page 9
Safety guide ....pages 6-8
Electrical basics ....pages 10-13
Low insulation ....pages 172-173

FAULT FINDER (DRY CYCLE AND TUMBLEDRIERS)

The lists below each of the main faults indicate the sequence in which they should be examined.

Symptom: No dry cycle

Check: Correct setting
Manufacturer's instructions
Electrical basics ....pages 10-13
Plumbing basics ....pages 26-29
Door locks ....pages 75-78
The wiring harness.............pages 142-143

Symptom: Poor drying of clothes

Check: Load size, manufacturer's instructions ....pages 154-157
Electrical basics ....pages 10-13
Plumbing basics ....pages 26-29
Drying components ....pages 144-153
Air ducts/motors, i.e. main wash, pump and fan ....pages 69-74, ...110-122, 144-167
Heaters (airflow type)..........pages 144-153
Temperature control...........pages 85-92
Timers ....pages 98-109
The wiring harness.............pages 142-143

Symptom: No heat at all on dry cycle

Check: Air ducts (blockages)...........page 161
Heaters (airflow type).........pages 144-146
Temperature control...........pages 85-92
Timers ....pages 98-109

Symptom: Heat too high

Check: Setting (see manufacturer's instructions)
Air ducts (blockages)...........page 161
Motors...............................pages 110-122
Temperature control...........pages 85-92
The wiring harness.............pages 142-143
Timers ....pages 98-109

Poor washing results

More often than not unsatisfactory washing is the result of incorrect operation of the machine by the user, rather than a straightforward mechanical or electrical fault of the machine. The commonest user faults are listed below.

Misuse of the controls

- To achieve good, consistent results with your washing, you must have a thorough understanding of your machine and its controls.
- Always remember, you tell the machine what to do. If in doubt, read the manufacturer's instructions.
- Check that the selected programme has the right water temperature and wash time for the fabrics in the load.

Incorrect quantity of detergent

- The amount of powder/liquid that you should use is usually displayed on the side of the pack. Remember that this is only a guide. Quantities have to be adjusted to load size, type and degree of soiling, and the hardness of the water supply.
- Make sure that the container used for measuring the powder/liquid is accurate. One cup measure of conventional powder weighs 85g (3oz). Use the scoop supplied with 'compact' powders; one scoop is normally equal to one cup of bulkier conventional powder.
- Make allowances for special types of soiling. The residue of ointments and thick creams, heavy perspiration and similar stains use up the suds activity very quickly. This can be compensated for by adding an extra half cupful (or scoop) of powder.
- A poor level of soil and stain removal, 'greying' whites or the appearance of 'greasy balls' on washed clothes is a clear indication of too little washing powder. It is never through overdosing.

Water supply

Washing powder is formulated to do several tasks: overcome water hardness, wet-out the fabrics, remove the dirt from the clothes and hold it in suspension away from the clothes.

It is obvious that the harder the water, the more the detergent has to work. Extra detergent or water softener should be added in the case of hard water and less in the case of soft water. The local area water authority should be able to advise you on the level of hardness of the water in your area.

Incorrect loading

- Overloading the washer prevents the clothes from moving freely inside the drum, resulting in inadequate dirt removal.
- Some programmes require reduced loads. If one of these programmes is used, the load must not exceed the manufacturer's instruction. If you are unsure, read the manufacturer's instructions.

Other factors

- Consider the age of the machine. Like any appliance, a washing machine has a restricted life span. The average life span of an automatic washing machine is around eight years according to manufacturers' figures.
- Make sure the machine is serviced regularly.
- Poor whiteness is a result of constant under-use of detergent and mixed loads. When the dirt is being only partially removed, there is a gradual build-up of deposits in the clothes. This can be avoided by using the correct quantity of washing powder and loading like fabrics together, that is, all whites, for example.
- Domestic changes, such as moving to a new area with a different water hardness, can affect the quality of the wash. The arrival of a new baby – and the concomitant lotions and nappy creams – may affect the wash.
- If the poor results are evident on only specific programmes rather than all programmes, loading may be incorrect. See the manufacturer's instructions.
- Most manufacturers recommend an occasional idle wash to keep the machine clean and free from deposits. This means a wash with no powder or clothes every month or so.
- Read the manufacturer's manual. Do not just assume that all machines are the same. Improvements on poor colours and whites will not happen magically; the process is gradual. Once a good whiteness has been achieved, correct washing and quantities of detergent are the only ways to maintain the standard.

POOR WASHING RESULTS WATCHPOINTS

1. **Wash clothes frequently.** Modern fabrics need frequent washing or dirt may become absorbed into the fibres.
2. **Use the right quantity of detergent.** Refer to the packet for the correct amount.
3. **Choose the recommended wash code and machine programme for the fabric,** to safeguard colour and finish, preserve shape and minimise creasing.
4. **Thorough rinsing is essential.** Some finishes, such as shower-proofing, lose their effectiveness if not well rinsed. Towelling fabrics, particularly nappies, may become harsh and scratchy. Always rinse at least twice.
5. **Treat stains quickly.** Give first-aid treatment immediately, whenever possible, by blotting with an absorbent tissue.

Chapter 4

Fault finding

Leaks

The leaks and all the methods of dealing with them described in this chapter relate equally to automatic washing machines, vented washerdriers and condenser washerdriers. Additional problems may be found with the extra hoses, connections and seals of condenser machines *(see Drying components pages144-153)*.
Specific problem areas on such systems are:

- Poor positioning of hoses resulting in their being chafed.
- Poor connection of rubber hoses to moulded plastic condenser units resulting in difficult-to-isolate faults.
- Lack of heat-resistant sealant (disturbed and not replaced during the course of earlier machine repairs) in areas such as heat duct to door seal inlet, fan chamber to both heat duct and condenser unit.

Ensure all connections are secure and correctly positioned and all the protective covers and sleeves are in good condition. Seal hose connections and condenser mouldings and ducts with the correct sealants *(see Useful tips, pages 168-171* and *Drying components, pages 144-153)*.

Using the flowchart

Bearing all this in mind, you can now easily and successfully follow the flowchart shown on the right. While it may, at first, seem blindingly obvious where the leak is coming from, it is still wise to follow the flowchart.

Box 1

The smallest of holes in the door boot (door seal) can be the cause of the biggest leaks. Check the seal and replace it if any holes are found or if it feels sticky or tacky *(see Door seals, pages 50-57)*. The contact between the door glass and the seal should be a good, secure one, with no scaling or fluff adhering to the glass. If either of these appears, remove by rubbing gently with a soft, non-abrasive scouring pad.

Box 2

Check the clamp band that secures the door seal to the washer for tightness. For specific details of different types of clamp bands *(see Door seals, pages 50-57)*.

Box 3

The dispenser hose is located at the base of the soap dispenser and forms the flexible connection with the tub. Inspect it thoroughly, as most of the water that the machine uses passes through this hose, so a fault can lead to some quite large leaks. This hose is usually a grommet type fitting and it should be checked for tightness, that is, it should not rotate. Scaling or powder marks running down the outer tub at the fitting point are indicators of a bad fit. Replace the hose if it feels sticky or tacky.

If this hose has leaked, the water will have contained detergent. If it has come into contact with the suspension legs, it may cause a loud squeaking or grinding noise just before and after the spin. This is because the suspension works its hardest at this time *(see Suspension, pages 93-97)*. This fault is more pronounced on Hoover automatics.

If there is a hose connected between the outer tub and the rear of the machine, this is an air vent tube. It cannot leak, as no water normally passes through it. However, it may have come into use if the machine overfoamed, overfilled or spun while still full of water. The hose is a grommet-type fitting and should be checked for perishing. Most modern machines use the soap dispenser as the air vent as well as the water inlet, thus eliminating an extra grommet fitting hose in manufacture.

Box 4

The sump hose is the flexible hose located at the bottom of the machine. Depending on the make and model of your appliance, it will be in one of two configurations: linking the pump and the tub or linking the filter and tub, with a separate hose linking the filter and pump, thus creating a trap for any foreign objects to prevent their reaching the pump.

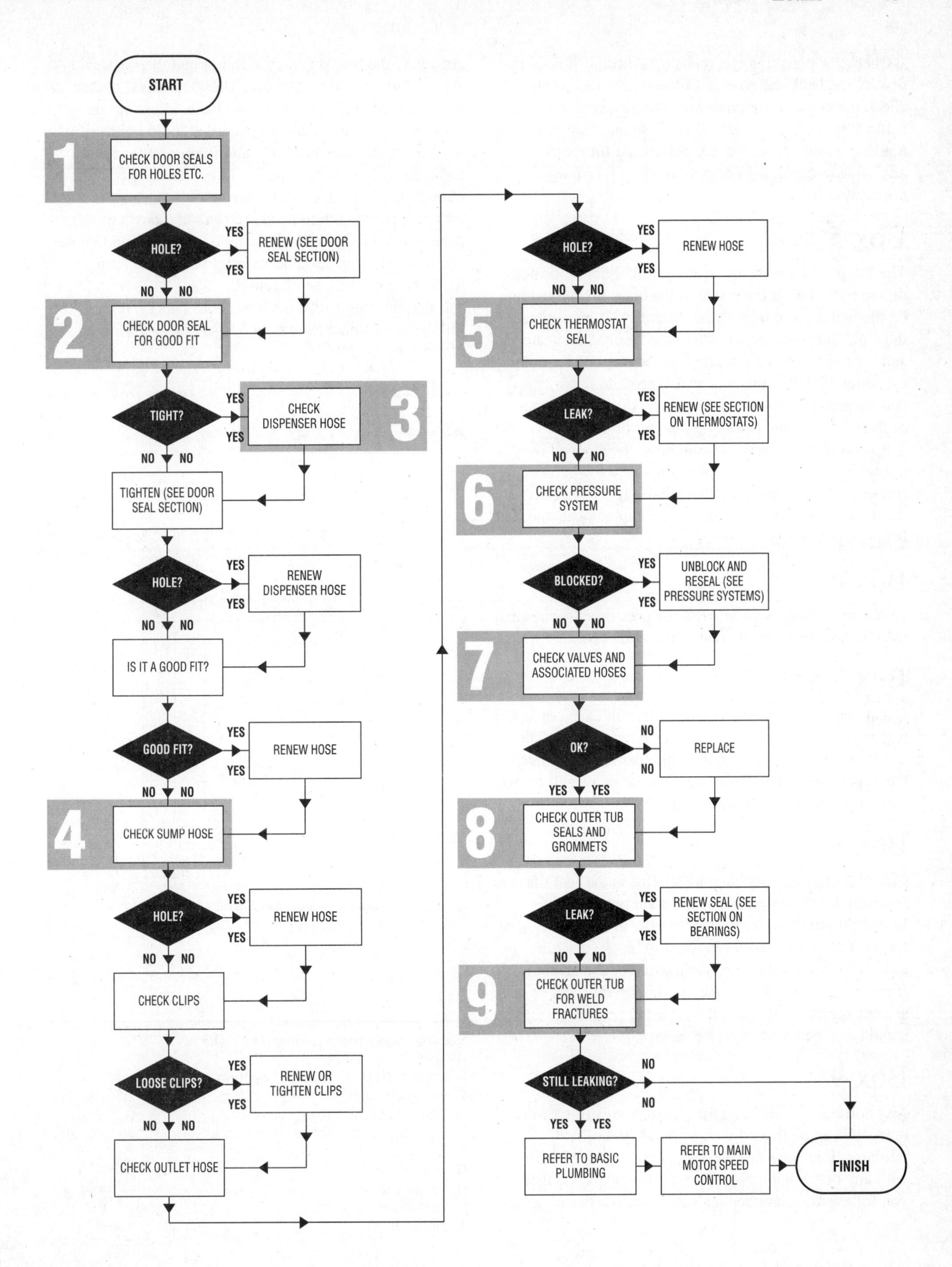
START
1
CHECK DOOR SEALS FOR HOLES ETC.
HOLE?
YES
RENEW (SEE DOOR SEAL SECTION)
NO
2
CHECK DOOR SEAL FOR GOOD FIT
TIGHT?
YES
3
CHECK DISPENSER HOSE
NO
TIGHTEN (SEE DOOR SEAL SECTION)
HOLE?
YES
RENEW DISPENSER HOSE
NO
IS IT A GOOD FIT?
GOOD FIT?
YES
RENEW HOSE
NO
4
CHECK SUMP HOSE
HOLE?
YES
RENEW HOSE
NO
CHECK CLIPS
LOOSE CLIPS?
YES
RENEW OR TIGHTEN CLIPS
NO
CHECK OUTLET HOSE
HOLE?
YES
RENEW HOSE
NO
5
CHECK THERMOSTAT SEAL
LEAK?
YES
RENEW (SEE SECTION ON THERMOSTATS)
NO
6
CHECK PRESSURE SYSTEM
BLOCKED?
YES
UNBLOCK AND RESEAL (SEE PRESSURE SYSTEMS)
NO
7
CHECK VALVES AND ASSOCIATED HOSES
OK?
NO
REPLACE
YES
8
CHECK OUTER TUB SEALS AND GROMMETS
LEAK?
YES
RENEW SEAL (SEE SECTION ON BEARINGS)
NO
9
CHECK OUTER TUB FOR WELD FRACTURES
STILL LEAKING?
NO
YES
REFER TO BASIC PLUMBING
REFER TO MAIN MOTOR SPEED CONTROL
FINISH

Check for perishing and replace the hose if it is tacky or sticky. Check the grommet fittings *(see Box 3)* and check the clips for tightness *(see Pumps, pages 69-74)*.

If a filter is fitted, check all hoses to and from the filter housing and the filter seal for defects. At this point, it is advisable to check the pump thoroughly *(see Pumps, pages 69-74)*.

Box 5

The thermostat and heater seal are usually both located on the back half or underside of the outer tub, depending on the make and model of the machine. An exception is the Hotpoint front loader, where the thermostat, heater and pressure vessel are located on the front of the outer tub, directly below the door seal. Access to these components is gained by the removal of the front panel of the machine *(see Door seals, pages 50-57)*.

A leaking heater seal can sometimes be stopped by tightening the centre nut. This increases the width of the rubber seal by squashing it 'vice like'. If the leak persists, the heater will have to be changed completely, as the seal is not sold as a separate item.

Box 6

Check the connections between the pressure switches and the tub (see Water level control, pages 63-68).

Box 7

Check the hoses and clips on the inlet valves in conjunction with the water inlet valves *(see pages 58-62)*. There is a very slight chance that the top of some valves may have split. This often becomes apparent as a small brown rust patch on the top of the valve.

Box 8

Check the tub seal and grommets. This is the seal that fits between the separate parts of the main tub assembly.
In addition to these seals, small rubber grommets may be found. These will have been fitted to block 'machine holes' that are used in the manufacture of the machine.
If these come out or leak, they should be replaced or have sealant applied to them. If the large tub seals leak, they should also be renewed *(see Bearings, pages 129-141)*.

Box 9

Any corrosion or flaking of the enamel covering of the outer tub can be treated with a good brand of rust inhibitor. Take care that it does not come into contact with any of the internal rubber hoses or seals and that you follow the manufacturer's instructions. Places to note are where the brackets for the motor and the suspension are welded on to the outer tub. These are stress points where the enamel may crack and flake, allowing rust to form. By the time a leak has started at these points, it is too late to save the outer tub, and it must be renewed completely if a lasting repair is to be made. Check plastic outer tubs closely for cracks. Generally speaking, it is very rare for the tub to need renewal on today's modern automatics. In addition, the cost of such items and all the relevant seals and parts necessary to complete such a repair may not be cost effective.

If the machine still leaks after these checks, refer to the following sections: *Plumbing basics (pages 26-29)* and *Motors (pages 110-122)*.

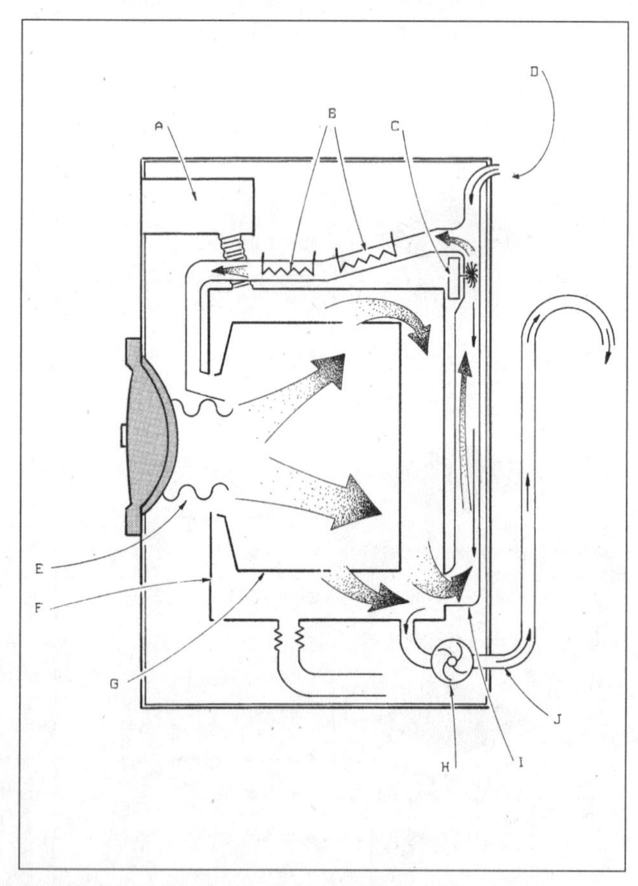

Simplified diagram of a combined washerdrier
A Soap dispenser
B Heaters housed in heater ducting
C Fan and motor for air circulation
D Cold water inlet to condenser unit
E Door seal with inlet for heater duct
F Outer tub unit
G Inner drum
H Outlet pump
I Condenser unit
J Outlet hose

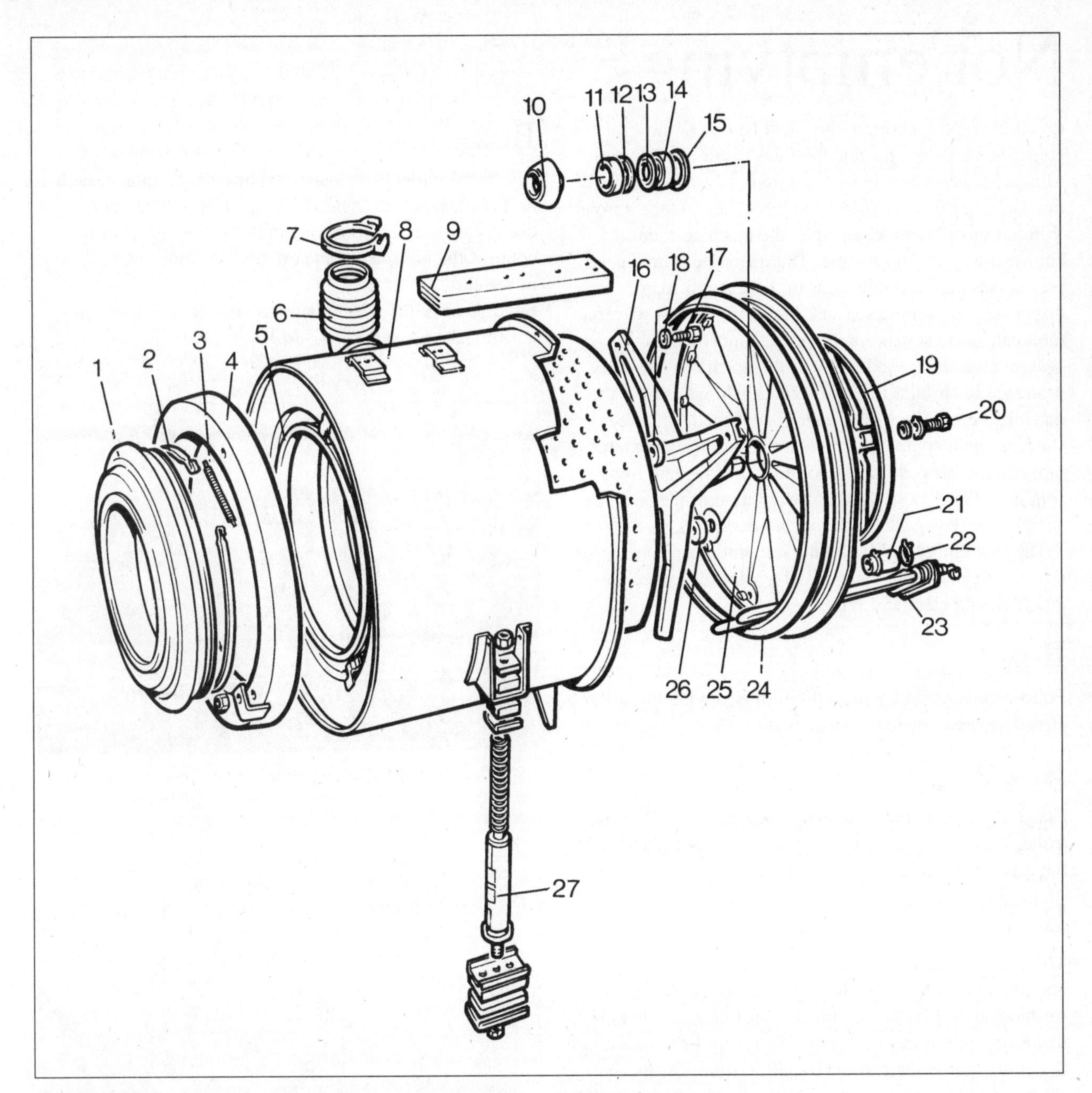

Exposed view of typical basic drum assembly. For additions to this basic arrangement, refer to accompanying schematic drawing and Drying components (pages 148-157).

1 Door seal (door boot)
2 Clamp band
3 Spring fastening for clamp band (could be bolt or rubber garter type)
4 Front tub weight (size and position may vary)
5 Tub lip on to which door seal locates
6 Dispenser hose
7 Dispenser hose clip
8 Outer tub (position of tub weights varies)
9 Top tub weights (locations vary)
10 Carbon face seal for main bearing
11 Front drum bearings (ball bearing type)
12 Front bearing spacer
13 Rear bearing spacer
14 Rear bearing (ball bearing type)
15 Securing clip (spring type)
16 Drum spider (detachable version)
17 Spider fixing bolt
18 Spider washer and spacer
19 Drum pulley
20 Drum pulley fixing bolt and locking tab
21 Pressure vessel hose
22 Pressure vessel hose clip (spring combination clips)
23 Heater
24 Rear tub seal (back half seal)
25 Back half casting
26 Thermostat grommet
27 Suspension unit (spring damper type)

Not emptying

Of all the faults reported, this must be one of the commonest. Often this and the 'leak' fault are the same. This is because there is no 'spin inhibit system' on many machines, particularly older models or ones made abroad. Without this system, even when the machine cannot empty, it will still try to spin. The increased drum speed pressurises the inner tub, causing leaks from soap dispensers, air vent hoses and door seals. In this way, one fault can cause several problems. On most recently designed machines and many of the machines currently produced in Britain, the level switch inhibits (stops) the machine before the spin if water is detected. If there is water in the machine, the pressure causes the switch to move to the off position, preventing the machine from spinning. For a more detailed description of the pressure switch, see *Water level control (pages 63-68).*

The 'not emptying' fault falls into three main categories. These are categorised as blockage, mechanical fault and electrical fault.

Box 1

Follow the emergency procedure for removing the water already trapped inside the machine.

Box 2

Check the outlet and sump hoses and the outlet filter if fitted. If there is a blockage or a kink, remove it and refit the pipe(s) and filter.

Box 3

The pump is located at the machine end of the outlet hose and junction of the sump hose. Check the small chamber for blockages. Check the impeller for free rotation and to make sure that it has not come adrift from its mounting to the pump motor shaft. If the impeller is adrift from the shaft, no water will be pumped although the motor itself will run. A quick way to check the connection of the shaft and impeller is to hold the shaft while trying to turn the impeller. If all is well, they will turn only in unison. Remember to turn anticlockwise, otherwise the impeller will unscrew from the shaft. If a fault is found at this point *(see Pumps, pages 69-74).*

Box 3a

If none of the faults in Box 3 are found and the bearings are not suspected, check the stator continuity of the pump windings *(see Electrical circuit testing, pages 23-25).*

Box 4

Check the outlet hose again. An internal blockage, such as a coin or button, can act like a valve and be very difficult to see. The best method of checking it is to connect the hose to a standard tap and observe the flow of water.

Finally check the wiring harness connections *(see The wiring harness, pages 142-143).*

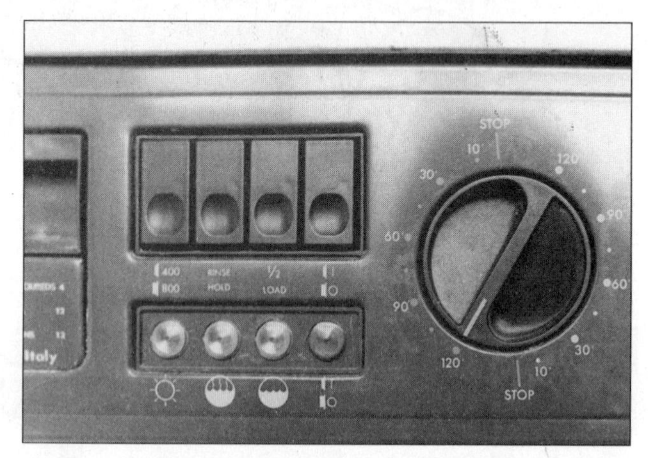

This vented combined machine has a rinse hold button second from the left. It may be correctly causing a spin inhibit; check the manufacturer's instructions for correct operation and use.

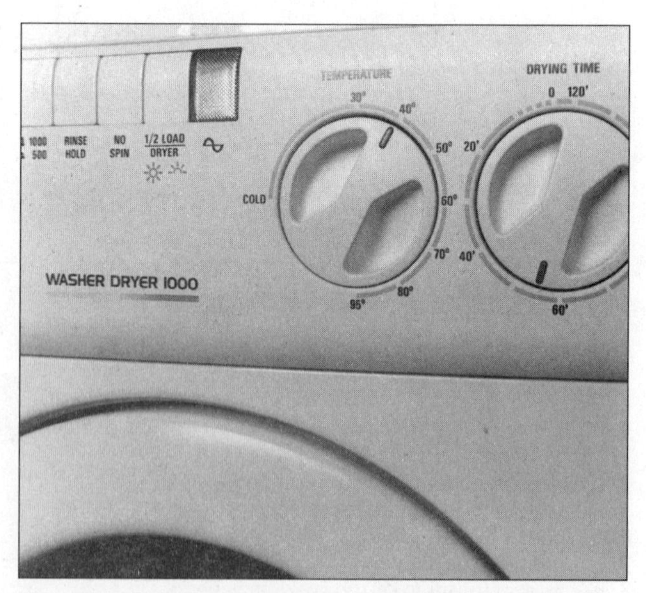

This combined condenser washerdrier clearly indicates the no spin option – third button from the left. However, errors in setting can still occur; always double-check.

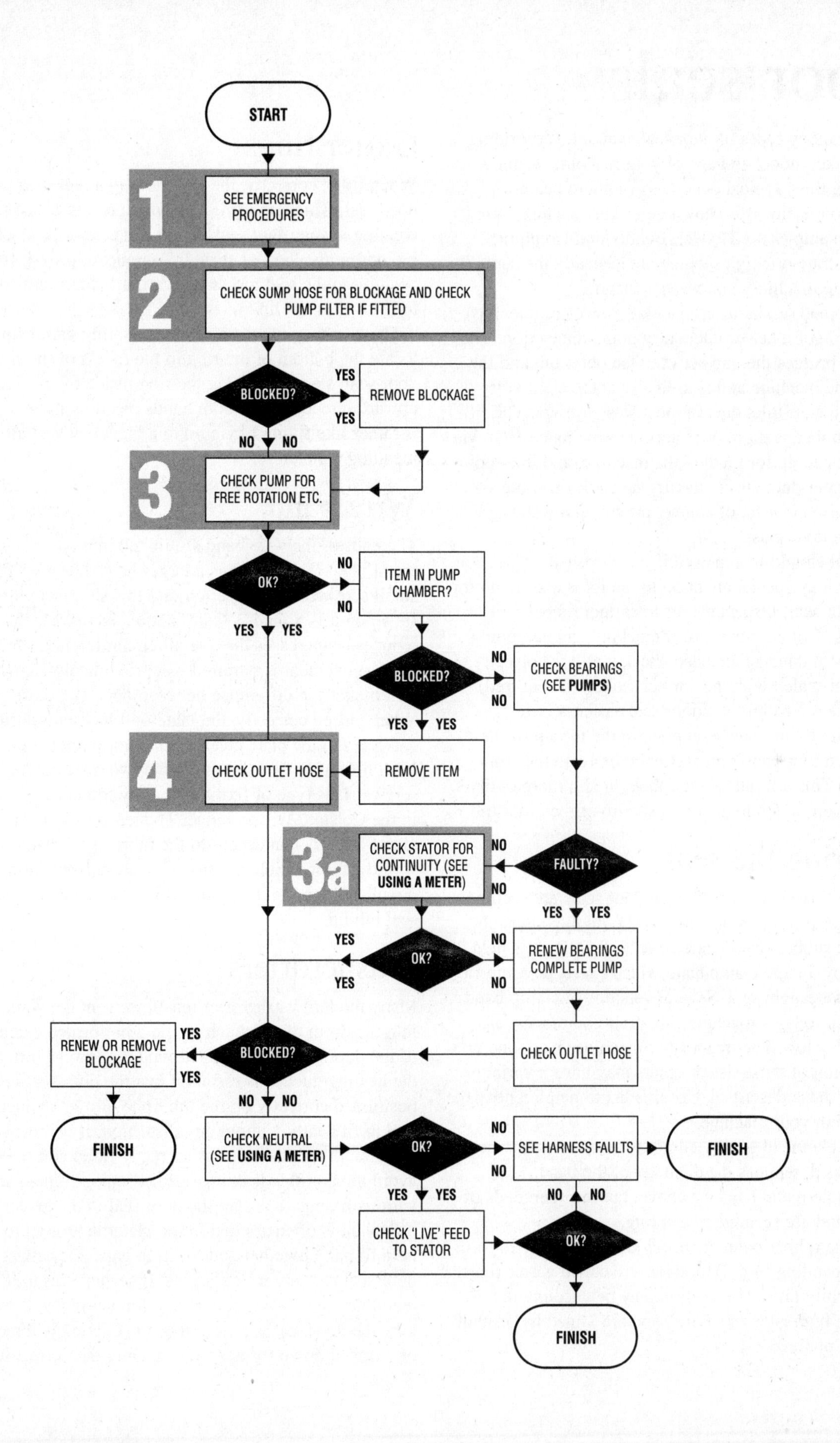
START
1
SEE EMERGENCY PROCEDURES
2
CHECK SUMP HOSE FOR BLOCKAGE AND CHECK PUMP FILTER IF FITTED
BLOCKED?
YES
REMOVE BLOCKAGE
YES
NO
NO
3
CHECK PUMP FOR FREE ROTATION ETC.
OK?
NO
ITEM IN PUMP CHAMBER?
NO
YES
YES
BLOCKED?
NO
CHECK BEARINGS (SEE PUMPS)
NO
YES
YES
4
CHECK OUTLET HOSE
REMOVE ITEM
3a
CHECK STATOR FOR CONTINUITY (SEE USING A METER)
NO
FAULTY?
NO
YES
YES
YES
OK?
NO
RENEW BEARINGS COMPLETE PUMP
YES
NO
RENEW OR REMOVE BLOCKAGE
YES
BLOCKED?
CHECK OUTLET HOSE
YES
NO
NO
FINISH
CHECK NEUTRAL (SEE USING A METER)
OK?
NO
SEE HARNESS FAULTS
FINISH
NO
YES
YES
NO
NO
CHECK 'LIVE' FEED TO STATOR
OK?
FINISH

Door seals

There are many types of door seal available, depending on the make, model and age of your machine. Removing and fitting three typical door seals (without heater ducting connections) is shown here. The machines used in these examples are Hoover, Bendix and Hotpoint. They illustrate principles which are basically the same for all washing machines and washerdriers.

A combined washerdrier machine, however, may have the addition of a heater duct entry point connection. The door seal bridges the gap between the outer tub and the shell of the machine and is used as a convenient entry point for the air inlet duct in most washerdriers. The flexible rubber seal provides access to the inner drum via the door opening for loading the machine, and inlet point for the heater duct during the drying cycle on some machines, while also, of course, providing a watertight seal to the door glass.

The seal should be renewed if it is perished or holed at any point. Pay special attention to the folds and mounting of the door seal. Inspect the air inlet duct rubber moulding, if fitted, for signs of cracking and hardening through heat damage and also check that it is correctly seated and sealed with the correct sealant (if required).

To make the fitting of a door seal easier, a little washing up liquid may be applied to the tub lip or the door seal tub lip moulding. (Do not apply on the front shell lip.) This will allow the rubber to slip more easily into position on the metal or plastic lip of the outer tub.

Securing the seal

There are three similar ways that door seals are secured to the outer tub. There is a formed lip on the outer tub. When the rubber seal is located on to this lip, it is held in position by a large clamp band which exerts pressure to create a watertight seal. Several versions of clamp bands are used in today's machines; the four commonest are described below. The majority of machines have one or a combination of those listed; others may have a variation of one of those described. Use this list to help identify the type used in your machine.

- A simple metal band secured by a bolt which, when tightened, reduces the diameter of the band.
- A simple metal band, as above, but the open ends of the band are secured by a spring.
- A large rubber band or spring joined together to form an expanding ring. Both types are called garter rings.
- A slightly unusual fastening may be encountered where both ends of a wire band are joined by a small metal plate.

Garter rings

When fitted correctly, the ring rests in a recess in the door seal, which, in turn, rests in the recess of the tub lip, creating a watertight seal. This band cannot be slackened by loosening a bolt or spring. To remove, prise it from its position in the seal recess with a flat-bladed screwdriver to lift it over the lip.

The best way to refit a rubber or spring garter ring is to locate the bottom of the ring in the recess of the fitted door seal, slowly working the ring inside the recess in an upward direction with both hands meeting at the top. This is rather like fitting a tyre on to a bicycle wheel after mending a puncture.

Wire band

The ends of the wire band are linked into two holes in a metal plate. The plate has a larger hole/slot in it. If a small screwdriver is inserted into this slot, forming a tight fit, when it is turned, a 'cam' action occurs which reduces or increases the overall circumference. Only a small movement is required – approximately a quarter turn makes the difference between open and closed. When closed correctly, the plate will lock into position.

Access to the plate can be gained through the door latch hole after first removing the two interlock fixing screws. This type of fixing is mainly found on machines in the Colston/Ariston range. The reason for this type of clamp band is that access to the more usual clamp band would be impossible because of a close front fitting circular concrete tub weight which surrounds the door seal tub lip.

Washerdriers

Many modern washerdrier machines vent the warm air into the drum of the machine via a preformed extension of the door seal, which fits around the lower portion of the heating duct. It is essential that the door seal is positioned correctly on the tub lip to allow the duct and seal to fit neatly without any distortion. It is most important that this union is correctly fitted and sealed to avoid the possible leak of water or vapour. There are various fixings to secure the door seal to the heater duct.

Sealant is often applied in the manufacturing process to help fit parts together and to fill in gaps left between the door seal and any irregularities. If sealant was used in the manufacture of your machine, you must renew it whenever the door seal is renewed or the ducting is removed or stripped down for any reason. Only the correct heat-

A

B

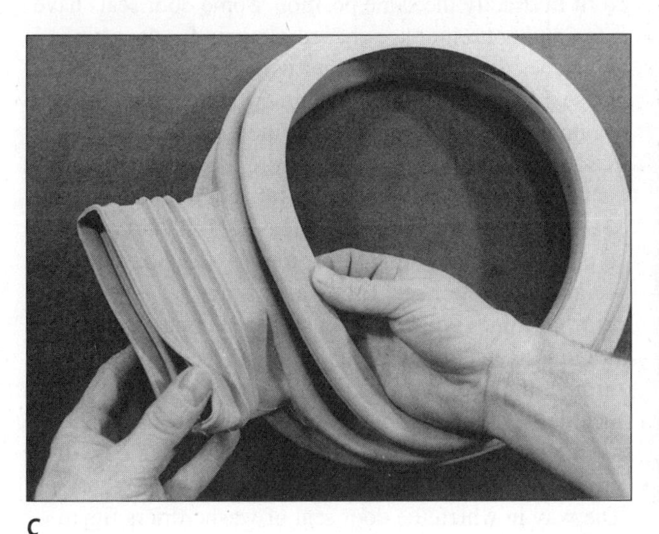

C

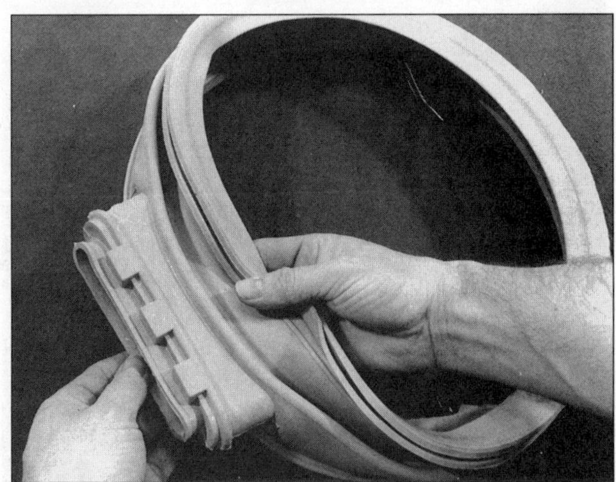

D

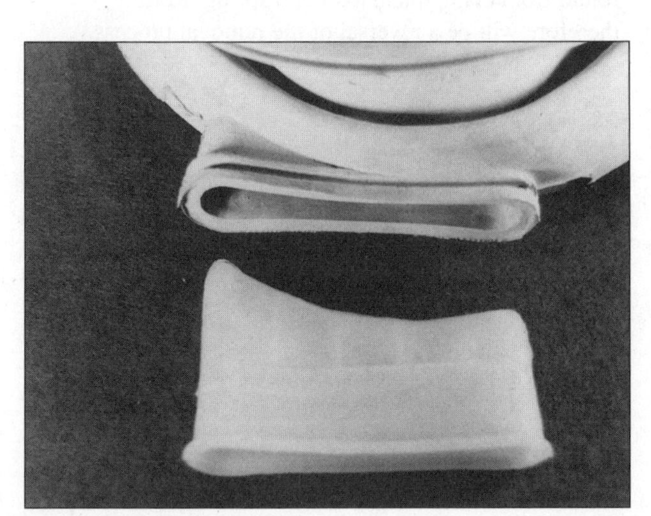

E

Washerdrier door seals with heater duct inlet protrusions
A Fagor
B Ariston
C Hoover
D Candy (with removable integral duct seal)
E Candy seal detailing duct seal

resistant sealant should be used *(see Useful tips, pages 168-171)*. Do not use the kind of general-purpose sealant designed for watertight seals on hoses and pressure systems, because this is not adequate.

Fitting the door seal to the shell of the machine is similar to the tub lip system. Three major variations are found. The most straightforward seal simply grips the shell lip with no additional support other than the elasticity of the door seal itself. For safety reasons modern machines now use various methods of secondary fixing to prevent the door seal being easily removed or dislodged from the shell lip. It is essential that such fixings are replaced if they have once been removed.

Not all washerdrier machines use the door seal for the heater duct entry point. This early Hotpoint machine has a normal door seal; air inlet is via a fixed outer tub fitment.

The second type of seal also uses the shell lip, with the aid of a clamp band. A recess is formed on the outer edge of the seal for a clamp band or clamp wire to be inserted. This ensures a firm grip on the shell lip. As with the inner tub lip, many variations can be found.

The third method involves a plastic flange screwed to the outside of the shell. The screws holding it pass through a recess in the outer front lip of the door seal and so secure it firmly to the front panel.

Variations of both of the later clamps can be found on modern machines. All are fitted as a safety measure to restrict access to the inner of the machine. If it has had to be removed for any reason, it is essential that any such fixing is replaced correctly. Before removing the old door seal examine it and note its correct positioning to aid the subsequent repair and renewal as the new seal will need to fit in exactly the same position. Some door seals have a top and bottom or pre-shaped sections for door hinges or catches and will fit no other way. It is easier to line up the seal before fitting rather than try to adjust it when the clamp bands have already been fitted.

Some machines have one or more tub weights mounted on the front section of the outer tub, encircling the door seal and tub clamp fitting, and leaving little or no access to the clamp or tub lip. Do not remove the front tub weight to aid door seal fitting as it is not necessary for most machines – Creda, Zanussi, Ariston and Bendix, for example. Zanussi and Creda seals can be changed through the door opening in the front panel of the machine, without removing the top at all. However, having said this, removing the top of the machine will provide more light, so you can remove it if necessary. The way in which the door seal of washerdriers fits to both outer tub and shell are identical to wash only machines, that is, lip and clamp. Fitting however will be found from clamp plate to clamp spring. Refitting therefore will be a reversal of the removal process.

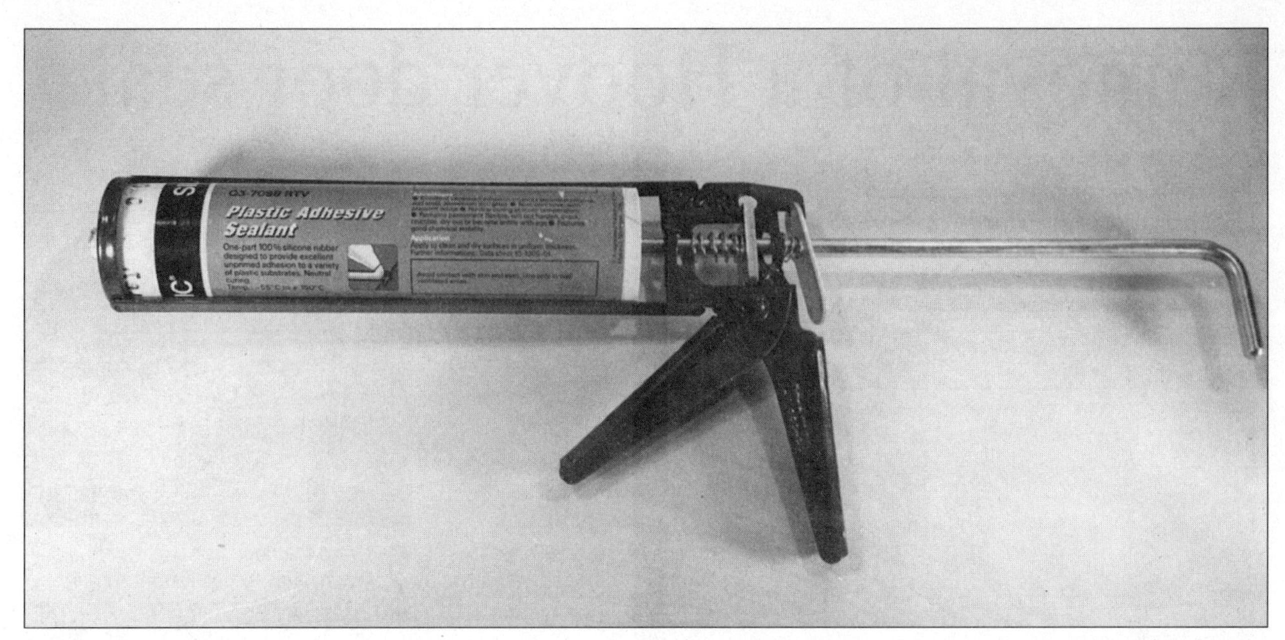

This type of sealant is sometimes used to secure the door seal to heater duct inlet. If it was used originally, it must be renewed if disturbed or when a new door seal is fitted.

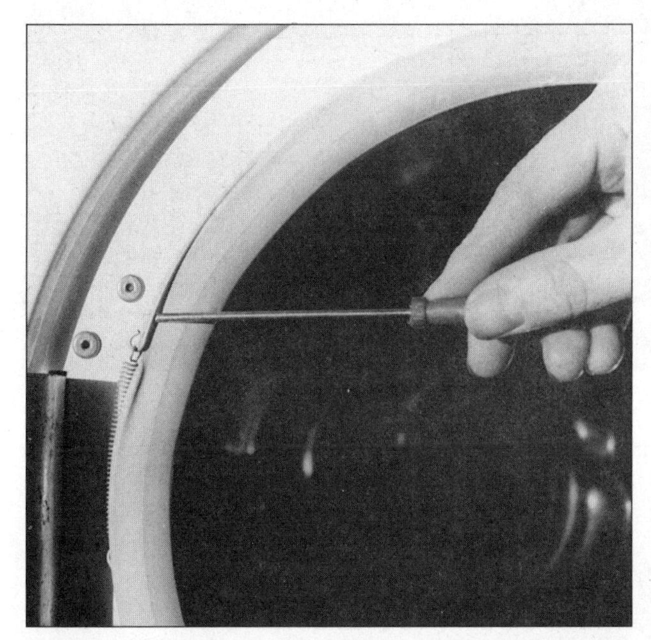

The front lip of this door seal is secured by a wire and spring type fixing. Carefully insert a flat-bladed screwdriver to ease the fixing from its position (the door has been removed for photographic purposes only).

When removed from the shell of the machine this Zanussi washerdrier tub assembly clearly shows the large tub weight completely encircling the door seal. This type of seal is held in place by a garter spring. Removal and refitting of both seal and securing band can be carried out with the unit in situ. The unit is shown removed from the machine for photographic purposes only.

Removal of a Hoover door seal

Inspect door seals regularly for signs of damage, cracking and hardening. Replace as necessary.

TOOLS AND MATERIALS

- ☐ Non-abrasive pad
- ☐ Door seal

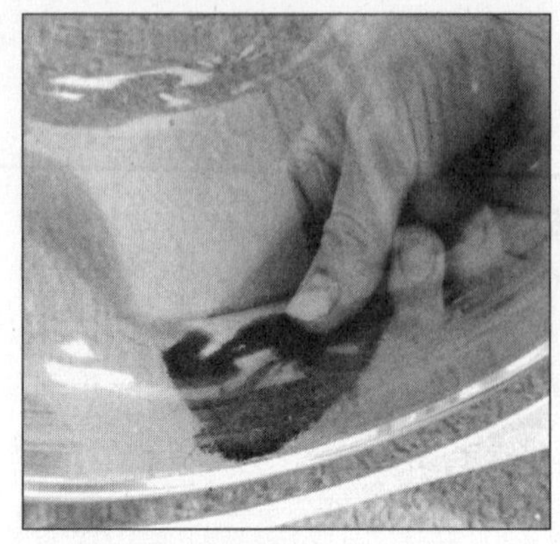

1 Check the inside of the door glass for scale deposit ridge and clean off with a non-abrasive pad.

2 Grasp the door seal firmly and pull downwards to free it from shell lip. Some machines may have clamp band on the front lip; in which case, remove this first.

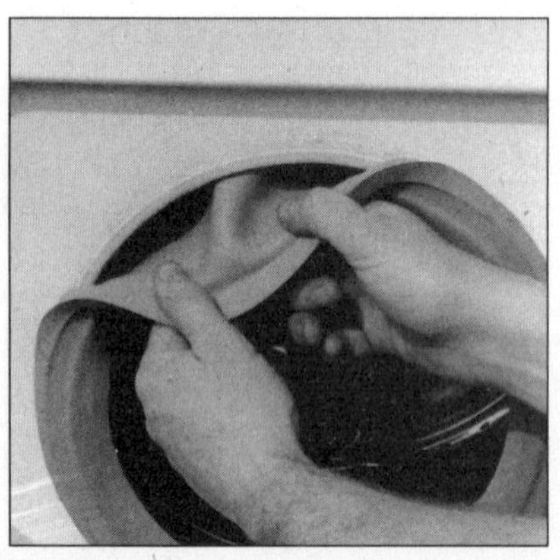

3 Once it has been freed from the lip, continue pulling downwards.

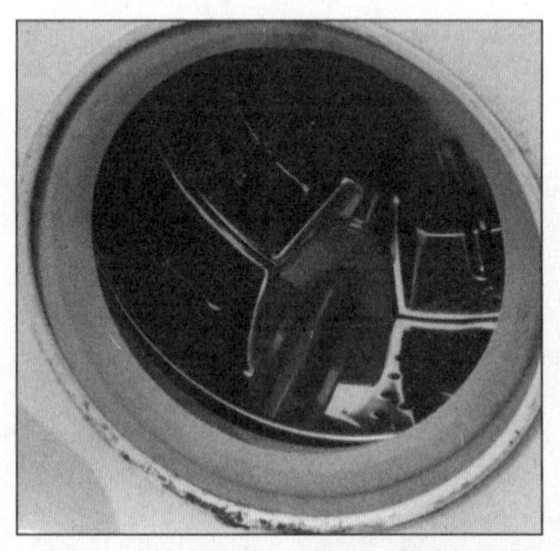

4 Free the complete door seal from the front lip and allow it to rest on the inner side of the front panel.

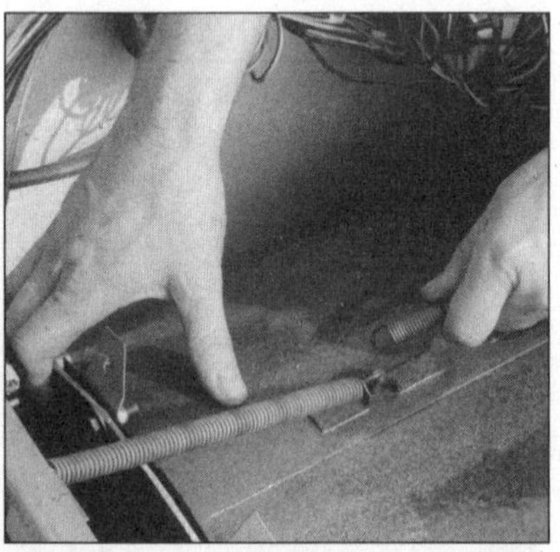

5 With the top removed, free the top support springs or tie and lean the outer tub unit back as far as possible.

6 With the position of the clamp band and bolt in view, remove the band. Free the seal from the clip.

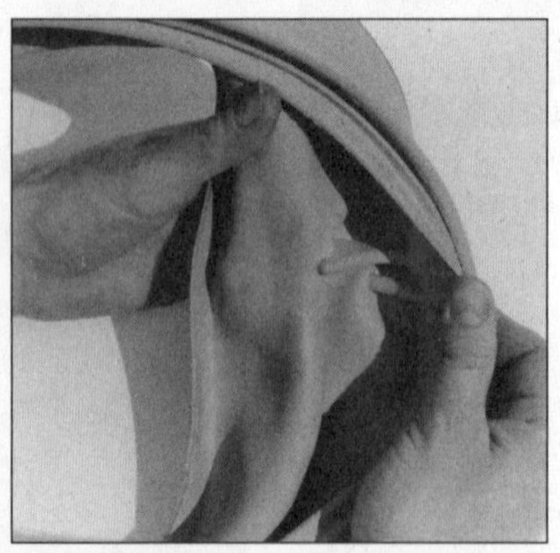

7 Fit the ridge at the 9 o'clock position when viewed from the front of the machine.

Removal of a Bendix door seal

Inspect door seals regularly for signs of damage, cracking and hardening. Replace as necessary.

TOOLS AND MATERIALS

- ☐ Non-abrasive pad
- ☐ Screwdriver
- ☐ Door seal

1 Remove the plastic flanges around the door seal front.

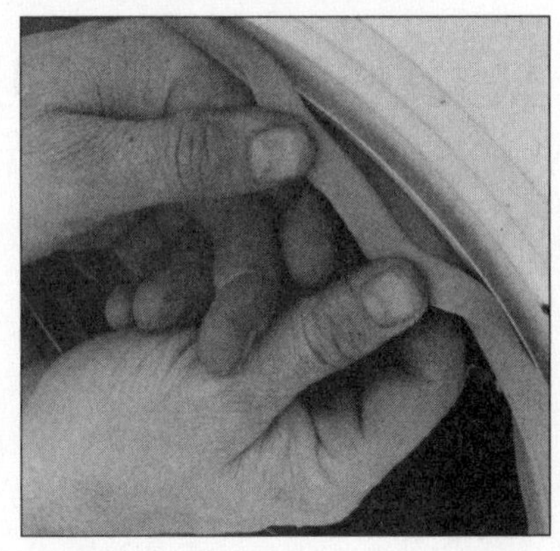

2 Remove the door seal from the front lip of the shell.

3 The position of the clamp band is obvious when the top of the machine has been removed.

4 Note the orientation of the band inside the machine. It must be refitted in the same position.

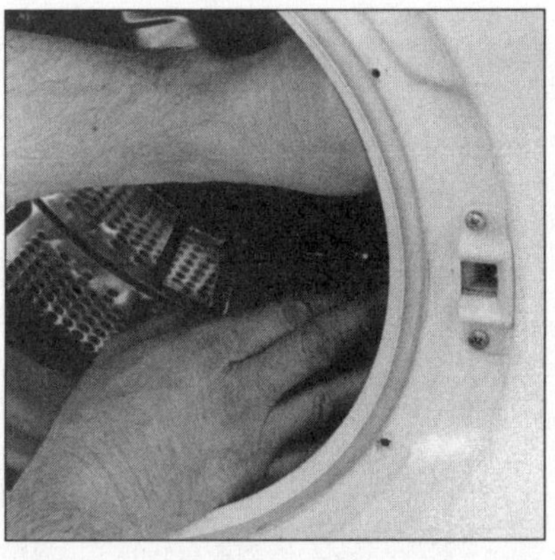

5 Slacken off the tub clamp band bolt and remove old door seal.

6 With a non-abrasive pad clean off any scale and/or deposit on the tub lip before fitting the new seal.

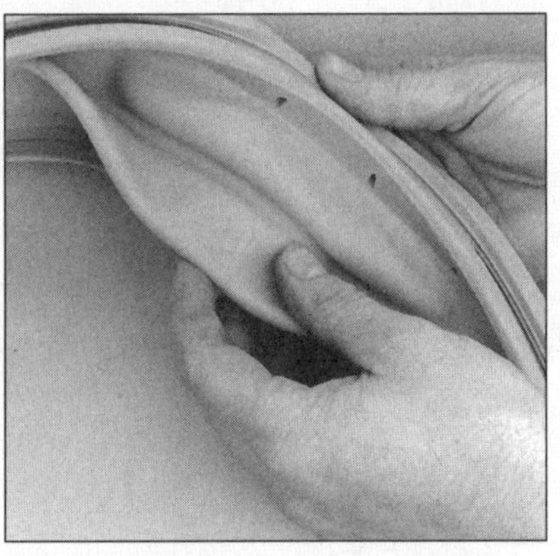

7 Check the new seal before fitting and smear the inner lip with a little washing up liquid to help slide it into position.

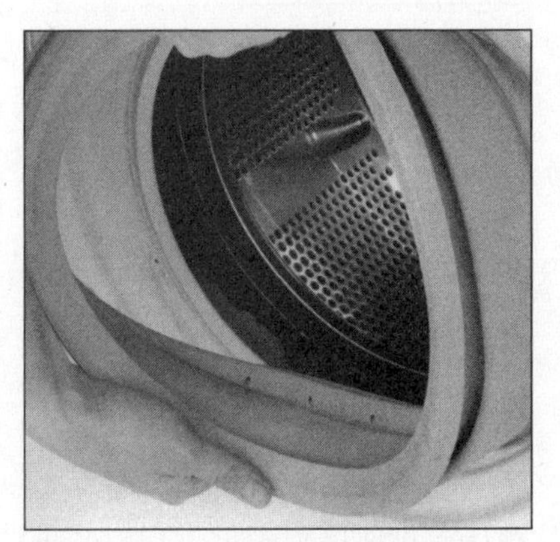

8 Ensure that the three drain holes on the door seal are fitted at the bottom of the tub lip.

Removal of a Hotpoint front loader door seal

Inspect door seals regularly for signs of damage, cracking and hardening. Replace as necessary.

TOOLS AND MATERIALS

- ☐ Screwdriver(s)
- ☐ Door seal
- ☐ Washing-up liquid

1 Remove the outer plastic surround screws and top and bottom section.

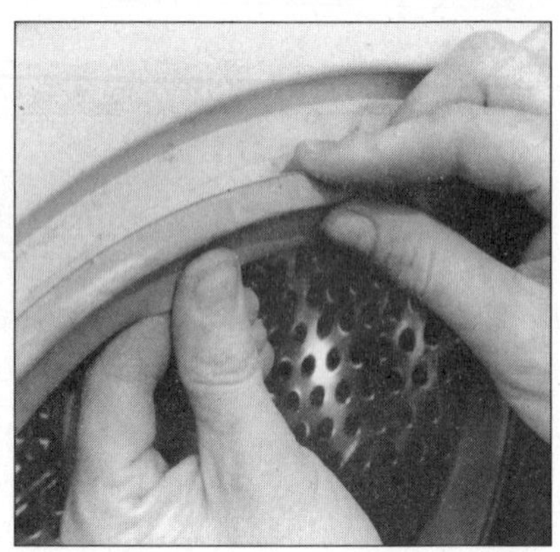

2 Pull the seal to release from shell lip, hinge and catch.

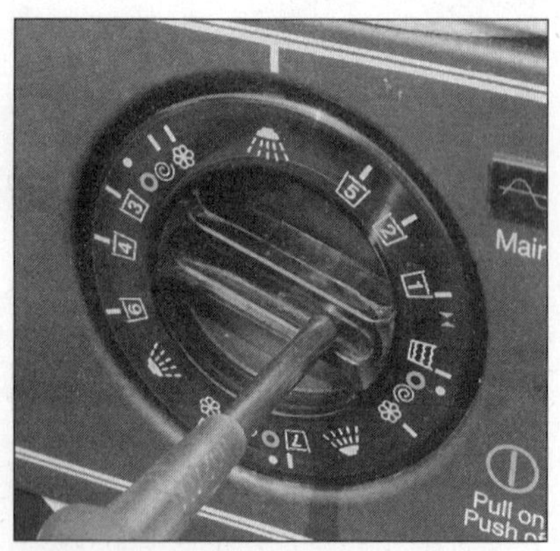

3 Unscrew the timer knob centre and remove the timer knob. Also remove the two front fascia fixing screws found behind the timer knob.

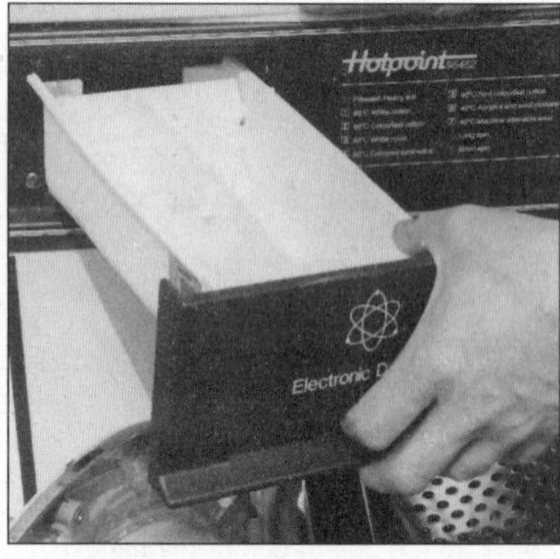

4 Pull out the soap dispenser drawer completely and remove the front fascia fixing screws.

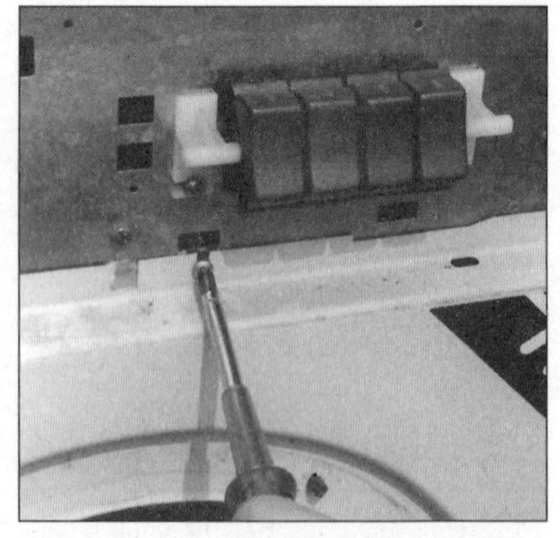

5 With the front fascia removed, remove the screws securing the front panel of the machine.

6 Four hexagonal headed screws secure the front panel under the bottom edge. (On later machines, three Philips headed screws will be found).

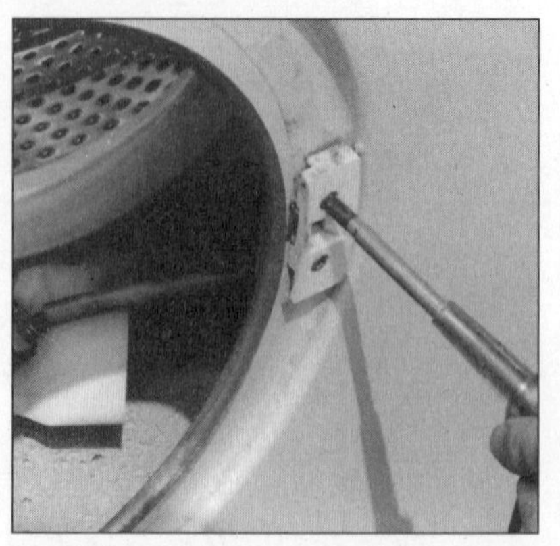

7 Remove the door switch assembly and pressure switch bracket.

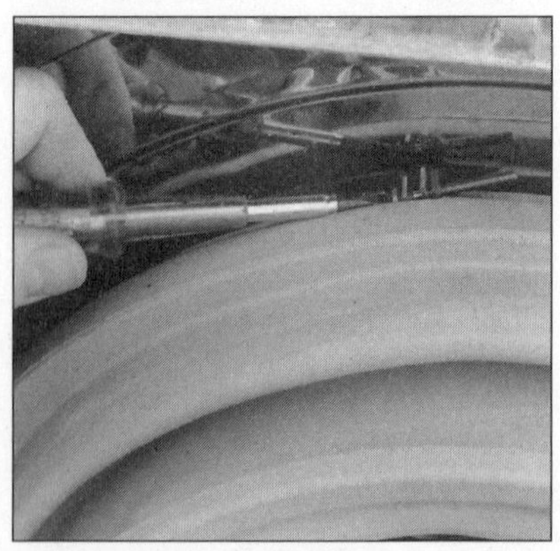

8 With the front panel removed, you can easily remove the clamp band.

9 Note the position of hinge and catch mouldings. Pull the door seal free from the tub lip.

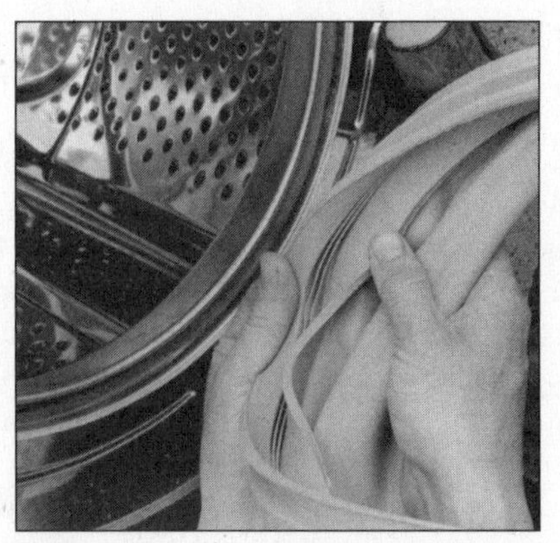

10 When you are fitting a new seal to the tub lip, the tub gap can be adjusted slightly. (Note: the inner lip of the door seal is ribbed.)

11 When the new seal is fitted in this position, check that the inner drum rotates without fouling the door seal inner. Adjust if necessary to obtain the smallest gap possible before refitting.

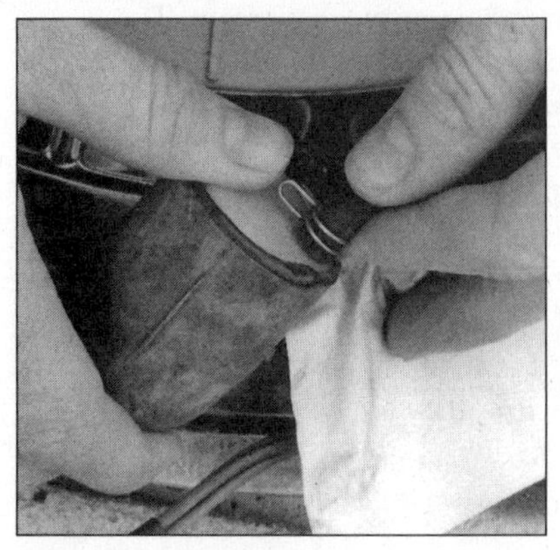

12 With new door seal in this position check that all of the leads, hoses and pressure vessel are correct before refitting the front panel.

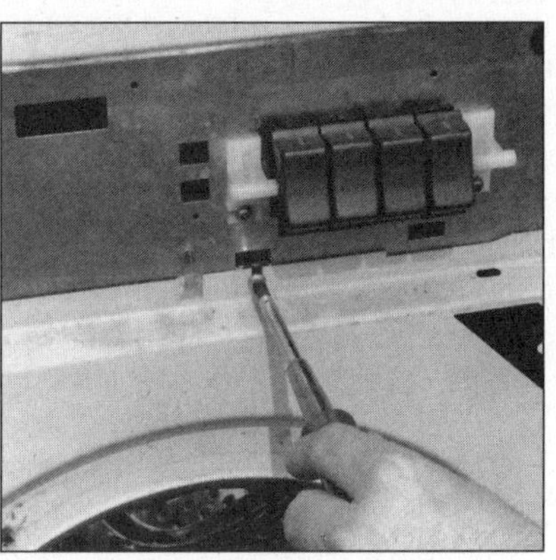

13 Refit the front panel, door catch and all front panel fixings.

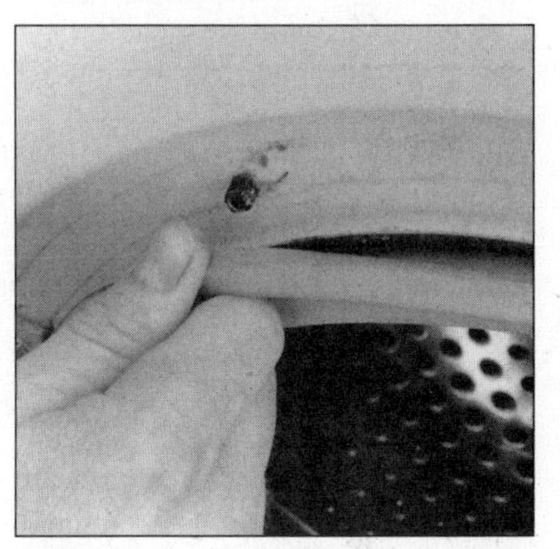

14 With the front panel secured, fit the door seal to the front panel lip.

15 Lubricate the hinge and catch points with a little washing-up liquid and ease into position.

16 Refit the plastic surround and ensure that the ends locate correctly. The plastic pips can be moved to aid fitting. If fitted correctly the new seal should not have undue kinks or twists.

Water inlet valves

Several of the most commonly reported faults are not taking powder, not filling at all, not filling in certain parts of the programme and, additionally, condenser water supply problems with some washerdriers only.

Many configurations of water valve can be found, from single hot or cold to much larger units consisting of three or more individually operated valves grouped together to control the flow to several outlets from one inlet. All inlet point threads are the same size, but outlet hose connections from the valve may differ. A wide variety of fixing brackets are also used, outlet angles can be in-line (classed as 180 degree valves) or angled downwards (classed as 90 degree valves). The general operation of the electro-mechanical action is described here and covers all such valves regardless of size or fixing styles.

A solenoid coil of some 3,000-5,000 ohms (3-5KΩ) resistance, when energized (supplied with power), creates a strong magnetic field at its centre. This field attracts a soft iron rod or plunger up into the coil, holding it in that position as long as power is supplied to the coil. When power is removed from the coil (de-energized), a spring at the top of the plunger recess returns to its resting position.

How the valve works

The two states of the water valve are shown here in detail. With no power supplied to the solenoid coil (A), the soft iron core (B) is pressed firmly on to the centre hole of the flexible diaphragm by spring (C). As chamber (D) is only at atmospheric pressure and the water is at least 0.3kg/cm^2 (4 psi) – somewhat higher – pressure is exerted on the top of the diaphragm, effectively closing it tight. The greater the water pressure, the greater the closing effect of the valve. Therefore, no water will flow.

The pressure on top of the diaphragm is via a small bleed hole (E). It is essential that this very small hole is not obstructed. Although very small, it is a major factor in the correct operation of these types of pressure-operated valves.

When power is supplied to the solenoid coil, the resulting magnetic attraction of the coil overcomes the power of the spring (C) and pulls the plunger up into the coil centre. This allows an imbalance of pressure to occur by exposing the centre hole of the diaphragm. The imbalance lifts the flexible diaphragm and allows water to flow into chamber (D), so water flows.

It is easier for the water to lift the diaphragm than to balance the pressure by flowing through the very small bleed hole. Any enlargement or blockage of this essential bleed hole will immediately render the valve inoperative.

There are several benefits to such valves. The higher the pressure supplied to it, the tighter the valve will close, the cost is relatively low, they are very reliable and are simple to change if faulty.

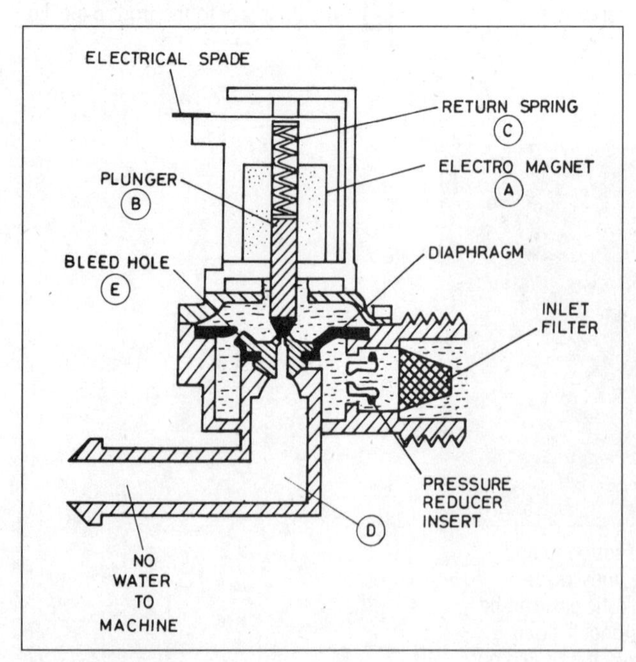

De-energized valve.

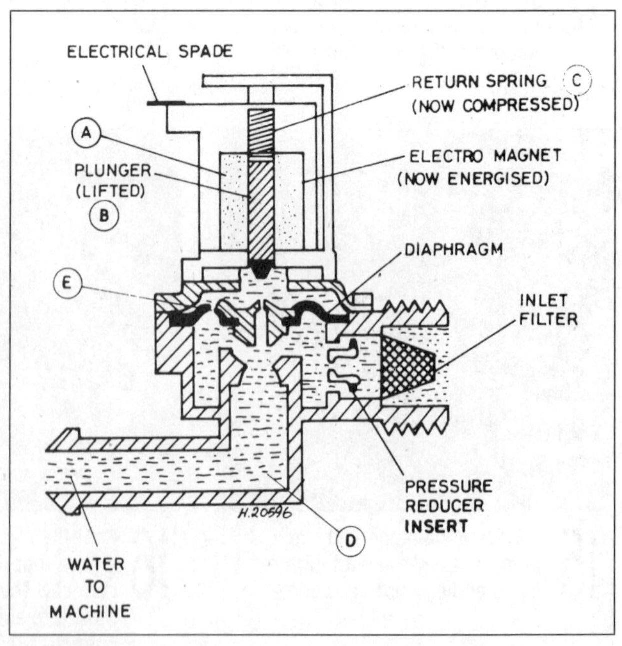

Energized valve.

Typical faults

- As with ordinary house taps, the valve seat may wear and allow a small trickle of water to pass, even when de-energized. If the taps are left turned on over a long period of time, this will cause the machine to fill when not in use, resulting in an overfill and possible flood.
- When de-energized, the valve may fail to allow the plunger to return to its normal resting/closed position. This problem will cause severe overfill and flooding. Turning off the machine will not stop the overfilling. Complete isolation of both power and water supply is required and renewal will be necessary.
- The valve fails to allow water to flow due to open circuit in coil winding *(see Electrical circuit testing, pages 23-25)*. The valve fails to allow water to flow due to a blocked filter on its inlet. Carefully remove and clean. Do not allow any particle – no matter how small – to escape past the filter as it could block the bleed hole.

Types of valve

Water valves come in many sizes and an assortment of shapes – single, double and triple valves or a combination of all three. On the double and triple valves, each solenoid operates one outlet from a common inlet. Unfortunately, a fault on one coil or one outlet will generally mean a complete renewal of the whole valve assembly, as individual spare parts are not available.

On condenser washerdriers with double or triple valves, a restrictor is fitted to one of the valve outlets. It is important that any replacement valve has this internal restrictor fitted. The valve with the restrictor is for water supply to the condenser unit, as it requires a slow trickle of water to operate correctly *(see Drying components, pages 144-153)*.

Condenser washerdriers and some condenser dry-only machines require a cold water supply during the dry cycle. Without it, the machine will not dry. The flow of water from the valve is restricted to slow it to around 35ml (1½ fl oz) per minute during the drying sequence. The water flow is restricted by a plastic insert in either the valve outlet to the condenser or at the receiving end within the condenser unit itself. The insert has small holes through which the water can pass at a given rate.

If the restrictor becomes blocked, the water flow will become too slow for the condensing action to take place. This is a fairly common fault, especially in machines with restrictors or spray points mounted in the condenser unit. Such problems are often related to water hardness.

If the restrictor is removed, cracked or dislodged, the flow of water may become too great, thus preventing

Single valve: Red for hot supply. White for cold supply.

Double valves, 90 degrees and 180 degrees outlet respectively. Typical use is one side for wash/rinse cycles, the other for restricted condenser supply.

Triple valve configuration. Typical use is pre-wash, main wash and restricted condenser supply.

Double valve clearly showing restrictor in left-hand outlet used for condenser unit supply.

drying. This results from a water build-up within the machine between the periodic pump action, or water droplets being picked up by the airflow and deposited on the clothes. As the machine uses separate valves, that is, one for normal cold inlet for pre-wash, wash and rinsing and the other restricted for dry cycle only, the washer-drier may fill and, of course, operate correctly for wash cycles, but fail to dry.

When obtaining replacement valves, make sure that if the original valve had a restrictor, the new one also has a corresponding restrictor fitted to the correct outlet. In common with all valves and hose connections, make sure that they are secure and watertight . This is especially important with machines fitted with restrictors at or within the condenser, as pressure can build up within the supply hose (valve to condenser unit) and if not securely fitted, may blow off or weep.

Further details of the condenser unit, its components and specific faults can be found in Drying components *(pages 144-153).*

Verifying a fault

In this theoretical instance, the machine has been loaded and a programme selected but it has failed to fill. Moving the timer/control to a pump out position confirms that power is being supplied and that the door interlock is working *(see Door locks, pages 75-78).*

Box 1

Reselect wash programme to confirm that machine was originally set and turned on correctly.

Box 2

With the machine correctly set, this confirms that, although the machine has electrical feed, no water is entering to begin the filling/washing action.

Box 3

This may seem too obvious to mention, but many an engineer has been called out to find the taps are in the off position. The usual response is that the taps are 'never turned off' and some other devious member of the family or innocent plumber has done the dirty deed! This comment raises the cardinal rule that all automatic washing machines should be turned off at their isolation taps when not in use. This may seem pointless, but the objective is simple. If an inlet pipe should split or an inlet valve fail to close correctly, a quite disastrous flood might occur if the taps are left on.

Box 4

By unscrewing the hose from the valve, you can easily check the water supply by turning the tap to which it is connected on and off, ensuring that the free end of the hose is held in a suitable container. Failure of water flow could be due to a faulty tap or tap shaft or an internal fault of the supply hose. Some manufacturers supply rubber inlet hose seals that have metal or gauze filters moulded into them. They recommend that the two washers supplied with the filters are fitted at the isolation tap end of the supply hose as a first line filter for the valves. Check if such filter washers were used during the original installation by unscrewing the supply hose from the isolation tap. Clean or renew them as required.

Box 5

Checking of the water valve inlet filter can be carried out while the hose is removed for Step 4. Take care not to allow any particles to escape past the fine mesh filter and into the valve. Carefully clean the filter of all scale and debris and refit.

The filter can be removed by gently gripping the centre with pliers and pulling it free of the main valve body.

Box 6

Ensure that the water supply to the valve is adequate to operate the valve *(see Plumbing basics, pages 26-29).*

Boxes 7 and 8

If the heater is on when there is no water in the machine, a pressure system fault is indicated and should be checked *(see Water level control, pages 63-68).* If the heater is in the off position when there is no water in the machine, the valve is suspect. The valve is easily changed by removing the fixing screws and detaching the internal hose(s) from the valve. Make a note of the wiring and hose connections that are on the valve, remove them and replace with a new valve assembly by simply reconnecting the hoses and wires in a reverse.

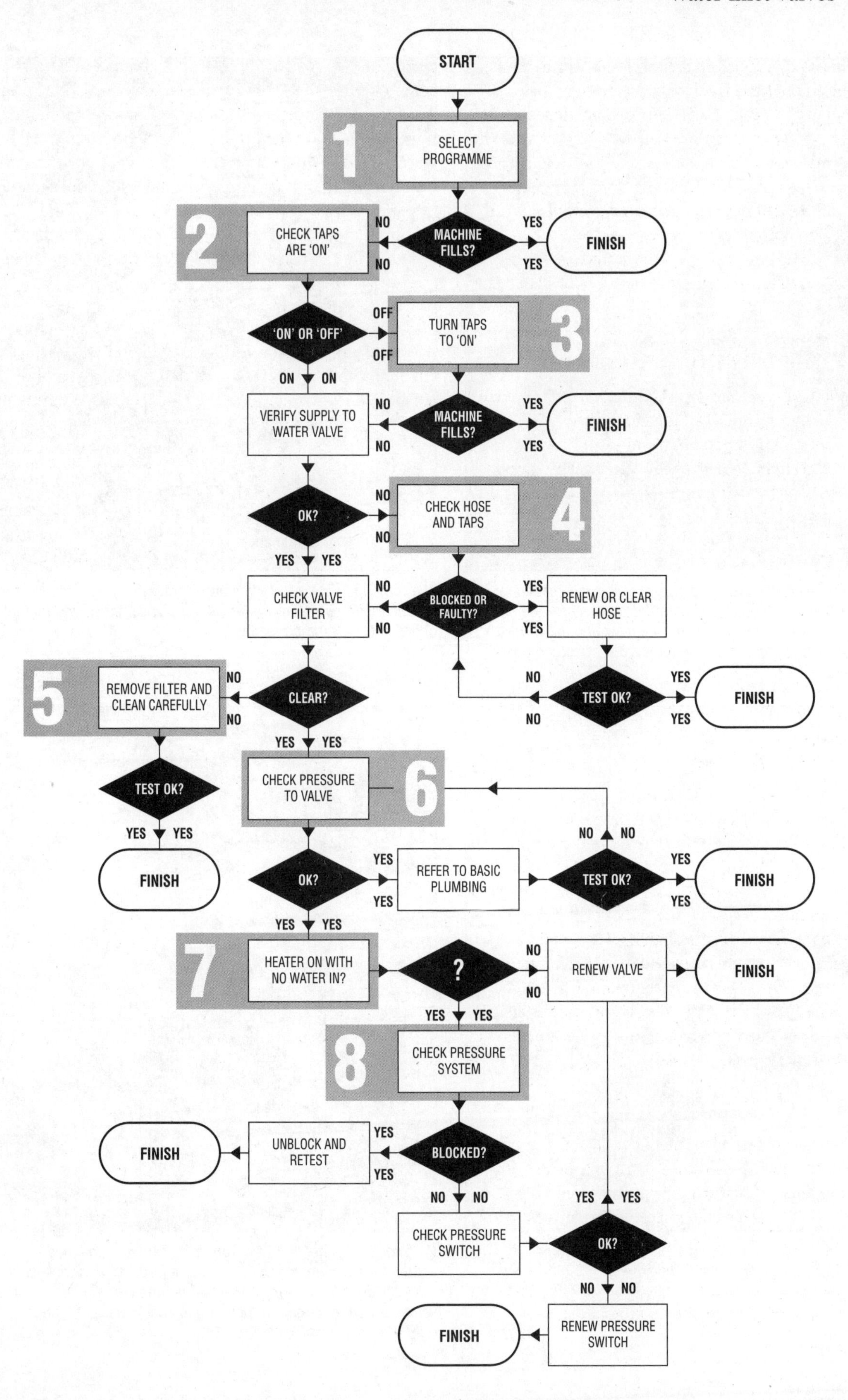
START
1
SELECT PROGRAMME
MACHINE FILLS?
YES
FINISH
NO
2
CHECK TAPS ARE 'ON'
'ON' OR 'OFF'
OFF
3
TURN TAPS TO 'ON'
ON
VERIFY SUPPLY TO WATER VALVE
MACHINE FILLS?
YES
FINISH
NO
OK?
NO
4
CHECK HOSE AND TAPS
YES
CHECK VALVE FILTER
BLOCKED OR FAULTY?
YES
RENEW OR CLEAR HOSE
NO
TEST OK?
YES
FINISH
NO
CLEAR?
NO
5
REMOVE FILTER AND CLEAN CAREFULLY
TEST OK?
YES
FINISH
YES
6
CHECK PRESSURE TO VALVE
OK?
YES
REFER TO BASIC PLUMBING
TEST OK?
YES
FINISH
NO
YES
7
HEATER ON WITH NO WATER IN?
?
NO
RENEW VALVE
FINISH
YES
8
CHECK PRESSURE SYSTEM
BLOCKED?
YES
UNBLOCK AND RETEST
FINISH
NO
CHECK PRESSURE SWITCH
OK?
YES
NO
RENEW PRESSURE SWITCH
FINISH

With the filter removed, clean and inspect it. If it is damaged, or if you suspect that dirt may have entered the valve, renew the unit. This procedure can be carried out without removing the valve(s) from the machine. Here, it is shown removed for photographic purposes only.

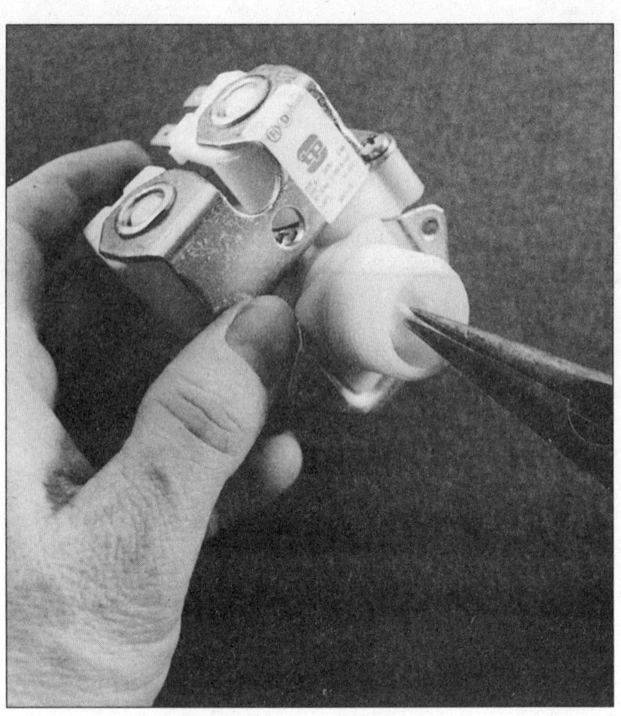

The valve inlet filter can be removed for cleaning by carefully lifting the filter with pliers. Ensure the filter is not damaged and that no debris gets past during removal.

Triple valve, generally used on cold supply, found on some automatic washing machines and dishwashers.

With the rear panel removed from this Hotpoint combined condenser machine, the single hot valve (top left) and triple cold valve (below it) can clearly be seen. The hose configuration from the valves to both detergent dispenser and conditioner unit (on the right) are clearly visible.

Water level control

Modern automatic washing and combined washerdrier machines have several fill levels each of which corresponds to the wash cycle selected. For example, there is a high level fill for delicate programmes and a lower level for more robust wash programmes. There may also be the facility for an intermediate level option if a half load selection is available. The amount of water used for a selected programme is governed by a pressure system.

Location

The pressure switch has no standard location, but is usually positioned at the top of the machine. It can be identified as a large circular switch with several wires and a plastic or rubber tube attached to it leading to a pressure vessel. There are several sorts of pressure vessel.

The pressure vessel may be an integral part of the plastic filter housing located behind the front face of the machine's shell. Alternatively, it may be an independent unit located to the rear or front of the outer tub. There are also pressure hoses that function in the same way as the rigid pressure vessel. These hoses are either grommet fitted to the lower part of the outer tub or directly moulded to the sump hose. Machines with moulded plastic outer tubs normally have provision for a rigid pressure vessel to be mounted on the lower section. The position and style of the switch and pressure vessels vary, but the basic way they operate does not.

How it works

The pressure switch does not come into contact with the water, but responds to air pressure trapped within the pressure vessel or pressure hose. When water enters the tub and the level rises, it traps a given amount of air in the pressure vessel. As the water in the tub rises, this increases the pressure of the trapped air within the pressure vessel. This pressure is then transferred to a pressure-sensitive switch via a small bore flexible tube.

The pressure switch is a large circular device that houses a thin rubber diaphragm which is expanded by the pressure exerted on it. The diaphragm rests alongside a bank of up to three switches, each of which is set to operate at a different level of pressure. Operating in this way, the switch is totally isolated from the water, ensuring maximum safety.

Possible faults

To create the highest pressure in the chamber of the pressure vessel, it must be positioned as low as possible in the machine. Unfortunately, any sediment collecting in the machine is liable to accumulate at this point and may easily block the entrance to the vessel. Similarly, because of its very small internal diameter, the pressure tube can also block. The pressure that this device creates is very small and can be easily blocked by a very small obstruction, such as a small lump of powder or sediment.

This Zanussi machine has a rigid pressure vessel (left-hand side of picture) connected to the sump hose.

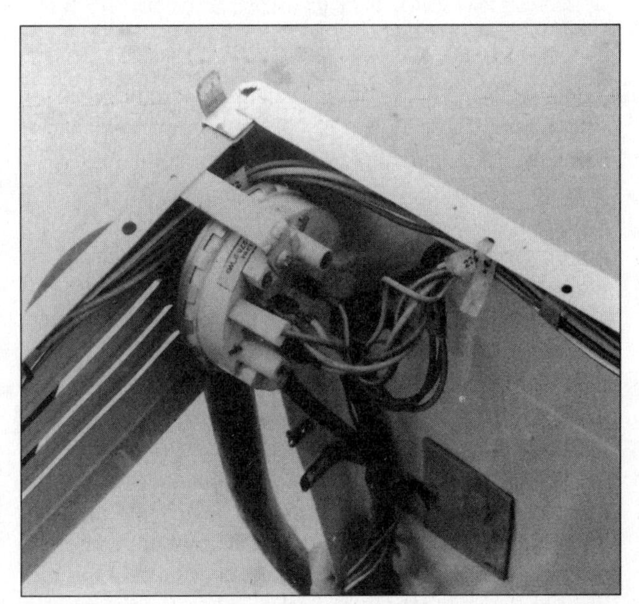

This shows the position of the pressure switch in a typical automatic washing machine. The positions may vary with makes, but all will be found as high in the machine as possible.

The pressure vessel used on machines such as Hoover, Creda and Servis.

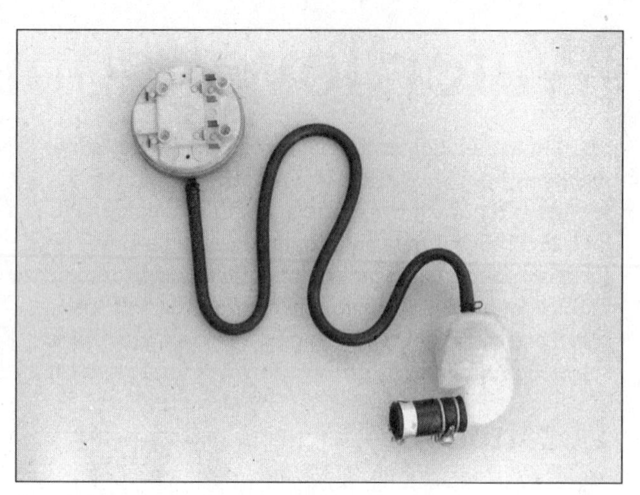

A typical pressure system.

Typical triple level switch; this is more common on machines with plastic tubs or plastic drum paddles.

Single level switch.

The seals and hoses of the system are also of great importance. These should be checked for air leaks and blockages. Any puncture or blockage creates a loss of pressure, resulting in the incorrect operation of the switches. In other words, if the air pressure in the pressure vessel leaks out, the vessel fills with water, indicating that the machine is 'empty', so the water valves are re-energized, 'filling' an already full machine. The result is obvious – and most unfortunate!

The problem just described assumes that the air is prevented from actuating the pressure switch. If a blockage occurs while the switch is pressurised, the machine works as normal until it empties. The next time a programme is started, the pressure switch will already be pressurised. Therefore the machine will not take in any water, but proceed to turn the heater on. Although most heaters are now fitted with a TOC (thermal overload cut-out), this may not act until some damage has been done to the clothes inside the drum.

Checking a pressure switch

If you blow into the switch via the pressure tube, you should be able to hear audible 'clicks' of the switches. This should also happen when the pressure is released. If your machine uses a single level of water, one click will be heard. Two levels of water will produce two clicks. If your machine has an economy button, a third faint click will also be heard. Do not blow too enthusiastically as this may damage the switch. Remember the pressure these switches operate on is very low.

Many machines have an overfill level detection system which will activate the outlet pump should any excess water enter the machine. Several systems use the third or fourth switch of the existing pressure switch bank, operated only by the increased pressure caused by the overfill.

Unfortunately, systems that use the same pressure vessel for both normal and abnormal water level detection may fail to detect overfilling if it is caused by a blocked pressure vessel or hose fault that allows the pressure to escape.

Systems using a separate pressure vessel and a separate pressure switch for detecting overfilling are much less prone to failures of this nature. Nevertheless, they still require cleaning and checking frequently.

Two main faults may occur within the pressure switch. When the diaphragm becomes 'holed' or porous, the switch can be operated and clicks heard, but it will click back again without being depressurised. Secondly, the contact points inside the switch may 'weld' themselves together. This is not unusual if an item such as a heater short circuits and 'blows' a fuse. It does alter the number of clicks heard, as one or more switches may be inoperative. Moving the unit can often free the contact points. However, this is not a lasting repair, as the switch inevitably fails again through contact point damage.

All these faults require replacement with a new switch. Give the make, model and serial number of the machine when ordering, as pressure switches are internally pre-set for specific machines although they look similar. Fitting is a simple direct exchange between the old and new.

The diagrams below illustrate the operation of a double level pressure switch.

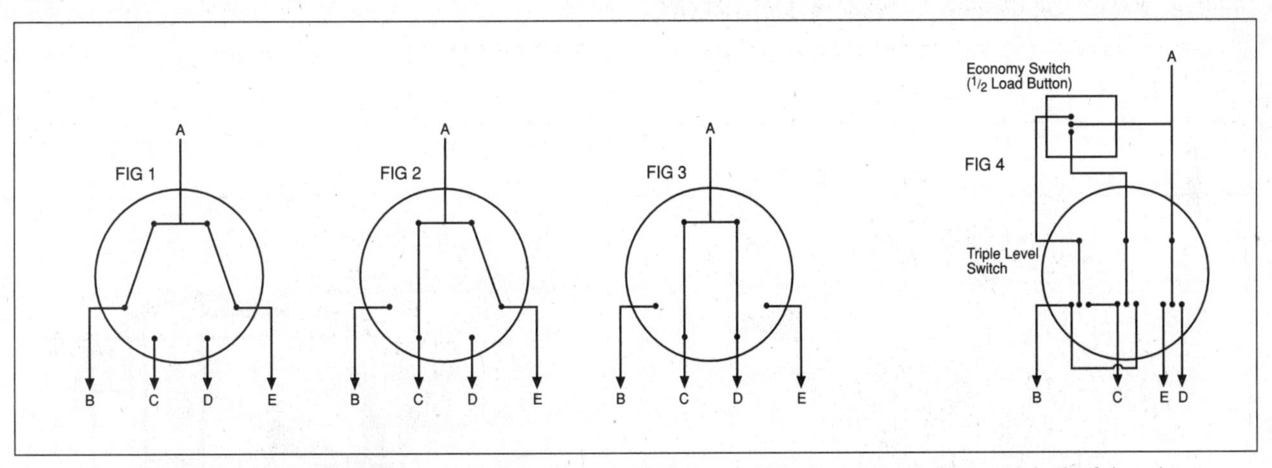

Fig. 1 shows the machine filling with water. B and E are hot and cold valves respectively; the machine is filling with both of the valves.
In Fig. 2 the lowest level of water is reached. The pressure breaks the connection with the hot valve (B), and remakes it with the heater switch (C). The cold valve (E) continues filling.
Fig. 3 shows the highest level, with the cold fill stopping, switching in the motor (D).
Fig. 4 illustrates the way a 3-level switch is used in conjunction with an economy switch, to give an alternative level as an economy feature.

Any loose connections on the pressure system will allow the pressure to drop. This causes overfilling. Ensure a good seal.

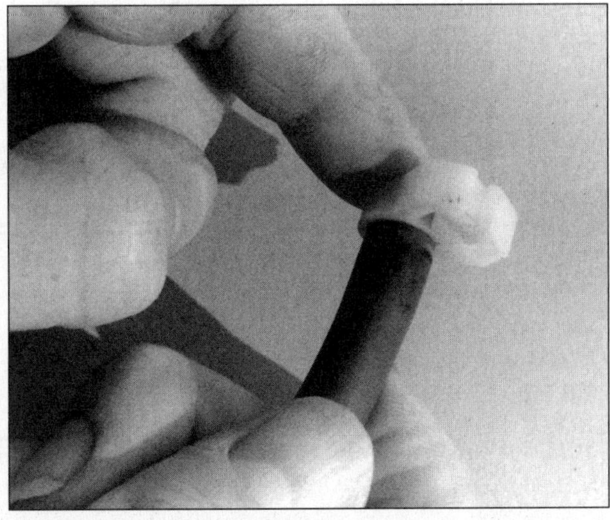

Check the pressure tube for chafing and small porous cracks. Renew if suspect.

Diagrams illustrating the theoretical operation of a single level pressure switch.
***A** The live supply.*
***B** The empty position of the pressure switch. With the switch in this position, power supplied to A via the programme timer is allowed to flow to the fill valve via B. When the pre-set level of water is reached, the diaphragm in the pressure switch pushes the contact arm across to contact C. Power to the fill valve is, therefore, stopped and transferred to connection C, which in turn may supply the circulation motor and heater, for instance.*

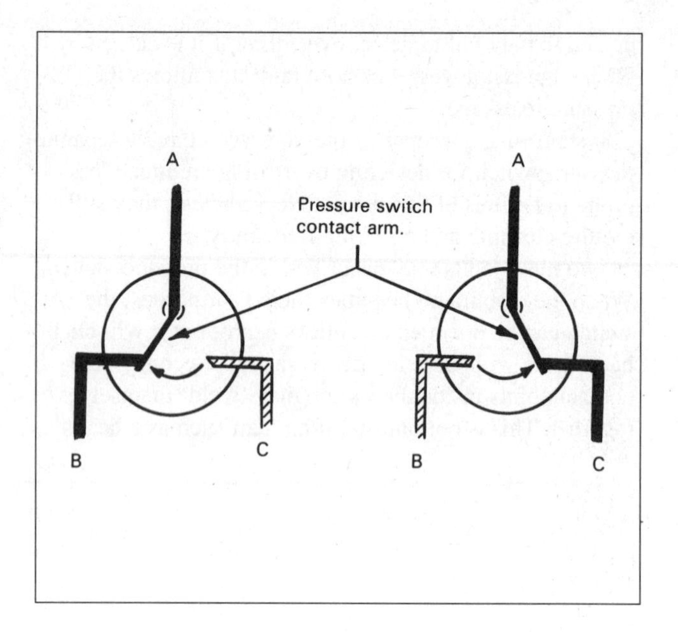

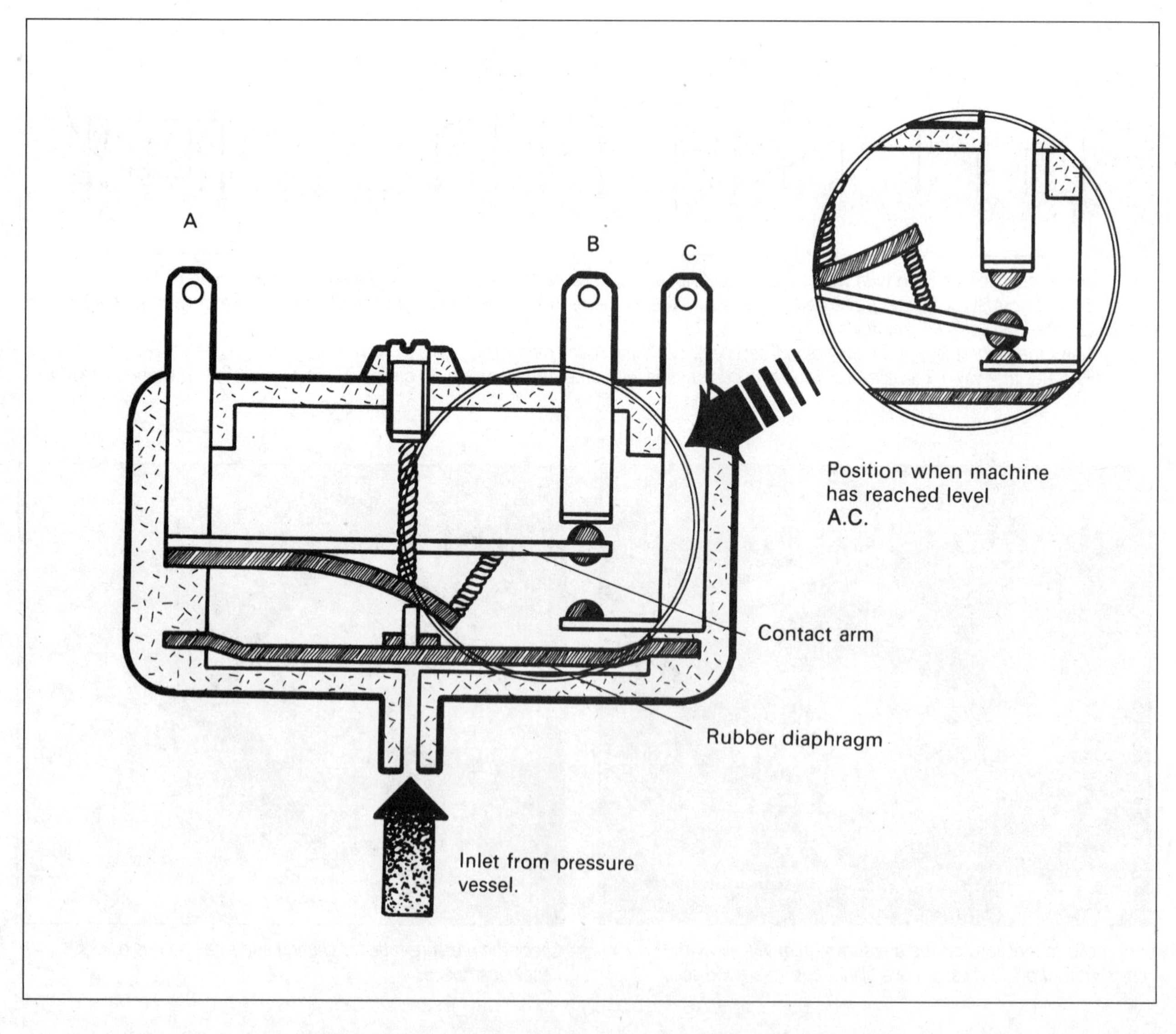

WATER LEVEL CONTROL WATCHPOINTS

1. **Check the pressure system at yearly or half yearly intervals,** depending on the water hardness in your area.
2. **Any hoses or tubes that have been disturbed must be resealed,** and any clips tightened.
3. **Blowing down the accessible end of the pressure tube may seem an easy solution to remove a blockage.** However, it may effect only a temporary cure.
4. **Always use a good low lather detergent specifically formulated for use in automatic machines.**
5. **A pressure switch failure should be suspected only when the system has been thoroughly cleaned, checked, sealed and re-tested.**

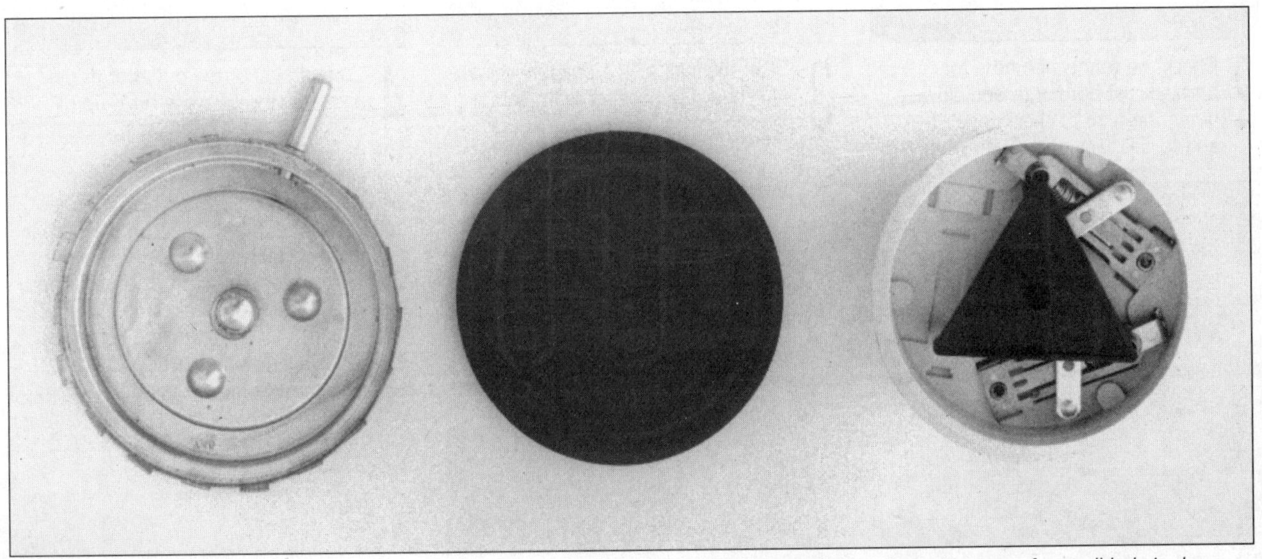

An internal view of a pressure switch showing the diaphragm and switches. This pressure switch is faulty because of a small hole in the internal diaphragm. It appears to operate correctly, although the pressure decreases during the wash and the machine overfills.
If the first functional test is rushed, this type of fault may be over-looked. The switch illustrated was stripped down only to confirm the fault. Such switches have to be replaced when faulty, because they cannot be repaired.

Clearing a pressure system blockage

TOOLS AND MATERIALS

- ☐ Screwdriver(s)

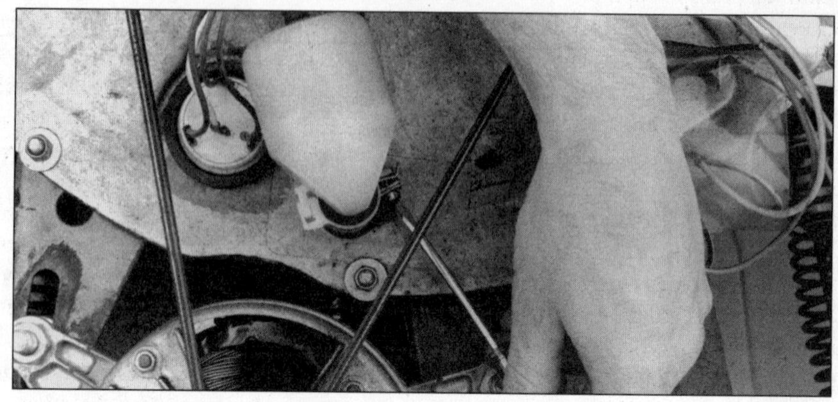

1 Note their positions and then remove all connections to the pressure switch. Remove the complete system from the machine.

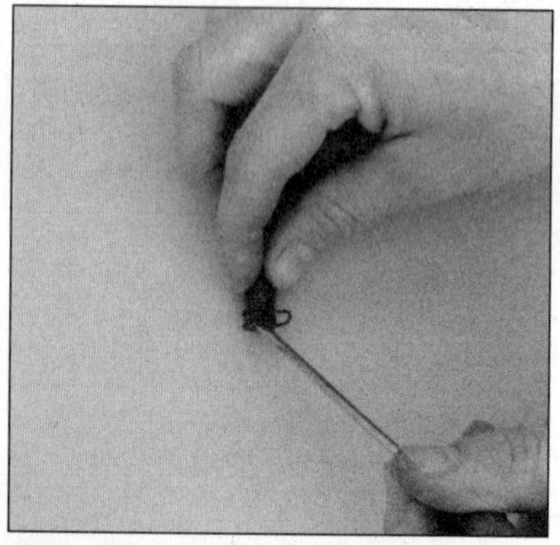

2 Check the connecting hose for blockages at both ends and blow down the tube to check for air leaks and to clear any obstructions.

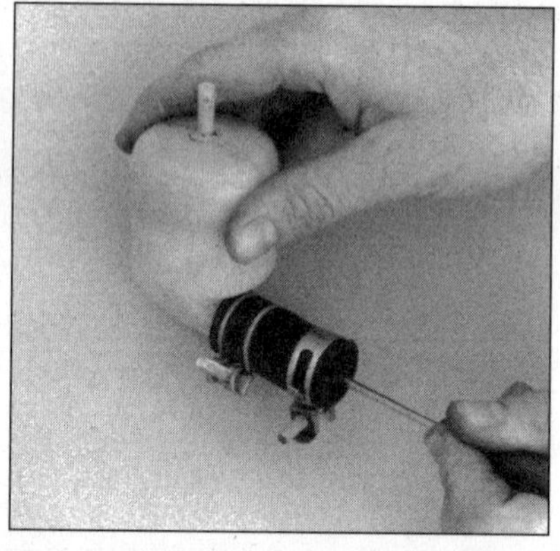

3 Clean the inlet to the pressure vessel and check the hose for leaks and/or perishing.

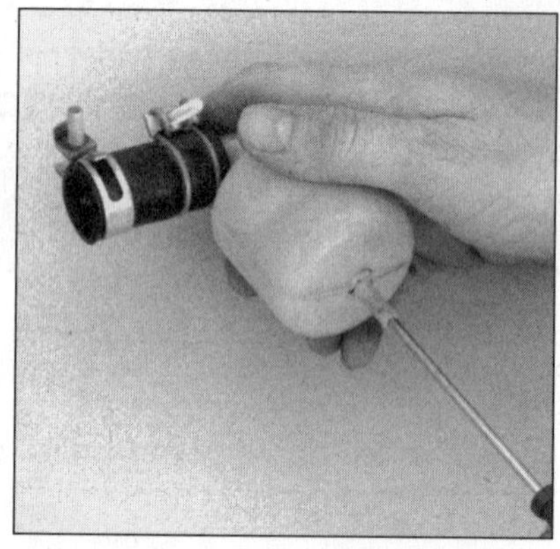

4 Carefully check the outlet of the pressure vessel for any build-up of sludge.

5 Wash out the pressure vessel thoroughly to remove any loose particles and sludge.

When refitting, check the pressure tube and any rubber hose connections for wear – rubbing on pulley, belt or clips. Follow a similar procedure for all types of pressure vessel. Take care to reseal all hose connections that are removed. Remember to wash away all loose particles, because even the smallest blockage in this system will cause trouble.

Pumps

The pump, a vital part of the correct functioning of the machine, is prone to a variety of faults. Leaks from the pump may not be apparent, but the resulting pool of water usually is. So here are a few points to look out for.

Firstly, check all clips on the hoses to and from the pump and tighten if they are loose.

If the leak remains, check the pump's shaft seal. This is the seal that forms a watertight barrier on the rotating shaft of the motor directly between the impeller and the front motor bearing. If fluff and lint collect between the seal and the impeller itself the rubber seal becomes distorted. To check this, remove the pump chamber securing clips or screws and, while securing the rotor of the pump motor, turn the impeller clockwise to undo it from the shaft (impeller and rotor are often left-hand threaded). Remove the pump chamber and clean away any objects adhering to the shaft. Then refit, ensuring the pump chamber seal is in position.

If the seal still leaks, it is because it is worn or has softened. This usually means complete pump renewal. This is not as costly as you may think, because many genuine and 'patterned' pumps of good quality are now available at very low cost. While this may seem drastic for such a small seal, remember that water containing detergent will have been entering the front pump bearing long before the leak was bad enough to see. As a result, the pump itself will probably be damaged and next in line to cause trouble.

The only instance where complete renewal can be avoided is with early Hoover machines. This type has a large, flat disc-like seal, which is the pump chamber seal, and the shaft seal in one. If the seal has leaked, the front bearing should be replaced at the same time as the seal. If the rotor shaft is even slightly worn, it is usually easier and quicker to fit a complete compatible pump to avoid any further trouble. In fact, this may prove to be the least expensive remedy.

Other leaks may be attributable to the pump. For example, the blades of the impeller may be damaged, broken off or badly worn away by a solid object, such as a coin, lodged in the pump, or tight bearings may cause slow running of the motor. Both these problems cause poor water discharge, that is, slow draining (detected by functional testing). This, in turn, may cause the machine to spin while some water still remains, thus causing other hoses and so on to leak or the machine to fill to too high a level, as some machines have a timed rinse fill action.

On other machines, slow drainage may mean that the machine does not spin at all. This is because the pressure switch detects the presence of water in the machine which the slow pump failed to discharge in its allotted time, so it does not allow a spin to take place by either missing out the spin completely or stopping the wash cycle at the spin position. This may occur on the intermediate spin half way through the rinse sequence on many machines.

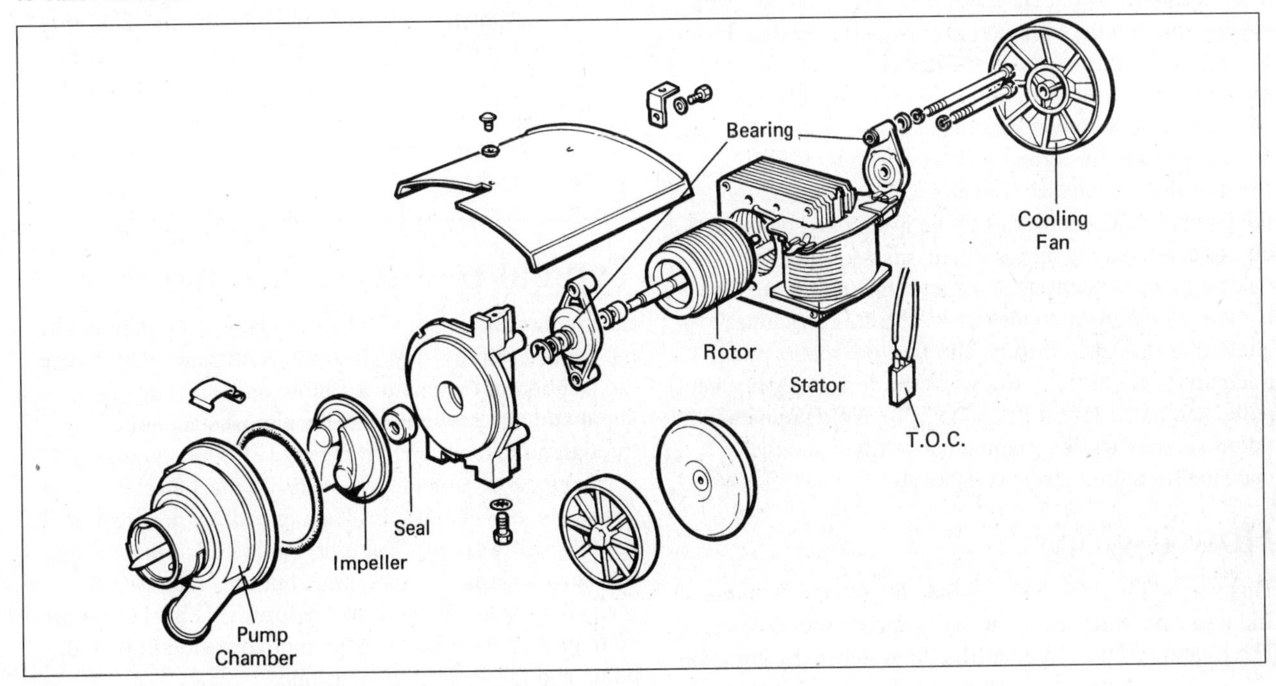

Exploded view of typical pump.

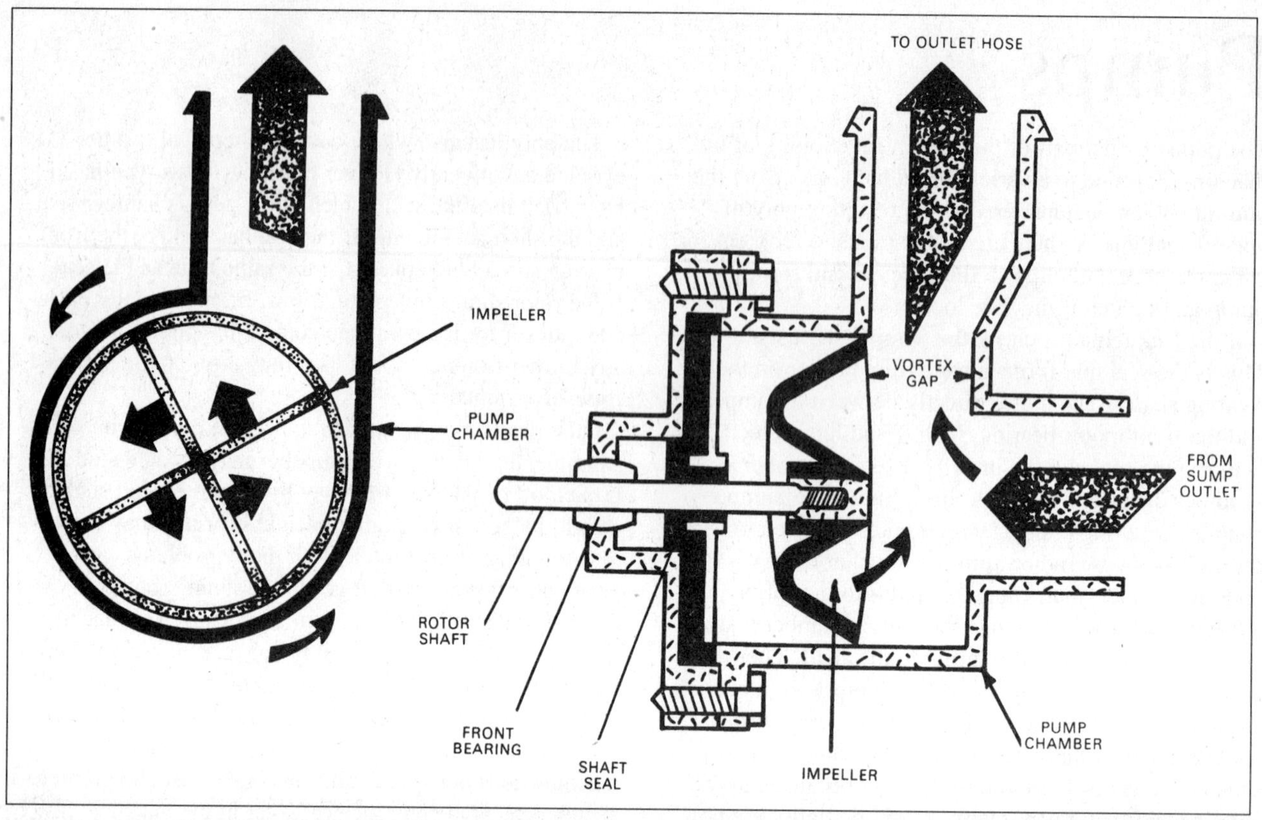

Vortex style pump. Dynamic force exerted on the water in the direction of the outlet.

Checking the impeller and bearings can be done at the same time as checking the seal.

Condenser driers

Similar faults on condenser driers may give rise to poor drying, through the failure to discharge the condensate and water used/produced in the drying process quickly enough. This leads to a build-up within the outer tub which is enough to wet the clothes but not enough to be visible. Ensure the correct pump-out rate by using a functional test sequence. Depending on make and model, the pump on a condenser machine is operated periodically or continuously during the drying sequence to discharge the condensate and water. Failure through wear is more common on condenser washerdrier machines although some manufacturers do fit more robust pumps to compensate for this. Always fit the correct replacement pump and make sure it has a TOC. To avoid premature failure, a good quality pump must be fitted and the machine should be serviced regularly.

How it works

This simple illustration shows the outlet pump chamber and impeller. Water from the sump enters from the front. The rotation of the impeller lifts the water in the direction of the narrower outlet hose. Some machines are fitted with a non-returnable valve system on the outlet hose – watch out for blockages at this point.

There are two types of impeller. One is simply a paddle type which is more prone to blockages. The other is like the one illustrated and is called a vortex pump. This type of impeller is more a flat etched disc that allows a gap between it and the pump chamber. This gap lets particles pass through more easily than the bladed impeller version. The vortex impeller applies lift to the water as shown in the diagram. The action is similar to the rotating vortex created when a bath or sink empties.

Typical pump replacement

Ensure that the machine is isolated before attempting any repair. The step-by-step illustrations on pages 73-74 show the location of the pump assembly on a Hoover automatic; the position of this is almost standard throughout the entire Hoover range of both automatic wash only and combined washerdrier machines.

The machine shown was leaking badly when inspected, but was still working. The user admitted that the machine had been leaking for some time, but now more water seemed to be leaking out than before. This is obvious in pictures 3 and 4 from the degree of corrosion to both the pump and the shell of the machine.

Access to typical sump hose filter or catch pot. As can be seen from the two photographs below, pump positions vary between makes but the basic operation is the same.

Pump mounted lower left-hand side Philips washerdrier connected to accessible front filter unit.

Pump mounted right-hand side directly to sump hose with internal filter accessible only during routine servicing.

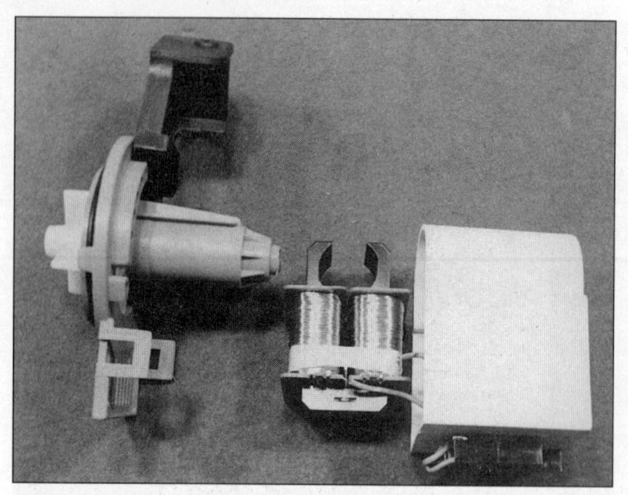

This new type of pump uses a permanent magnet rotor (see Motors, pages 112-125) and avoids many of the problems associated with early pumps. It is becoming popular with many manufacturers.

Typical electric pump with sump and outlet hoses of the type found on many of the leading makes. The main differences between electric pumps are the pump chamber mouldings. All pumps empty at a rate of about 36 litres (8 gallons) per minute.

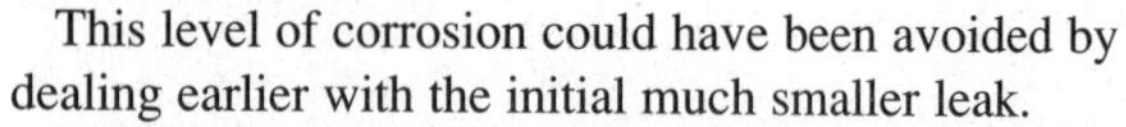

This level of corrosion could have been avoided by dealing earlier with the initial much smaller leak.

Pictures 5 and 6 show variations of pumps that may be encountered on this type of machine, and pictures 7 to 12 show the further stripdown of the pump. In this case, it was thought best to renew the pump completely because of the severity of water and detergent damage that had been done to both the bearings and metal laminations of the stator. Again, this could have been avoided if the leak had been dealt with rather sooner.

The shell and mounts were coated with anti-rust compound before fitting a complete pump of the type shown in picture 6. Care must be taken that the anti-corrosion liquid does not come into contact with any rubber hoses or seals.

After replacing all hoses and connections – a simple reversal of the removal procedure – the machine was moved back into its correct working position and reconnected to the water and power supplies. All connections and work was double-checked before trying a short rinse programme to ensure that the new pump functioned correctly, the repositioned clips were watertight and that no other leaks were present, This was followed by a full functional test via an RCD protected supply.

The user was advised of the unnecessary danger and, in this case, damage caused by using the machine while ignoring a very obvious fault.

Anti-corrosion compound

The anti-corrosion coat may be one of several types, all available from DIY car centres and hardware shops. Follow the manufacturer's instructions, taking care not to allow any contact with rubber hoses or plastics. When using anti-corrosion gel or rubber sealant indoors, protect the floor from spillages and ensure adequate ventilation.

Pump replacement

TOOLS AND MATERIALS

- ☐ Screwdrivers
- ☐ Spanners
- ☐ Anti-rust compound
- ☐ Pump

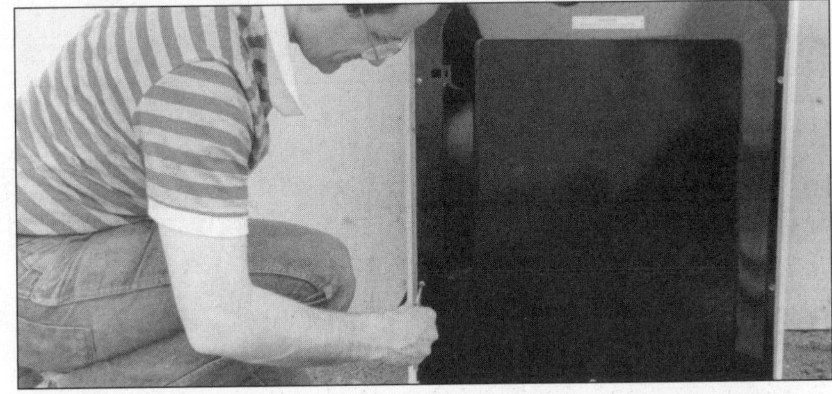

1 Isolate the machine and remove the rear panel. Note the location of the pump – back, right-hand corner.

2 Protect the machine face and carefully turn the machine over. The position of the pump is now clearly visible.

3 Note the positions of the hose clips and connections and remove them. Corrosion may be found on the mounting. Treat with anti-rust compound before fitting the new pump.

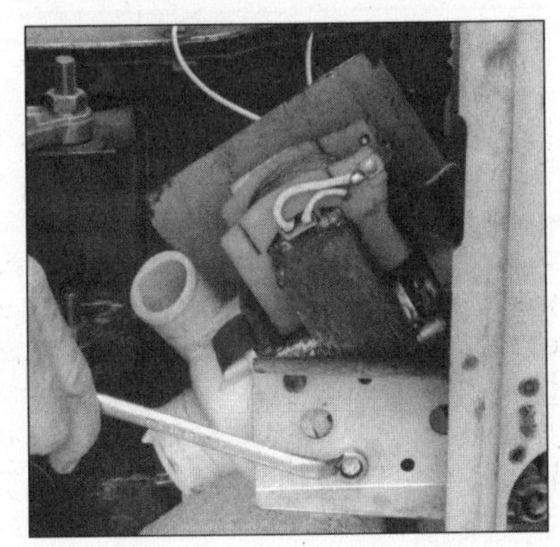

4 Support the pump while you remove the securing bolts. Withdraw the pump from the machine.

5 Two types of Hoover pump – different styles of impeller and stator – are shown.

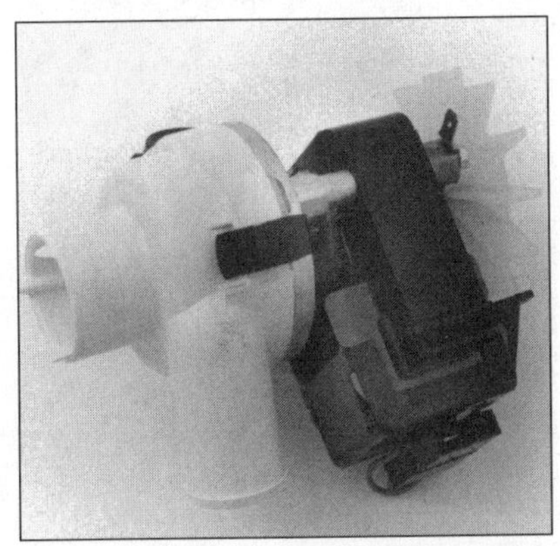

6 This type of pattern pump fits Hoover, Creda, Servis and Hotpoint. Many other styles are available for other machines.

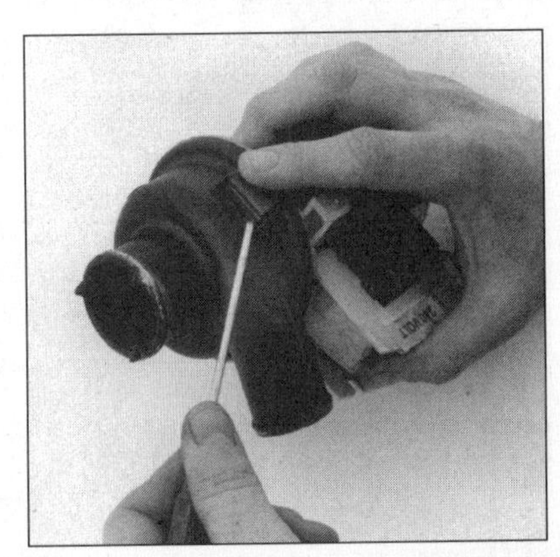

7 Lever the small clips loose using a small screwdriver. Hold the clips lightly to prevent them springing off. Screws may be found in place of clips.

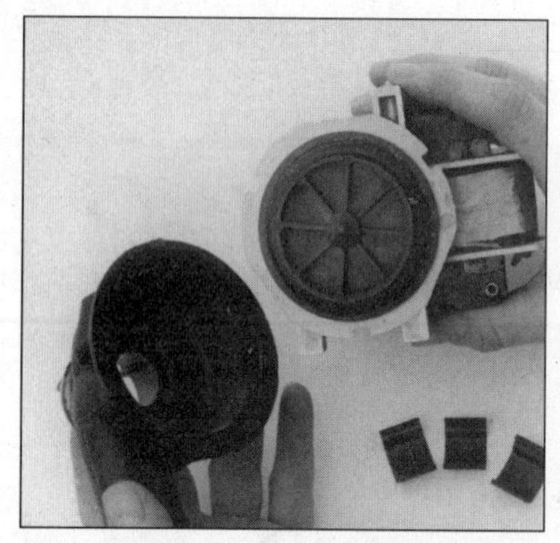

8 Note the pump chamber position and remove to expose the impeller.

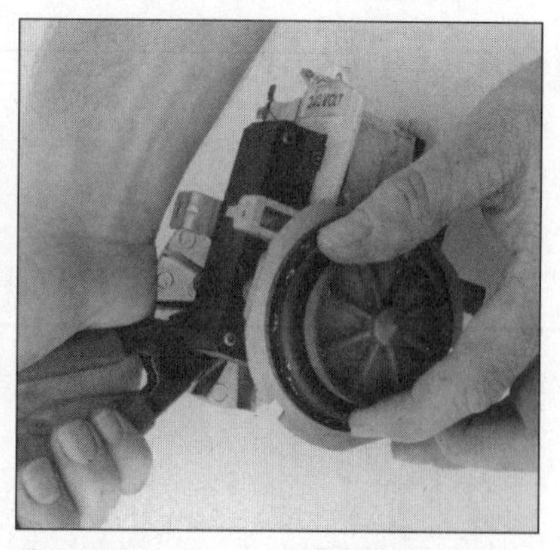

9 Holding the rear shaft securely, turn the impeller clockwise and remove (left-hand thread).

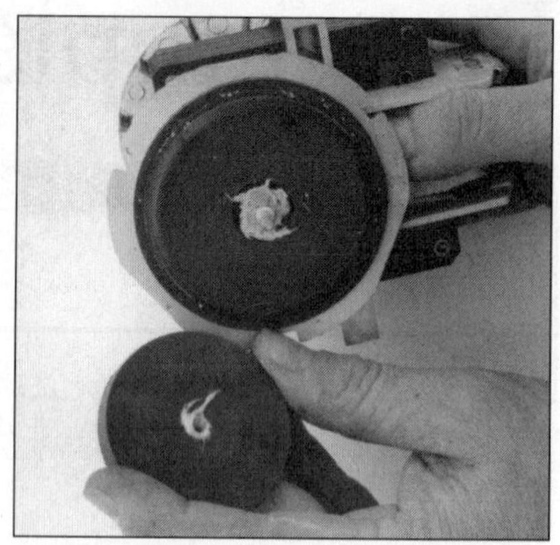

10 The rear seal is now exposed. In this case, it is badly worn by a build-up of lint on the shaft.

11 With the seal removed, check the front bearing for wear and damage. Check the rear bearing.

12 This rear view of the seal shows the extensive wear.

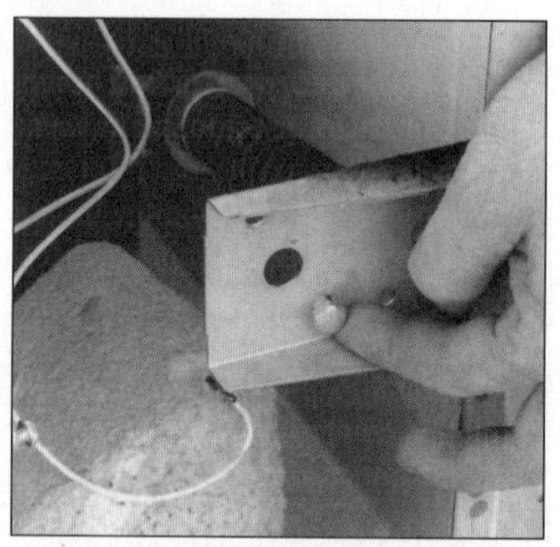

13 When fitting a new pump, check the bolt hole sizes of the mounting plate. Sometimes they may need enlarging.

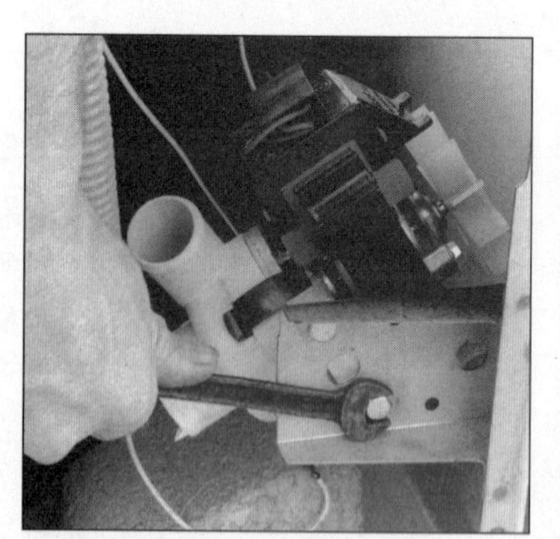

14 Secure the pump firmly into position.

15 Reconnect the terminals. If the pump has an earth tag, ensure that it is a good fit, as with all connections.

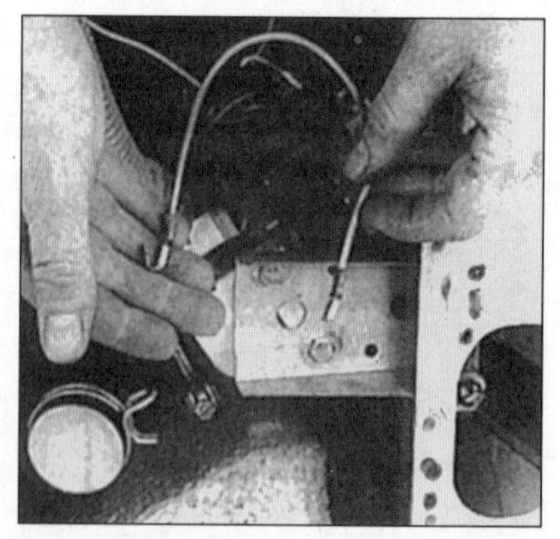

16 If a pump has a plastic mounting plate, the metal stator laminations must be linked to the machine earth path. Make a short lead to connect the earth tag on the pump to the fixing bolt and secure. Refit the hoses, clips and panels before testing.

Door locks

The electrical switch behind or near the main door latching device of washing machines is called an interlock. It is designed to delay opening the door for up to two minutes after the machine has stopped. The delay time differs between the makes and models of different machines.

Types of interlock

A machine with a push/pull timer knob action may also have a manual interlock for double protection. The manual interlock system is quite straightforward, bolting or unbolting the door with a push/pull action of the timer knob via a latching mechanism. The mechanical interlock acts in much the same way as the electrical version in preventing the door from opening if the machine is turned on, except that there is no time delay. Mechanical interlocks are always in addition to electrical interlocks, which incorporate a delay to entry.

A more recent version of interlock incorporates a pressure switch type of system. This does not allow the door to open if any water remains in the machine. A mechanical operation, it works even when the machine is unplugged. The door can only be opened when the water has been drained out. Bear this in mind if the pump fails or there is a blockage. The system is easily recognisable by the pressure tube leading to the door interlock. Such systems may use a separate pressure vessel or a 'T'- junction arrangement from the water level pressure switch tube.

Whichever system is used, faults similar to those described in Water level control *(pages 63-68)*, such as blockages and air leaks, may be encountered.

Washerdriers

Some washerdrier machines include a simple micro-switch attached to the door interlock and forming one unit. The micro-switch is in circuit during the dry cycle and operates directly via the door latch and slide action of the interlock. These micro-switches and interlock combinations are used on machines when the two-minute delay facility during the dry cycle is not used. This arrangement allows the door to be opened if required, but the micro-switch opens circuit and interrupts the cycle.

Take care with combined washerdriers as the door glass can reach extremely high temperatures. Keep children and pets out of reach when the machine is in use.

Dry-only machines

On basic tumble-dry-only machines, a simple micro-switch activated by the door latch pecker or hinge movement is used. No delay action is available and the simple opening and closing of the door stops and restarts the programme. This system is described in detail later in this chapter.

How interlocks work

Manufacturers all have their own versions of interlocks, so it is impossible to illustrate all the different types. Here, the Klixon (3DB) switch is used to illustrate the internal workings and theory.

Fig. A (page 77) shows the state of the interlock before power at L. The door bolt is disengaged and the bi-metal

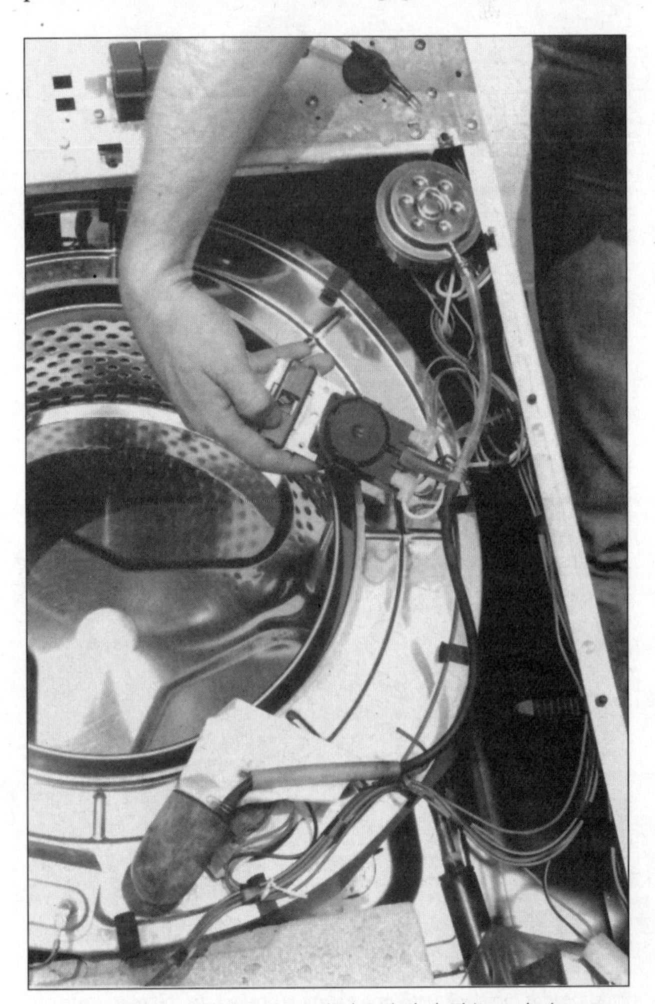

New-style interlock with pressure lock included. This works in conjunction with the pressure switch, preventing the door from being opened if any water remains in the machine.

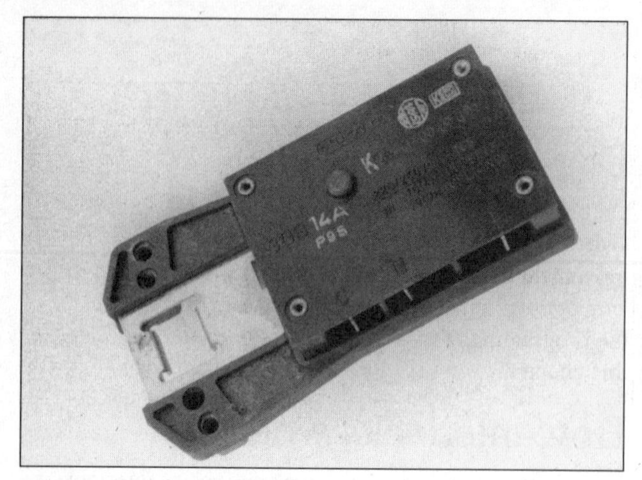

Zanussi 3DB style interlock.

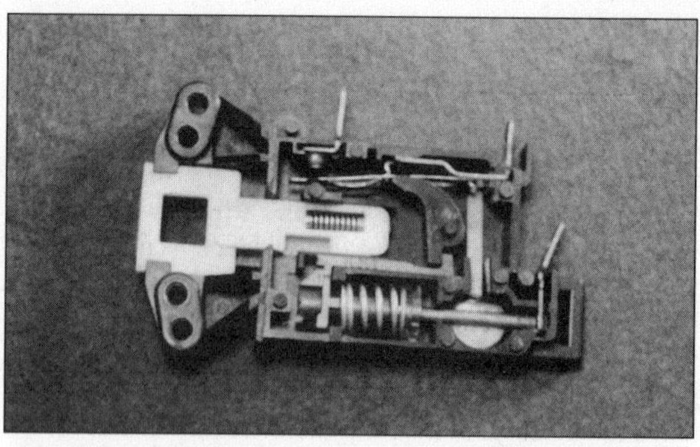

Internal view of interlock that uses an oil-filled piston instead of a bi-metal strip. The piston can be seen in the lower section of the interlock, running through its large return spring on the left and supported on the right by the heater pod.

Typical dry-only door safety switch. Note the operating arm. Because the switch carries all the power it is susceptible to overheating problems due to poor connections or contacts.

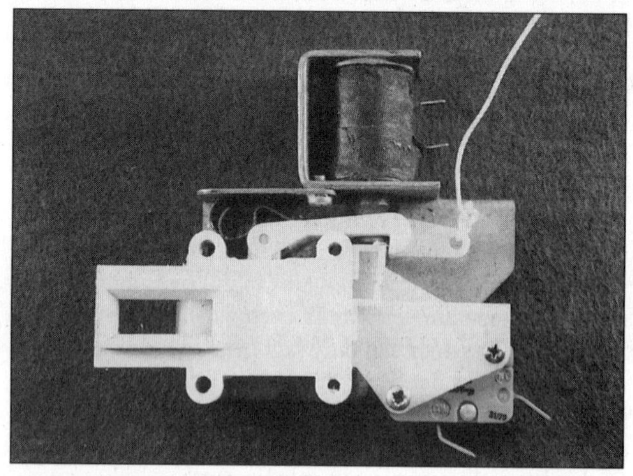

Solenoid operated interlock as used on many computer-controlled machines. The micro-switch is clearly visible. So, too, is the cord for manual activation of the interlock in the event of power failure or a fault.

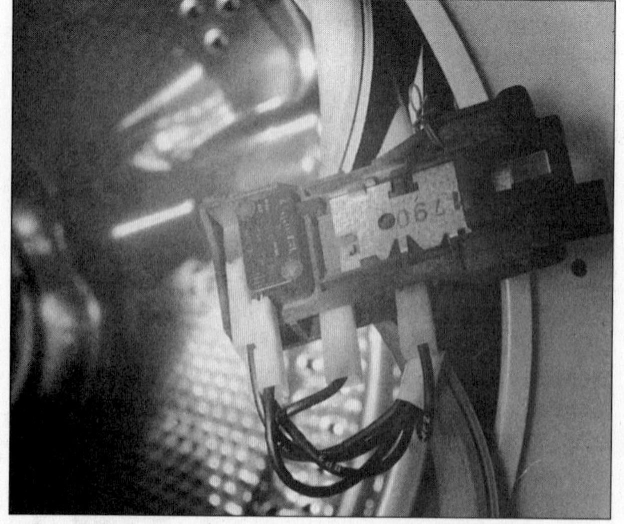

Washerdrier interlock with integral micro-switch to the rear. Note how access is gained via the door seal front lip by removal of the front clamp band.

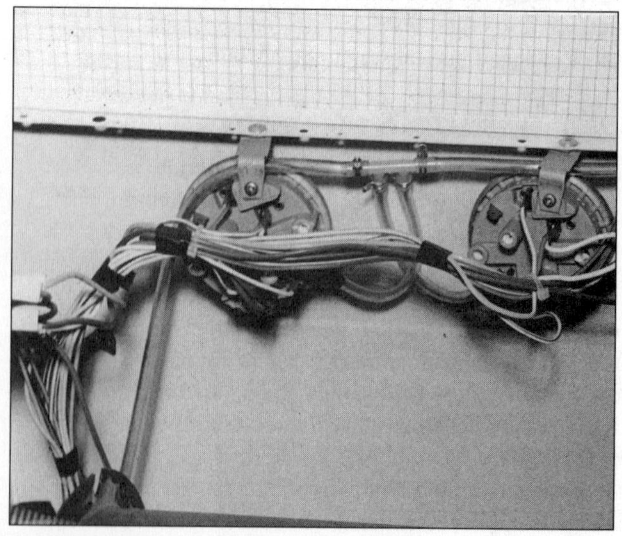

Double pressure switch system with 'T' junction and tube to the right leading down to a pressure-operated interlock.

strip is in its 'rest' position. Therefore, there is no connection at point Y, so no power is transmitted to X.

Fig. B shows the state of the interlock when power is applied to point L. The heater is activated, heating the bi-metal strip. This bends, engaging the door bolt and making the connection at Y, allowing power to flow to X. When the power is disconnected, the heater is allowed to cool and the bi-metal strip then bends back to its 'rest' position. This action can take up to two minutes, creating the delay after the machine has stopped. The delay time (cooling of heater and bi-metal) varies according to the ambient temperature, style of interlock and position of the appliance.

Some interlocks use a small oil-filled piston arrangement in place of a bi-metal strip. When heated, the oil expands and the resulting piston movement is used to actuate the interlock. As with the bi-metal system, cooling allows the piston to retract and de-latch the interlock.

The heater referred to in this chapter is not the large heater in the drum, but a minute one used only to heat the bi-metal strip.

Modern machines tend to link the interlock to all the other functions, so if the interlock fails, power to the rest of the machine is severed and the machine becomes inoperable. On older machines, interlock failure results only in no motor action throughout a normal programme.

A two tag interlock (1 DB) is called a straight through interlock because, although locking occurs, switching does not.

Computer-controlled interlocks

Computer-controlled machines may have electro-mechanical interlocks similar to those described above. However, they may be completely different both in the way the door is opened and in the way that access is restricted via the door until the programme has finished and the machine is completely empty of water. The basic operation is described here, but this does not refer to any specific make as the various manufacturers incorporate subtle differences.

Door opening is by means of a push button action that actuates a simple electrical switch. This makes contact and supplies power to a solenoid-operated catch mounted within the machine in place of the door interlock. With power supplied to the solenoid, the door latch is released and the door opens. To avoid continuous supply to the solenoid, a micro-switch is incorporated within the unit to open circuit the solenoid coil as soon as the door latch pecker is released.

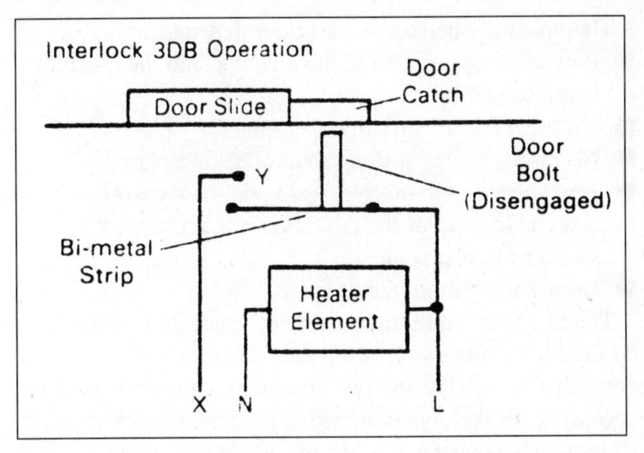

Fig A
This shows the state of the interlock before power at (L). The door bolt is disengaged and the bi-metal strip is in its 'rest' position. Therefore, there is no connection at point (Y), so no power is transmitted to (X).

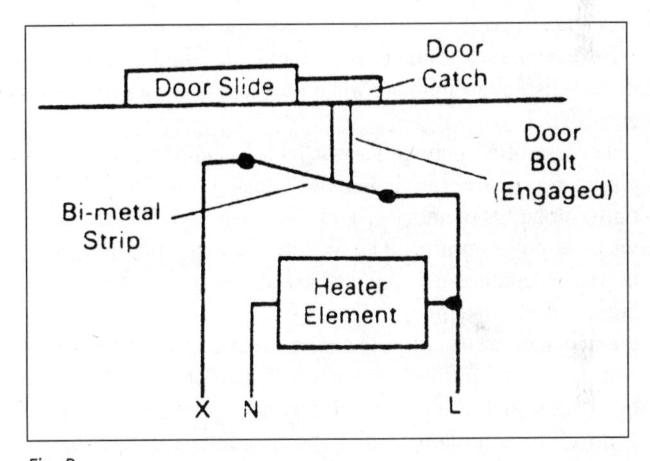

Fig. B
This shows the state of the interlock when power is applied to point (L). The heater is activated, heating the bi-metal strip. This bends, engaging the door bolt and making the connection at (Y). This then allows power to flow to (X).

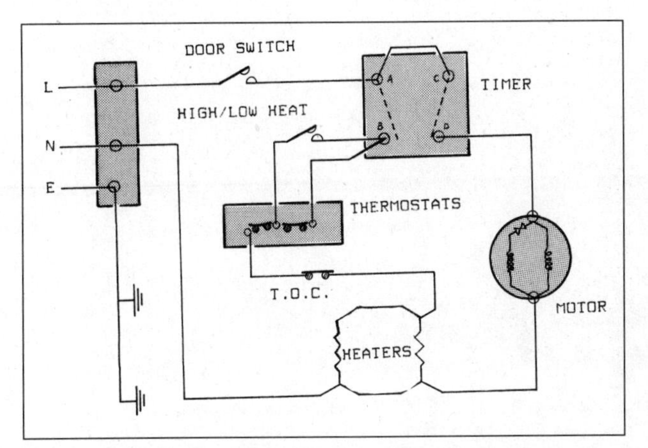

Schematic wiring diagram of a simple tumble dry only machine.

The opening operation of the door depends on:

- Power being supplied to the machine and the machine being turned on.
- No programme currently operating.
- No motor action or drum rotation taking place.
- Any water in the machine being below the lowest detectable level of the pressure switch, that is, all switches in rest position.
- Door open button being pressed.

The last four requirements are continuously monitored by the micro-processor within the electronically-controlled timer. For the first condition, a power supply is necessary for the door solenoid to operate. However, many machines have a means of de-latching the door mechanism manually in the event of a power failure or when faults occur; read the manufacturer's instructions. Before carrying out mechanical actuation of the latch, ensure the machine is unplugged (isolated) and that the water level is below the door level *(see Emergency procedures, page 9, and also photograph on page 75).*

Several makes of machine use an LED (light emitting diode) display to indicate when door opening can be activated.

The use of a micro-processor within such machines allows for greater interaction and sensing to be carried out. The larger memory size of the chips allows more variable programming. The way in which the programme is written enables it to react to variations within the circuit of the machine. This, in turn, gives rise to a greater number of criteria being monitored to ensure compliance with safety and so on. The micro-processor board also includes a clock chip which can help in controlling programme times and functions accurately. The timing function may also be used to time a delay to the door open switch if required by the manufacturer of the machine.

Dry only machines

Most tumble dry only machines do not incorporate a time delay system, so access to the load compartment is not restricted by time. However, a micro-switch is actuated when the door is opened to open circuit and sever power to the motor, stopping drum rotation. Because rotation is slow on dry-only machines, the drum quickly stops when the motor is turned off by the micro-switch.

On some machines, only the motor is open circuited by the opening of the door, leaving the heater still in circuit. The motor used to rotate the drum usually rotates the fan which circulates the air. Consequently, in such machines, if only the motor is stopped when the door is opened, the heater temperature will quickly rise and open circuit its TOC. If the door is left open for long periods, cycling of the TOC will occur with the possibility of damage to both TOC and heater.

This will only occur if the timer is still on the heat section and will only cycle until the remaining heat time runs out, that is, completes its pre-set time. Continuously opening and closing the door is not recommended. To avoid such problems, many machines now use the door micro-switch to open circuit supply to all components. Make yourself aware of the system used in your machine.

Computer-controlled tumble dry only machines use a system similar to that described above. Dry programmes on such machines, once interrupted, will not restart simply on the door's closing but require the start button to be pressed again.

Similar systems can be found on higher specification mechanically timed machines. These machines use a relay start once normal operation has been interrupted by the door being opened. The remaining drying time cannot commence until the start button is pressed. This system is safer than the simple micro-switch only operation, which restarts the drum rotating as soon as the door is closed.

Heaters

Although they vary in shape and size, the heating elements used in washing machines, washerdriers and tumbledriers are usually one of two basic types. Exposed single wire elements are used exclusively in dry only machines. Metal sheathed elements are found in washer-driers for heating the airflow (mounted in a special duct on combined washerdriers) and in washerdriers and washing machines for heating the wash water (immersion type). The solid/sheathed element is also used in many dry only machines to heat the airflow.

The first type is simply an exposed length of conductor, which heats up when a current is passed through it because of the resistance properties of wire. Because it is an exposed conductor it must be housed and supported in a way that avoids accidental contact, which could cause electric shock or burns. It must also be housed so that it can dissipate the heat generated and be supported by heat-resistant insulators at regular intervals.

Sheathed elements

Metal covered or shrouded elements are found in all types of machine: washing machines, vented washer-driers, condenser washerdriers and dry only. The way in which the element is insulated and supported by its solid outer sheath allows it to be shaped into any one of thousands of configurations to suit any situation. However, this should be done only at manufacture and no modification or bending should be made to old or new elements.

When a current is passed along the conductor the resistance causes it to heat up. However, the conductor in this type of element is housed within a tube and surrounded by magnesium oxide, an insulating material. Heat can be transferred to the outer sheath, but the current cannot. Outer sheaths are made from various types of metal to suit specific requirements and conditions. The heating portion of some elements is designed to be used submerged in water, whereas others are used within the machine to heat the airflow during the drying cycle.

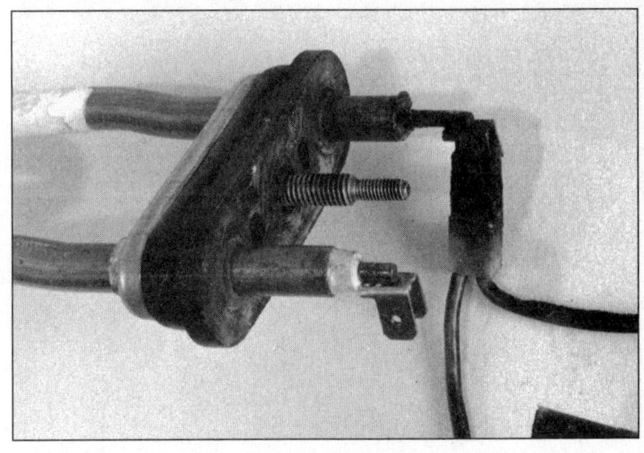

Overheated terminal due to loose connection on live supply.

The designs of both exposed and sheathed elements vary enormously from product to product, but the way in which they work is always the same. As with motors, a TOC is normally present to avoid overheating. The TOC is a common area of fault and replacements must be

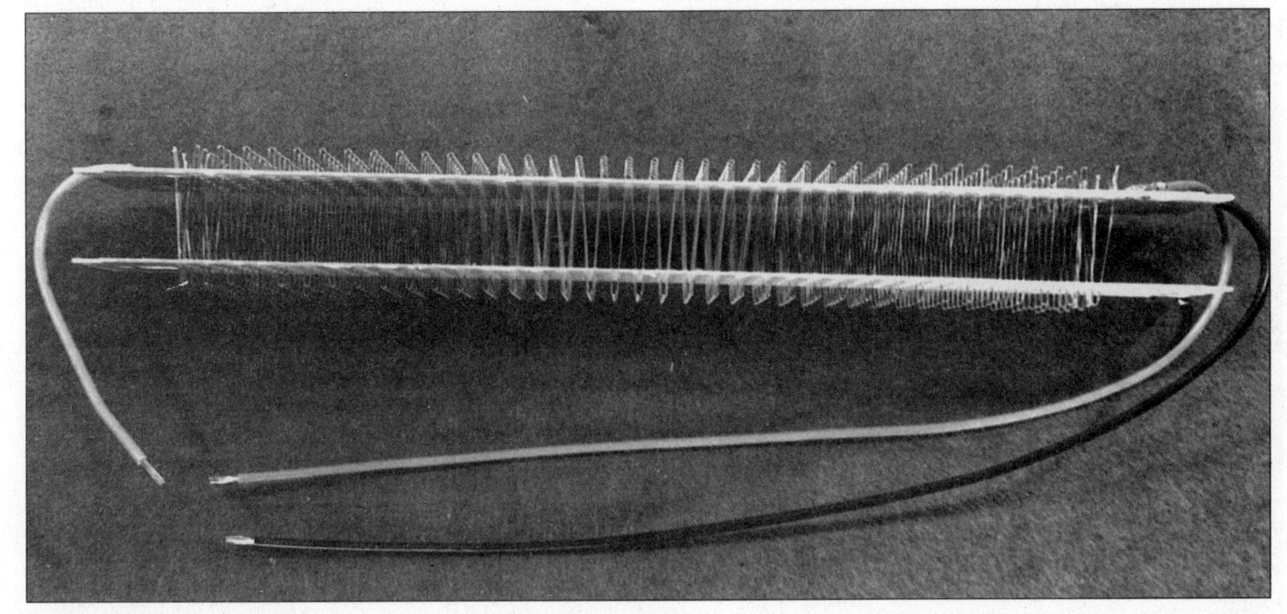

One variation of exposed element used in many tumble dry only machines.

The wash water heater is clearly visible on the lower right-hand side of the outer tub back half of this condenser washerdrier machine.

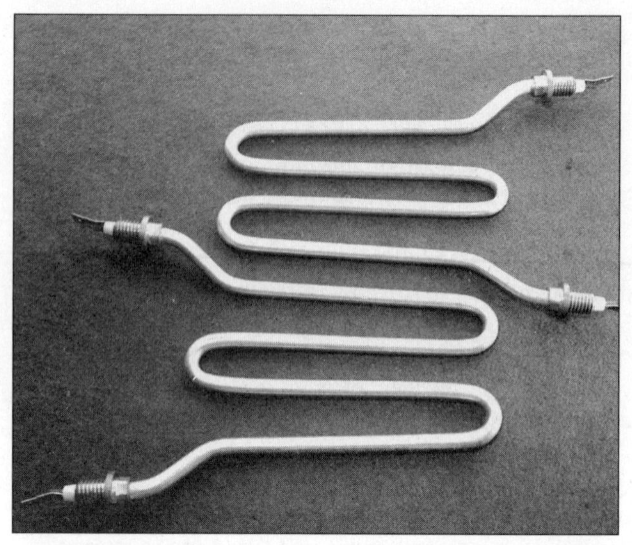

Sheathed airflow heaters are used in combined machines. The Newpol type is shown.

identical to the original. To avoid overheating, make sure there is a good airflow over the element, that is, check that fluff does not block the air intake.

Common faults

One of the commonest faults with both exposed and sheathed heaters is an open circuit: no current flows through the heater, so no heat is produced. This is often simply the result of a broken or loose connection to one of the heater terminals. This overheats, leaving an obvious discolouration of the connection or terminal, resulting in a break of the circuit at that point. Alternatively, the break in the circuit may occur within the element itself. Heaters can very easily be tested for continuity *(see Electrical circuit testing, pages 22-25).*

Another fault that sometimes occurs is low insulation *(see Low insulation, pages 172-175).* Short circuiting of the heater accompanies the low insulation fault. This is caused by either a complete breakdown of insulation of sheathed elements, or the breaking of exposed elements or their mounts, allowing the exposed element to touch the earthed metal heat shield or housing. This results in the appliance 'blowing' fuses or earth tripping if an RCD is in circuit.

If any of these faults occurs, a complete replacement of the component(s) is required.

Heater duct of a Philips washerdrier machine (condenser type).

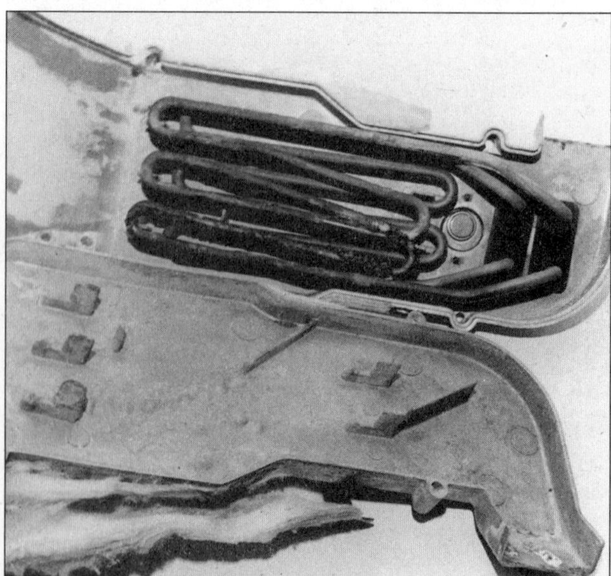

Typical configuration of airflow heaters housed in the ducting of a combined machine.

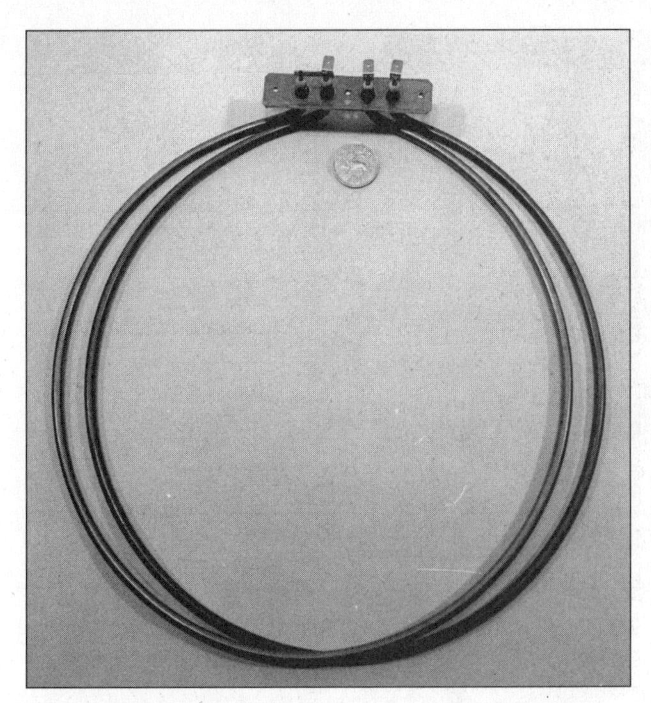

Sheathed elements are also found in tumble dry only machines. They are normally in a large circular single or double configuration, as shown. They may be mounted to the front or rear of the machine depending on make and model.

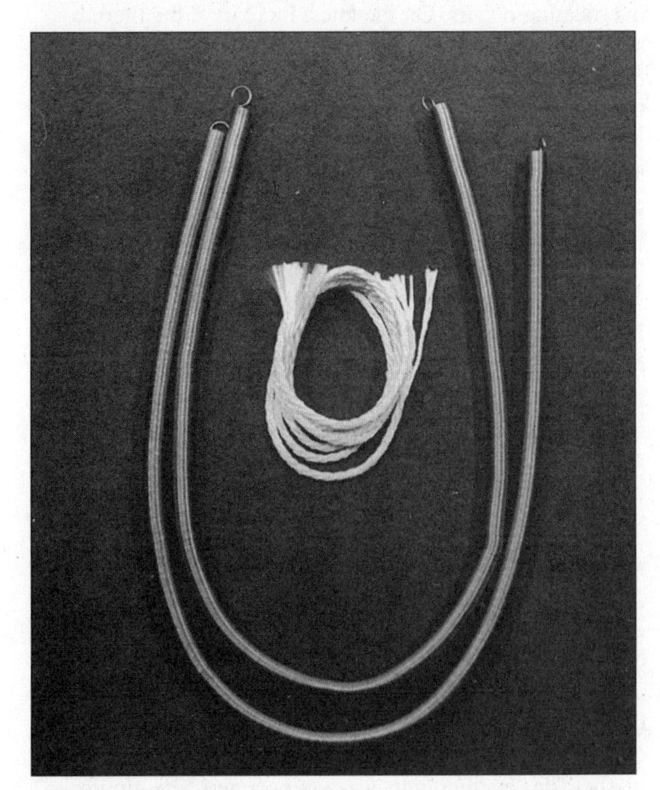

Spiral element kit for Hoover tumble dry only machine. Note the new glass fibre support cores. When fitting, measure the fixing length required, using a piece of string, and pre-stretch a new element before fitting it to the machine. Keep stretching evenly along the whole length, otherwise hot spots will form.

Exposed heater element (spiral type) supported on ceramic mounts (type 1).

Broken exposed elements that touch an earthed metal heat shield will 'blow' wired fuses only if the remaining element circuit is short and resistance low. It is possible for the remaining length of element to work using the earth as its neutral return. This type of fault results in poor drying because of the reduced size and, therefore, reduced heat output of the element. Check exposed elements and their mounts closely for this problem. If this fault occurs on an RCD protected circuit, the RCD will trip, preventing further use of the machine until the fault has been rectified.

The location

The wash water heater is usually located in the lower part of the tub assembly. It may be fitted through the back half of the tub or through an aperture in the tub base itself. An exception is the Hotpoint front loader range, including combined machines, where the heater is located on the front of the outer tub, directly below the door seal. Access is gained by removing the front panel of the machine *(see Door seals, pages 50-57)*.

Removing and refitting a heater

Make a careful note of all the connections before removing them. The heater can be withdrawn from its position if you slacken, but do not remove, the centre nut and tap it to release the tension. Gently ease the rubber grommet free from its position with a large, flat-bladed

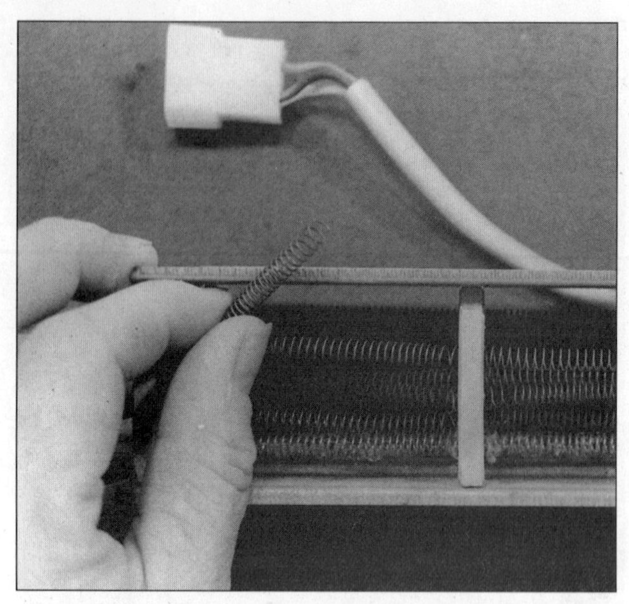

This Zanussi unit displays a typical broken element fault. A complete unit replacement is required.

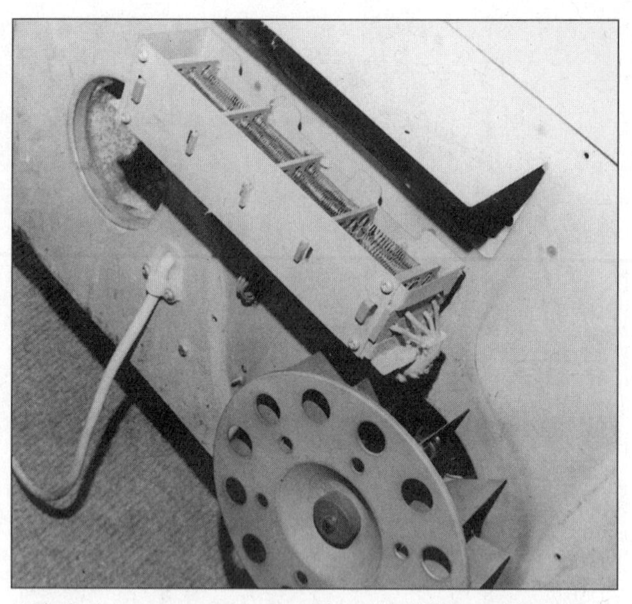

With the large rear cover removed from this Philips tumble-dry-only machine, the heater assembly and fan are easily accessible for cleaning and checking.

screwdriver. Fitting a new heater is a reversal of this procedure. Apply a little sealant to both surfaces of the grommet fitting. Do not overtighten the centre nut, because this distorts the metal plate.

The inner of the outer tub on most automatic washing machines has a raised flange or cover plate that engages the curved section at the end of the heater. It is important that this is located correctly when a heater is being fitted. Check that it is located and held correctly by pressing firmly but carefully downwards on the terminals of the heater while slowly rotating the inner drum. If the heater is not located correctly, it will pivot on the grommet mounting and allow the element to come into contact with the drum, resulting in a grating noise and vibration. If you hear such a noise during this simple test, slacken the heater clamp centre nut, remove the heater completely and relocate it correctly. Then try again. Do not exert excessive pressure on the terminals of the heater; try to press down on the exposed outer sheath.

Machines with plastic or nylon outer tubs are fitted with overheat protectors which are linked in line with the live feed to the heater. They are essential for safety. If a pressure switch or pressure system fails, it is possible for the heater to be engaged with no water in the tub. This would, of course, be most unwelcome in a machine with a metal tub, although only minor damage would be caused to the clothes. In a plastic/nylon tub, the result would be extremely dangerous. Under no circumstances should the overheat protection device be removed or bypassed.

The overheat protector on early Philips machines was an integral part of the heater and was similar to a capillary thermostat switch. However, on later machines, a simple thermostat is used to open circuit the heater if overheating occurs. On the latest machines, the thermostat is connected in the live supply between the door interlock and the pressure switch. If overheating occurs, the thermostat goes open circuit and cuts power to all other components apart from the door interlock. Hotpoint uses a separate thermal fuse heater protector for boil dry protection of plastic outer tub machines. This and the later Philips thermostat are available separately.

If such items are faulty, the result is a failure to heat the wash water or a failure to move through the programme. If any type of protector is found to be open circuit, the cause must be identified and rectified before the protector is renewed. Thoroughly check the pressure system and switch. For more detailed information on thermal fuses and protection devices *(see Temperature control, pages 85-92)*.

Airflow heaters – metal sheathed

Sheathed elements similar to the water heaters described above are found in both washerdrier heating units and tumble dry only machines. Because space in the combined washerdrier is restricted, the elements used in washerdrier heating units are compact and very similar to their water heating counterparts. On the other hand, dedicated dry only machines that use solid sheathed elements have much longer, often circular elements.

This wash water element shows signs of scaling and fluff contamination.

The faults likely to occur are open circuit, low insulation and short circuit. Details of the various heating units on both washerdrier machines and dry only appliances are given elsewhere *(see Chapter 5, Drying, pages 144-167).*

Airflow heaters – exposed

This type of heating element is used predominantly in tumble dry only machines. It is easily distinguished from the sheathed type of heater. Although several variations and configurations of exposed element are used, they all fall into two basic categories. Simple spiral (spring-like) elements are stretched and supported on ceramic holders either in a large circular configuration or a compact box mount system. Single strand elements are woven into or on to a ceramic or mica support.

Whatever type is used, it must be correctly supported, free from fluff build-up and have all ceramic or mica supports intact. Check closely for any sagging of the element between the supports, because this could lead to short circuit between elements on compact heaters and to shorting to earth on larger circular heaters. Take care when cleaning or removing any fluff because both the supports and the elements become brittle with use and are easily damaged. Two elements are usually used to allow a high/low heat selection.

Do not make twisted joints to repair a broken element. A faulty element should always be renewed, along with any broken, cracked or charred support or insulation. Compact heater units are available only as complete units *(see Dry only machines, pages 154-167).* The large spiral elements like the ones shown on page 82 can be renewed individually *(see Dry only machines, pages 154-167).*

On machines with large spiral elements, a support cord runs through the centre of each element. This is to prevent the element from shorting to the metal heat shield in the event of its stretching or breaking. The support cord must be intact and correctly positioned so that it can support the element if breaking or sagging occurs.

The support cord on early machines was often made of asbestos and, with use, the cord itself became powdery and broke. If you encounter this type of support cord during repair or servicing be extremely careful as the dust can be harmful if inhaled. Dampen the work area to reduce dust, and dispose of the waste carefully. Glass fibre is now used for support cords because it is less prone to deterioration. It should be used to replace the old type of asbestos element support.

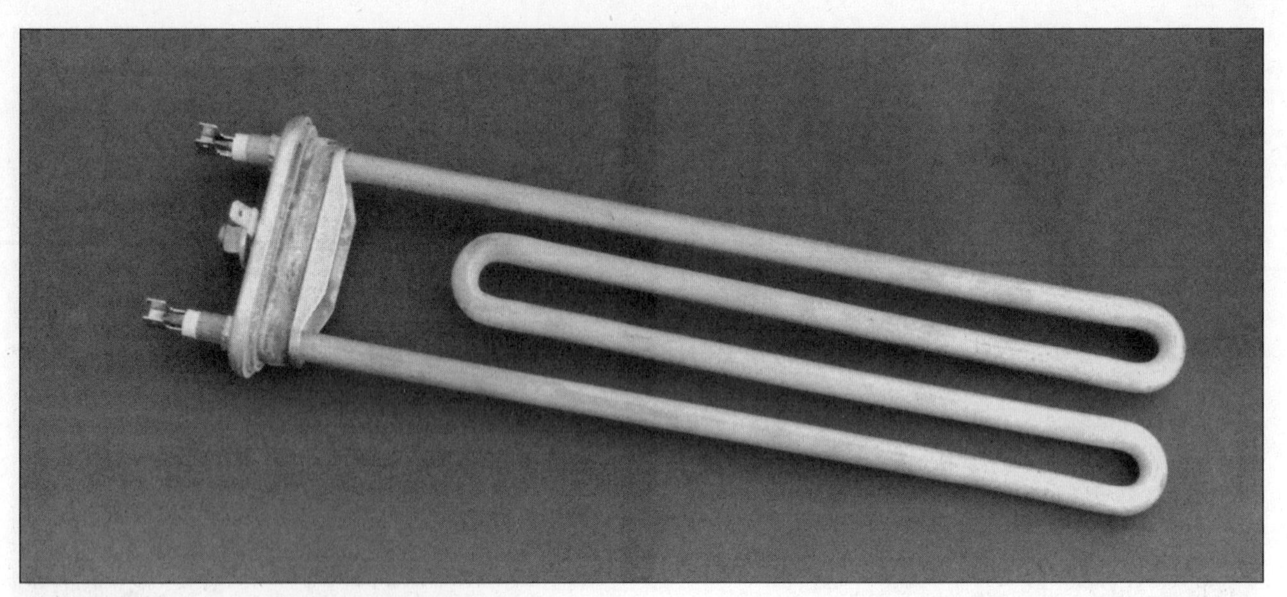

Normal style heater. (Do not fit to plastic tub machines).

Heater with overheat detector (arrowed). This is found on machines with a plastic outer tub.

Temperature control

A thermostat is an automatic device for monitoring temperature – water temperature, the direct heat of the heating element, in which case, it acts like a TOC, *(see Heaters pages 79-84)*, or airflow temperature of combined or dry only machines. The thermostat, sometimes simply called stat, will either 'make' or 'break' a circuit at a pre-determined temperature. Temperature ratings of fixed thermostats are usually marked around the metal perimeter on the back of the stat and are marked NO or NC, that is, normally open contact (closing and making a circuit at a given temperature) or normally closed (opening at given temperature). Some thermostats contain both variants.

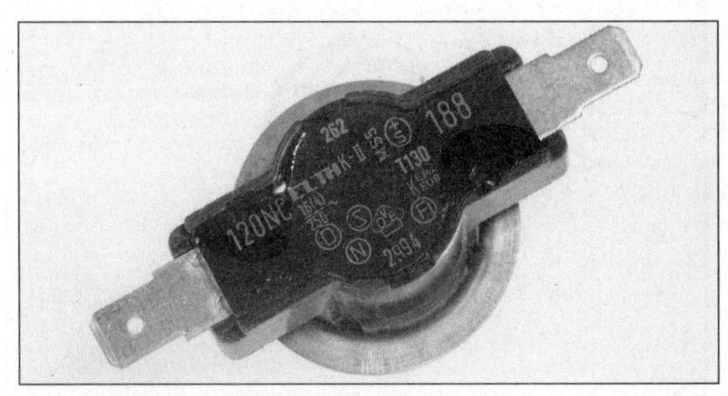

Single thermostat.

Location

The location of each thermostat depends on the job it has to do. The wash temperature thermostat is usually located on the back half or underside of the outer tub, depending on the particular make and model of the machine.

Exceptions to this are Hotpoint and Fagor front loading machines, where the thermostat, heater and pressure vessel are located on the front of the outer tub, directly below the door seal. Access to these components is gained by removing the front panel of the machine *(see Door seals, pages 50-57)*.

Thermostats used to control and monitor drying temperatures are located at a strategic point of the airflow. Thermostats used for overheat protection are located in or near the heater ducting (on combined washerdrier machines) or heat shields (on dry only machines).

Fixed thermostats of the types used in tumble dry only machines for temperature control and overheat safety (TOC).

How a fixed thermostat works

Fig. A (overleaf) shows a typical fixed (non-variable) thermostat, which may have one, two or three settings. Fig. B illustrates how a three-position thermostat works. The power enters the switch at X, but cannot proceed because there is no contact. As the temperature rises, and each pre-set temperature is reached, the bi-metal disc set to that temperature bends, making one of the three possible contacts.

Double thermostat.

Removing and fitting a standard thermostat

Make a careful note of the connections on the back and then disconnect the wires from the rear of the thermostat. Insert a small screwdriver at point (Z), and prise the thermostat away from its grommet fitting. Be careful if

Variable thermostat showing the switches, capillary tube and pod.

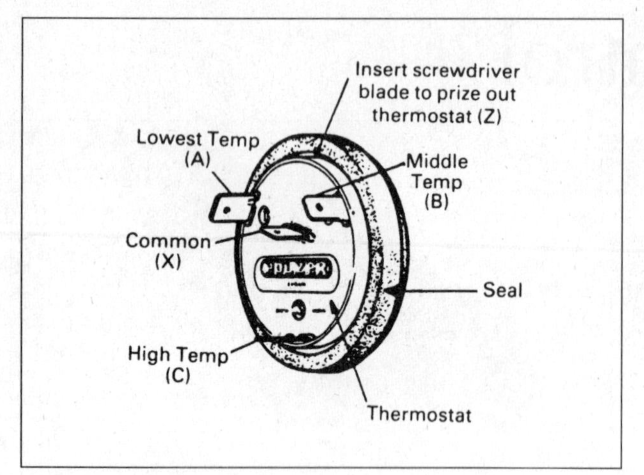

Fig. A

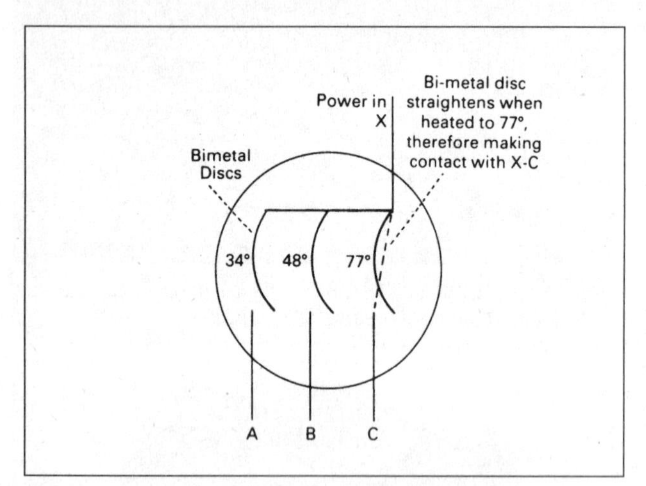

Fig. B

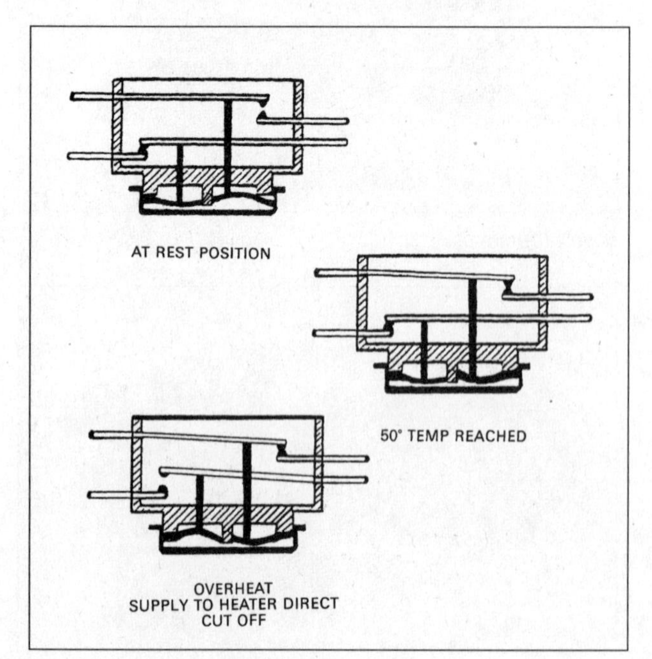

Fig. C - A typical fixed thermostat.

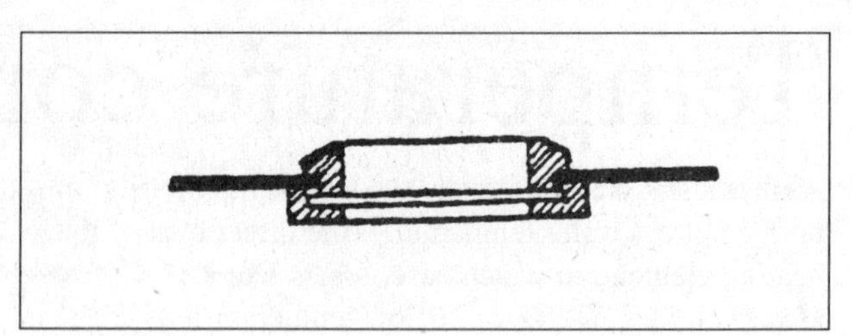

Fig. D

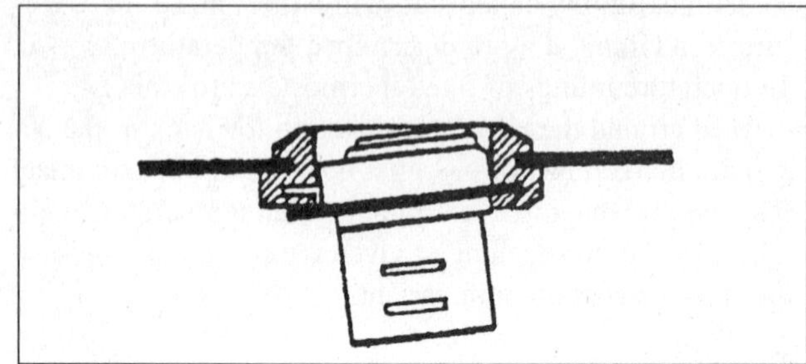

Fig. E

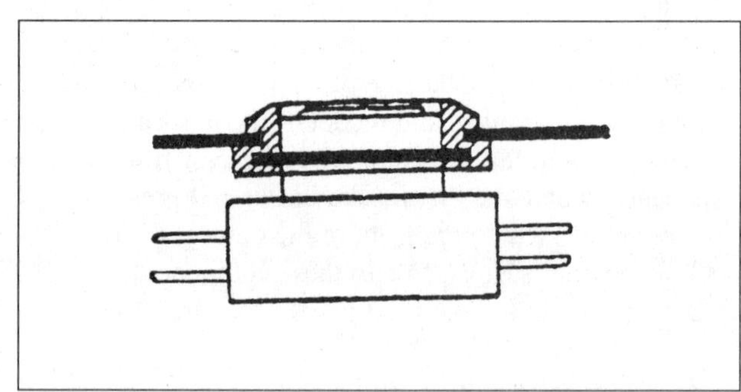

Fig. F

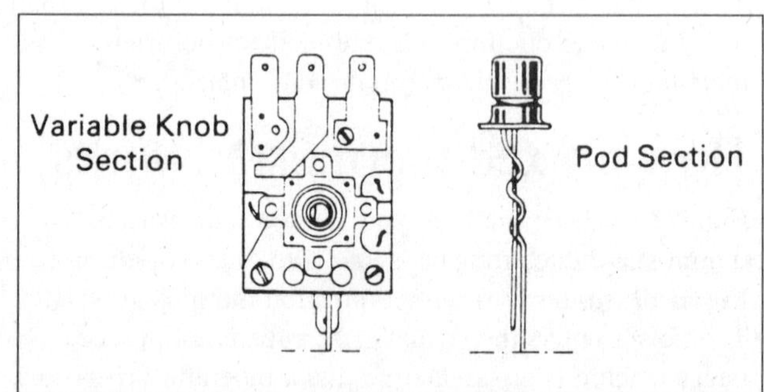

Fig. G - Typical pod type thermostat.

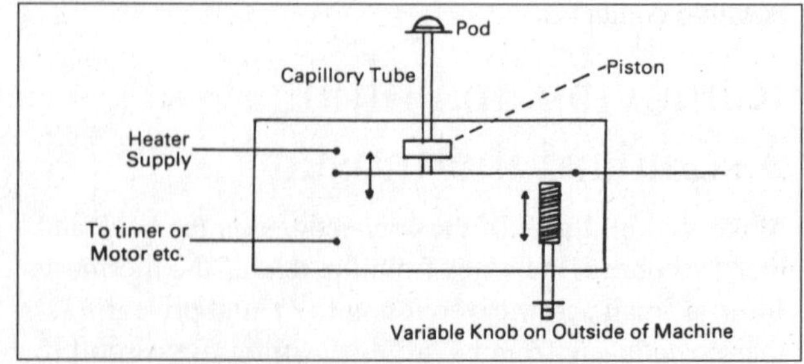

Fig. H - Internal workings of pod type thermostat.

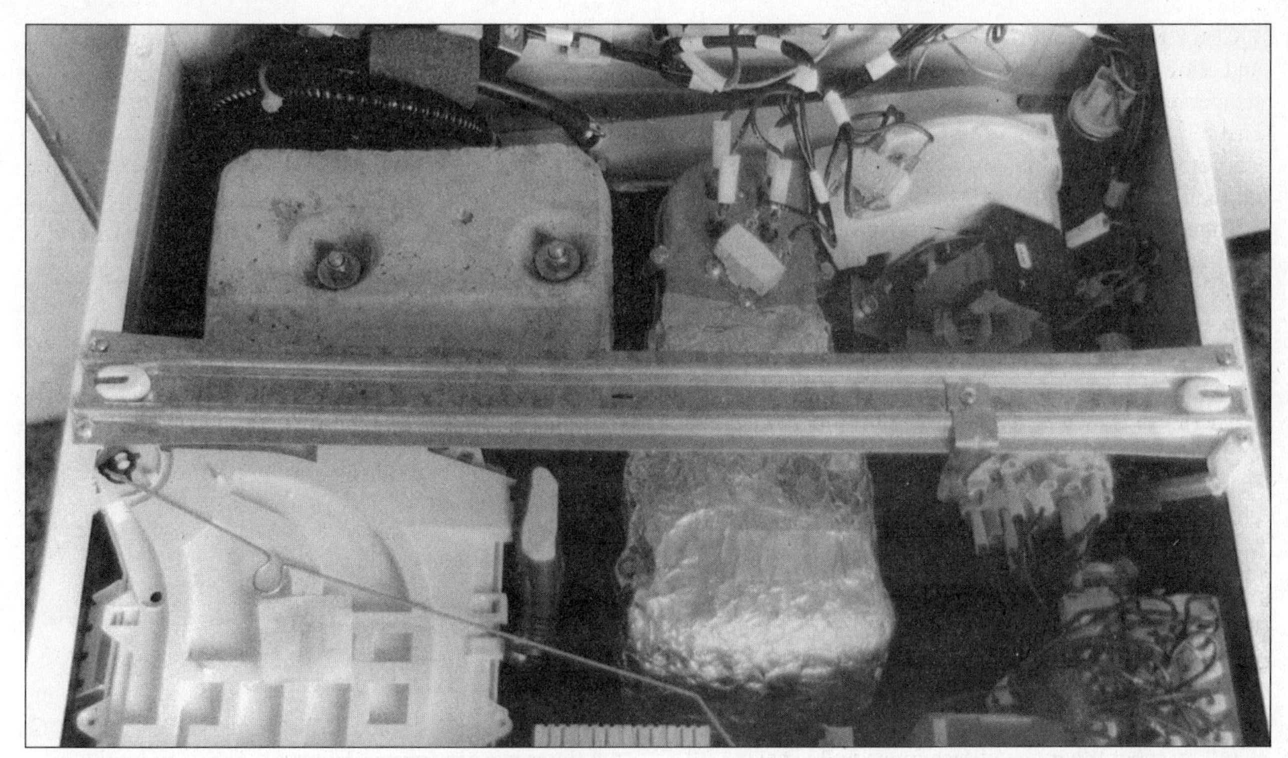

Indicated in this condenser washerdrier are (A) the variable temperature wash stat control, (B) dual fixed temperature air duct stat and (C) fixed temperature stat being used as a TOC to open circuit the air duct heater in the event of overheating.

sealant has previously been used, as this will have glued the stat into position. Before fitting a new thermostat, smear a little sealant on the grommet to aid correct fitting and avoid leaks. Fitting the new stat is a simple reversal of the removal process.

As always, before removing or repairing any component , isolate the machine from the main electrical supply by switching off and removing the plug from the wall socket.

Fig. C shows an alternative style of fixed thermostat – in this instance a 50°C (NO) normally open contact and an 85°C (NC) normally closed contact. The latter is a safety thermostat which operates if overheating occurs within the machine. The diagrams illustrate the position of the thermostat at rest. Bi-metal discs are mounted directly behind the metal front cover of the stat and are pre-set to distort at given temperatures – in this particular instance 50°C and 85°C. They are linked to contact switches by push rods. Any distortions of the discs either 'make' or 'break' the corresponding contacts as shown.

When removed from the machine, the thermostat's operation can be tested by placing the metal cover in contact with a heat source, such as a radiator or hot water, that matches or slightly exceeds the required temperature. Allow a little time for the heat to warm the stat and bi-metal discs. Testing for closing or opening of the thermostat can now be carried out *(see Electrical circuit testing, pages 23-25)*.

The fixed wash temperature stat (D) and pod for the variable stat (E) in position on the rear of the outer tub.

Check the temperature with a household thermometer and allow a few degrees either way of the marked temperature on the outer rim of the stat. Remember to check if the stat is normally NO or NC. Check that the stat returns to its normal position when cool, as indicated on the rim, that is, either NO or NC.

To refit, locate the metal lip in the grommet recess (D) and, with the aid of a flat-bladed screwdriver, ease the outer lip over the metal lip of the stat (C). Sealant will help locate and seal the thermostat into position, (E). Ensure that the thermostat is securely located into the grommet and that the outer lip is not trapped.

Thermostats may also be held in position by metal clips or clamps. Again, make sure of a good seal and check that the clips or clamps do not trap or touch any wires or connectors.

The variable thermostat

Fig G shows a pod type thermostat, found on machines that have a variable wash temperature control. Fig H is a schematic diagram of the internal workings. It consists of an oil- or gas-filled pod connected to the switch by a capillary tube. When the oil/gas in the pod is heated it expands within the sealed system and pushes a diaphragm. The diaphragm acts on the switch thus 'breaking' one circuit and 'making' the other. When the oil/gas cools it contracts, pulling the switch in the opposite direction. The switch is then in its original position and the process repeats if necessary.

Removing and fitting a pod type thermostat

The pod at the base of the capillary tube must be eased from its rubber grommet very gently. While doing this, take great care not to kink or pull on the capillary tube itself. It is extremely important that the capillary tube does not come into contact with any electrical contacts, such as the heater terminals, or with any moving parts, such as the main drive belt. After you have fitted it, check the entire length of the tube for any possible contact with these sorts of item. Also, a coiled section of at least two large turns should be left at a convenient position to absorb the movement of the tub assembly.

Testing

The standard thermostat can be subjected to a known temperature, such as that of a radiator or a kettle, and be checked with a small test meter for continuity. This process is described in Electrical circuit testing *(pages 23-25)*. The pod thermostat can be tested as above, ensuring that only the pod itself is immersed in water.

The state of the thermostat should be determined while at room temperature. On pod thermostats, the lowest and highest setting should be selected. During testing, ensure that the switch actuates both on rise and fall of temperature (see Electrical circuit testing, pages 22-25).

Thermostat operation flowchart

Box 1

The machine is turned on.

Box 2

The timer impulses, fills the machine with cold water and turns the heater on.

Boxes 3 and 4

The thermostat 'waits' until the water has heated to 40°C.

Boxes 5 and 6

When the thermostat closes (the water has reached 40°C), the timer washes for two minutes. (At this point the heater is still engaged).

Boxes 7 and 8

The above operation is repeated, again with the heater engaged. When the two minute wash has ended, the water will be at 45°C due to the extra four minutes heating.

Boxes 9, 10 and 11

The timer moves to the next position, which disengages the heater. It is then ready for the programme to continue as required.

Box 12

For the purpose of this flowchart, the washer will end here as we are only concerned with the operation of the thermostat at this time.

This is only used as an example to illustrate the use of the thermostat, and does not actually represent the way in which a wash is formed. For further information regarding the timer, see Timers *(pages 98-109)*.

This way of using pre-set thermostats gives a greater variation in wash temperatures. A combination of pre-set and variable thermostats, to protect the cooler washes, is common. In other words, if the variable thermostat has accidentally been left at 90°C but a delicate wash has been selected, the pre-set stat would override the variable stat, giving some protection to the wash.

Thermostat Flowchart

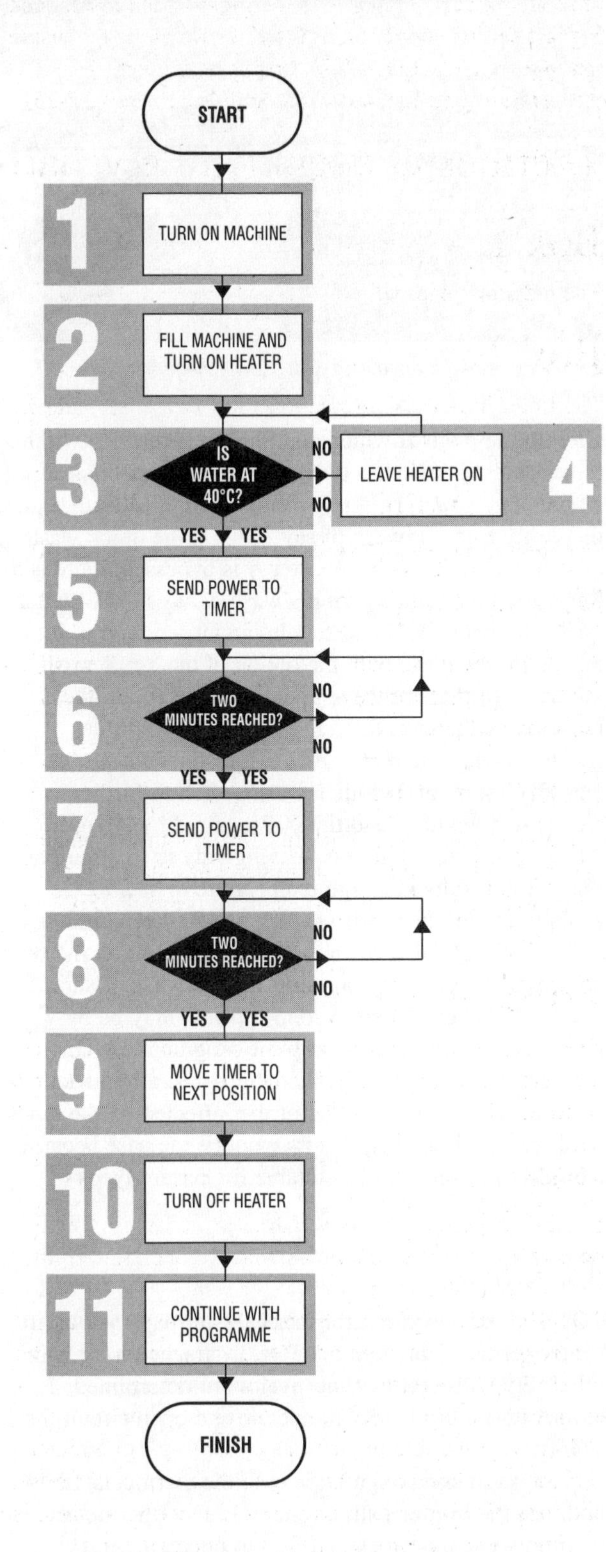

Thermistors

A thermistor is a solid state device used in place of a fixed or variable thermostat. Its particular properties allow it to be used as an infinitely variable temperature sensor that has no moving parts. It cannot go out of calibration, that is, give incorrect temperature resistance values. Occasionally, however, thermistors go 'open circuit', and sometimes connections to and from them may short circuit. Both are faults which give rise to temperature sensing problems.

Location

Like all temperature sensing devices, a thermistor must come into direct or indirect contact with the substance (air/water) or item that needs monitoring. Its location is similar to the other thermostats but methods of fixing differ.

How it works

Unlike other temperature control devices, the thermistor cannot work alone. It is an electrical resistor, and its resistance varies in relation to its temperature. There are two ways in which it varies, depending on the manufacturer and the requirements of the finished product. Thermistors may be positive or negative temperature co-efficient. In simple terms, this means a positive co-efficient thermistor's resistance increases as its temperature increases and, conversely, a negative co-efficient thermistor's resistance decreases as its temperature increases. Thermistors are therefore rated as PTC or NTC respectively.

The NTC thermistor is the version most often used in temperature sensing circuitry in automatic washing machines, such as Hoover, Servis and Hotpoint. A theoretical operation of this is described here. It is essential that only the correct variation of thermistor, which conforms to the rating requirements of the specific make of machine and its circuitry, is used.

The variation in resistance to temperature change forms part of an electronic circuit. The output of this controls the advancement of the selected wash programme in mechanically and electronically controlled machines. On electronically controlled machines – those without mechanical timers – the resistance of the thermistor is monitored directly by the main programme circuit board or sub-module *(see Timers, pages 98-109)*.

However, thermistors are also found on machines with mechanical timers/programmers and these work as follows. The resistance of the thermistor forms part of a temperature control circuit. There may be a separate module solely for this purpose or it may form part of the

Typical solid state thermistor.

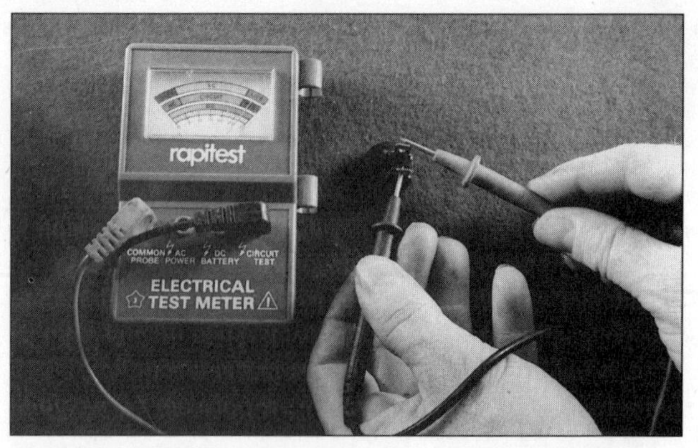

Continuity testing a thermistor with a low-voltage battery test meter.

motor control module, as in some Hotpoint machines.

The output voltage at D and C controls a triac (an electrical component within the circuit). The triac, in turn, switches the thermostop coil on a mechanical timer or impulse to the control panel of a computer-controlled machine *(see Timers, pages 98-109)*. Being an electronic circuit in either a mechanical or a computer-controlled machine, the operating voltage within this portion of circuit will be low (5 volts DC). Therefore, any electrical testing of the thermistor must be with a low-voltage test meter, because voltages of over 9 volts will damage the thermistor or module circuitry.

The two resistors ra and rb are of the same value. A 5v DC voltage supplied to point A will take one of two routes, depending on the resistance opposing it – ACB or ADB. Route ACB has four resistors within it which can be individually switched in and out of the circuit in accordance with the programme selected and temperature required for the particular wash cycle. The diode at point E eliminates reverse supply to the triac.

The route ADB contains the thermistor in its second leg – DB. If we assume that the water within the machine is cold, then the thermistor resistance will be high. This allows a current flow from D to C, thus energising the thermostop or holding the programme on the heat cycle until the pre-determined temperature (governed by the switchable resistors) is attained.

The release of the thermostop or impulse of programme is as follows. As the water temperature increases, the resistance of the thermistor decreases (NTC). At some point, the resistance in both sides of the circuit will be equal and at this point no current will flow between D and C and the triac will switch off. This, in turn, will release the thermostop on mechanical timers or allow impulse to the next stage of the programme on electronically controlled machines.

Variations in temperature are gained by switching in or out the required resistors in the CB leg of the circuit, thus altering the point at which equilibrium is reached within the circuit. All switches open, that is, all resistors in circuit – *rc* + *rd* + *re* + *rf* – results in high resistance and, therefore, a cooler wash of say 30°C. If the quick wash switch is closed, that is, resistor *rf* is bypassed, there is a lower total resistance giving a wash of say 40°C. Switch 1 closed, that is, two resistors in circuit – *rc* and *re* – results in say 50°C, with the option of the quick wash switch to further reduce temperature (and time) if required by the user. Switch 2 closed, that is three resistors – *rc, rd and rf* – in circuit results in say 60°C with the option of the quick wash switch to further reduce temperature (and time) if required by the user. All switches closed, that is, leaving only *rc* in circuit at this point, results in a high temperature of say 90°C, again with the quick wash switch option if required.

The option to alter the temperature selected by the set programme, that is, by pressing a quick wash or short wash option button on the control panel, may be by-passed itself by the timer on some programmes. This means that although in the theoretical operation detailed above, each setting could be further affected by the quick wash switch, in reality, this may not be the case because a quick-wash may not be suitable for certain types of wash loads.

TOCs

TOC is an abbreviation for thermal overload cut-out. In simple terms, if the item the TOC is attached to or is in proximity with gets too hot (over a pre-determined temperature), the TOC will operate and open circuit the supply.

It works in a very similar way to the thermostat; they both use the bi-metal strip system. In fact, thermostats are used on some machines as TOCs to open circuit the heater, for instance, if an overheat fault occurs. However, the term TOC generally relates to the smaller devices embedded within or on top of motor winding coil of all

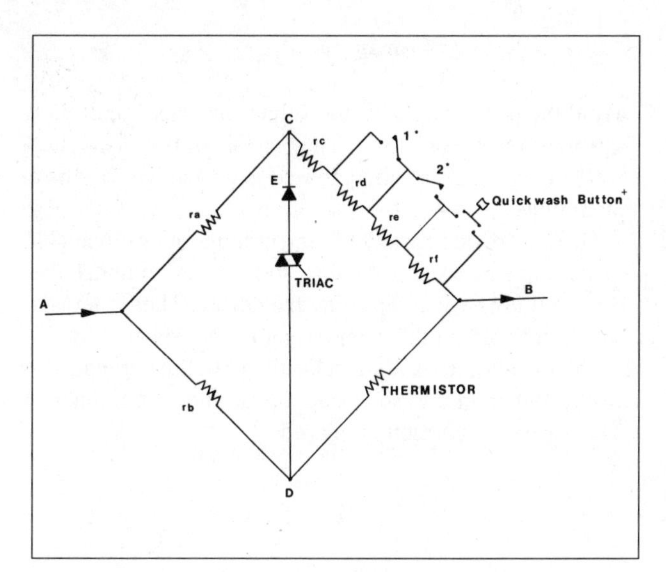

** = Switches controlled by timer wash selection. These switches would be electronic not mechanical on computer-controlled machines.*
+ = Button on front of machine, that is, user variation option.

types and above the heaters of many makes of tumble-driers.

There are several variations in style and size, each of which is matched to its particular use and position within the appliance. It is essential that only the correct style and temperature rating is used when renewing a TOC. Those used to protect motors, pumps, wash motors, tumbledrier motors and so on are rarely renewable and usually form part of the original winding or moulding.

If a TOC has gone open circuit, it will have done so for a reason. Therefore, this safety item must not be by-passed. Always ensure that any replacement component contains a TOC. Some pattern spares may miss out this fundamental safety device in order to cut production costs, so take note that such an omission could be unsafe.

Although designs and ratings of TOCs vary, there are two basic types of operation. The self-setting TOC is like a thermostat and resets when a normal working temperature returns. This may result in a cycling of the fault. For example, if a tumbledrier outlet or filter is blocked, the heater element overheats because of the reduced airflow. A safety TOC open circuits the heater supply. Unfortunately, on all but the most recent machines, when the heater cools, the TOC also cools and so the power is then returned to the heater and the cycle may carry on until either the pre-set drying time expires or the element or TOC fails.

The second type is the manual reset TOC. Its action is identical to the self-setting TOC except for one main difference. Once tripped, it cannot reset itself and has to be reset manually, usually by simply pressing a button or rod. This must be done with the machine isolated – unplugged – and only after the cause of the tripping has been eliminated.

Some washerdrier machines incorporate both a self-setting TOC, for low temperature fault levels, and a manual reset, if a cycling fault should occur or if the self-setting component should fail (in the closed position). Tripping of both self-setting and manual TOCs is common in both washerdriers and tumbledriers if there is a build-up of fluff or the air flow is otherwise restricted. Keep filters clear, and investigate any tripping of the safety devices immediately, in order to rectify the cause. A delay of up to four minutes (sometimes more) is possible before manual TOCs can be reset.

Tumbledriers manufactured from 1990 onwards incorporate a manual reset TOC as standard. Machines that have this facility have a reset button accessible on the outer casing, although the precise position varies from make to make. Before resetting, ensure all possible causes of faults have been eliminated or rectified.

Bladed self-setting TOC in position between the spiral elements of a Hoover tumble dry only machine (centre).

The button of the manual reset TOC is located on the rear of this tumble dry only machine. Manual reset TOCs can also be found in some combined machines.

Thermal fuses

Many small appliances now have this type of overheat protection device and it may also be found in some makes of washerdriers and tumbledriers. It is essentially a fail-safe device which, when actuated by a pre-determined temperature, goes open circuit. Once open circuit, it cannot self-set or be manually reset. The thermal fuse has to be renewed once the device has gone open circuit. Renew only when the fault that caused it to operate has been corrected.

It is a solid-state device and contains no moving parts or contacts which may in themselves fail. Thermal fuses avoid the possibility of fault cycling, are cheap and, in most instances, easy to fit. They are small but easily recognisable and are often housed in protective sheaths or mouldings.

There is a wide variety of temperature ratings available to suit the various applications, so take care to match the original rating when replacing the device. The rating is usually printed on the outer casing of the device.

The metal outer is not insulated and will be at mains voltage when in use, so ensure that all mounts, fixings and covers are correctly replaced.

Suspension

A great deal of vibration is produced during the normal washing and spin drying operations of all front loading machines. The level of vibration increases during the spin sequences, especially if the wash load is out of balance or severely under-loaded. If the outer tub unit were fixed rigidly to the shell or outer casing of the machine, both internal components and the area surrounding the machine would be damaged as a result of the excessive movement of the free-standing machine.

To avoid transferring this vibration, the inner tub unit of front-loading washing machines and washerdriers is supported within the shell of the machine by vibration absorbing supports – suspension. Because of the confines of the shell of a modern machine a limited amount of movement is allowed, but any excessive vibration and movement of the tub unit is removed or damped by the action of the suspension system supporting the unit.

The suspension is the system that controls all the movement of the tub unit when it is in use. Without the suspension, or when it is damaged, the whole tub unit moves violently when in use.

Large concrete or metal counterweights are used to improve balance during the spin cycle. The position of weights varies between makes. Ensure they are securely anchored and cannot work loose. Loose weights give rise to noises similar to bearing failure. If loose weights are left unattended, serious outer tub damage results. Shown here is a Candy outer tub unit with both top and lower front weights.

Transit fixings

To prevent damage to components within the machine when it is being delivered, some sort of packing will be fixed to the machine to stop any movement of the tub unit. It is essential that this packing, or transit fixings as they are commonly known, is removed. There are as many different types of transit packing as there are machines, so read the manufacturer's instructions to find out how to remove the packing from your new machine. It is advisable to retain the instructions and the packing in case you need to transport your washing machine again, when moving house, for example. If you do have to refit the transit packing, put a conspicuous sticker on the door to remind you to remove it before using the machine.

Types of suspension

The spring type suspension, which is simply large strong support springs, was used only on early slow spin machines.

Slide and damper types

Slide and spring damper suspension supports the tub from beneath with only small springs or straps at the top for holding the tub unit in mid fore and aft position.

The main faults to check for are those of guide wear. This allows the shaft to jump out of position and the top rubbers to soften or wear, resulting in a phenomenon called 'twisted tub'. The reason for this is that the suspension on one side of the tub is not correctly positioned, so that one side of the tub bangs on the side of the shell and causes damage.

A noise fault may also become apparent at the top of the suspension as a result of soapy water seeping into the suspension via the dispenser or dispenser hose. This is best removed by a spray of lubricant/moisture repellent and an application of Molycote to the top bush and slides. The top and the guides of the suspension are the only parts that should be lubricated in this way.

Lay the machine on its face when you are fitting top rubbers; the suspension should be held tightly with grips at the top end only. The top nut can then be undone. Do not hold the bottom of the shaft as any marks will quickly wear the plastic guides. When refitting, thoroughly clean the metal shaft and apply a smear of Molycote lubricant paste to the shaft and upper shaped washers. Apply the same paste to the plastic slides before refitting.

View of slide and spring suspension. This is prone to softening of the rubber mount at the top of the suspension leg. The result is a common condition known as 'twisted tub'.

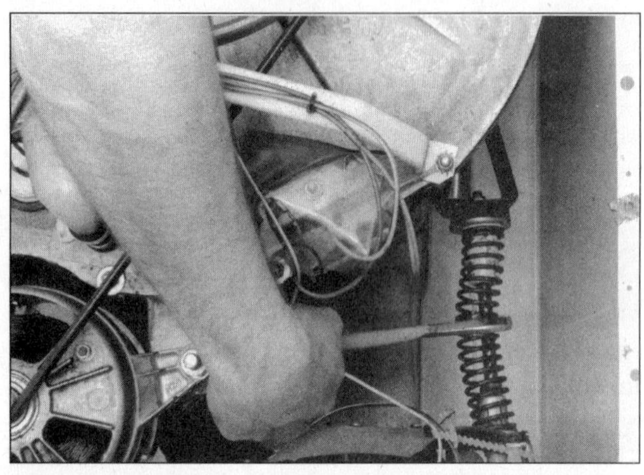

To renew the top rubber, the whole unit will have to be withdrawn. Grip the shaft through the spring at the top only, using adjustable pliers inserted through the spring.

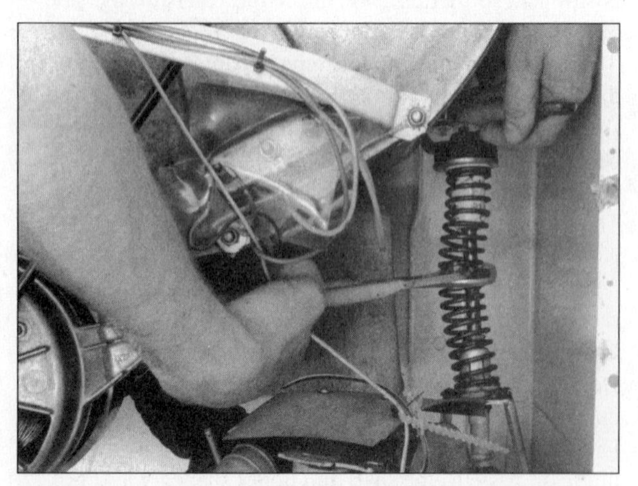

While you grip the metal shaft tightly, the securing nut can be removed (right-hand thread). Note the correct assembly of parts, and pull spring downwards to remove top bush. Do not be tempted to renew one side only – you must renew both sides.

The friction damper system

The friction damper system is not unlike the disc braking system on a motor car. Two support rubbers with pads made from asbestos or a similar material are mounted in two spring steel arms. These rest either side of the flat plate attached to the outer tub. When the outer tub moves, the action is slowed down or damped by the friction of the pads against the plate. This is a very cheap and extremely effective form of suspension.

When this type of damper is worn, the tub moves excessively and may squeak. The noise is caused by the rubber pad mounts coming into contact with the moving plate because the friction material is worn. This is easily overcome by renewing the pads.

Isolate the machine (switch off, plug out and water off) and lay it on its back or side so that the steel spring arms can be opened. Then the pads can be prised from their ball and socket joints.

If the pads on this system become glazed and/or shiny on their contact faces, they will make a chattering sound. It may be possible to avoid renewal by slightly roughening the faces with sandpaper to remove the glazing and then refitting them. However, if this does not work, the pads will have to be renewed.

Never put oil or grease on friction damper systems. Do not inhale the dust from the friction pads as it can be harmful to the lungs. Dampen with water during removal and cleaning to avoid airborne dust particles. Do not blow them clean. Dispose of old pads safely and wash your hands after contact.

The damper and spring system

The damper and spring system is similar to the shock absorbers on a car and it does the same job. If the smaller version of the system is used, the tub does not actually rest on the damper. It is hung from springs at the top of the tub, using the dampers at the bottom for shock absorption only. In the larger systems, however, the tub is held only by much larger dampers at the bottom of the tub with retaining straps/springs at the top to limit the amount of movement fore and aft.

Faults found with this type of solid damper have exactly the same symptoms as those of the friction damper system. The only remedy is the complete renewal of the faulty damper.

Isolate the machine and lay it on its face. Access to the dampers can be gained by removing the back panel and unbolting the damper from the shell and tub mounts.

The spring only system

The spring only system may also be found. The spring or the mounts can be changed separately if required or replaced as a complete unit. Note that the left and right springs of all systems are usually of different ratings, so be sure to specify the required side when obtaining a replacement.

Where two small springs are used for fore and aft support, it is quite possible that they may become dislodged during repair. It is essential that they are refitted correctly. Examine the correct positions and make notes of all springs and so on before you start.

Combination systems

It must be stressed that any combination of these systems may be found. A damper system may complement a spring system, or a spring system may complement a friction damper system. It is advisable to read the information on all types thoroughly before starting any repair on the suspension system.

Out of balance

The purpose of suspension in the automatic washing machine, whether wash only or combined washerdrier, is to damp the oscillations of the spring-mounted tub and drum unit. Counterweights of concrete or, in some instances, iron are used to help in the overall balance of the unit and to add weight for the further elimination of movement during use.

The suspension system has to work at its hardest during the distribute (pre-spin) and spin cycles. Under normal load conditions the simple suspension and counterweight system works extremely efficiently and well and reduces tub oscillation as long as all components and connections are secure and in good order. However, if a severe 'out of balance' condition occurs, excessive vibration may result. This will damage both the outer cabinet and internal parts because the suspension cannot cope with such adverse oscillations of the tub unit within the confines of the shell/cabinet.

Such an out of balance condition may result from a mechanical fault, such as worn suspension, underloading, such as one bath mat, overloading, such as a large duvet or heavy blanket, or by washing unsuitable items, such as trainers.

All but the earliest front loading automatic machines have a pre-spin speed – often called distribute – to balance out the wash load by rotating it at a pre-set drum speed. The centrifugal force created by the pre-set speed, usually around 83 rpm on the drum, arranges the wash load evenly over the inner surface of the drum before acceleration into the spin. However, this process can fail for several reasons.

- A balled load, that is, the knotting together of items. This usually results from poor loading – bundling all items into the machine together instead of separately.

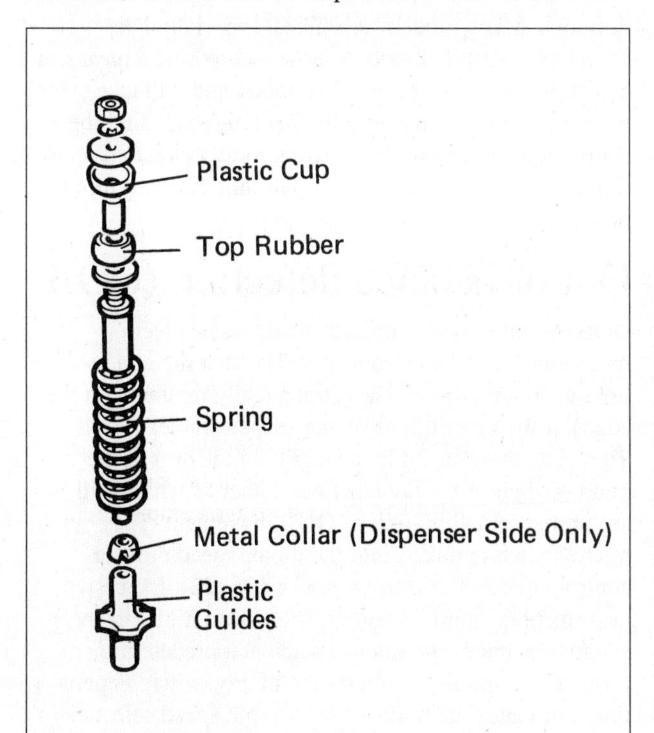

Spring-only unit Hoover type.

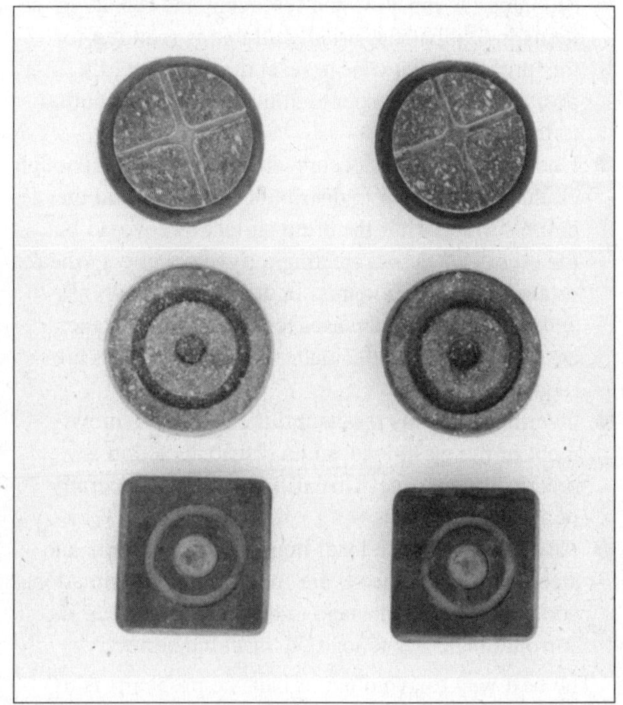

Friction pads from various machines: top – pad and mount for Zanussi, centre – pad for Indesit, bottom – square pad for Candy.

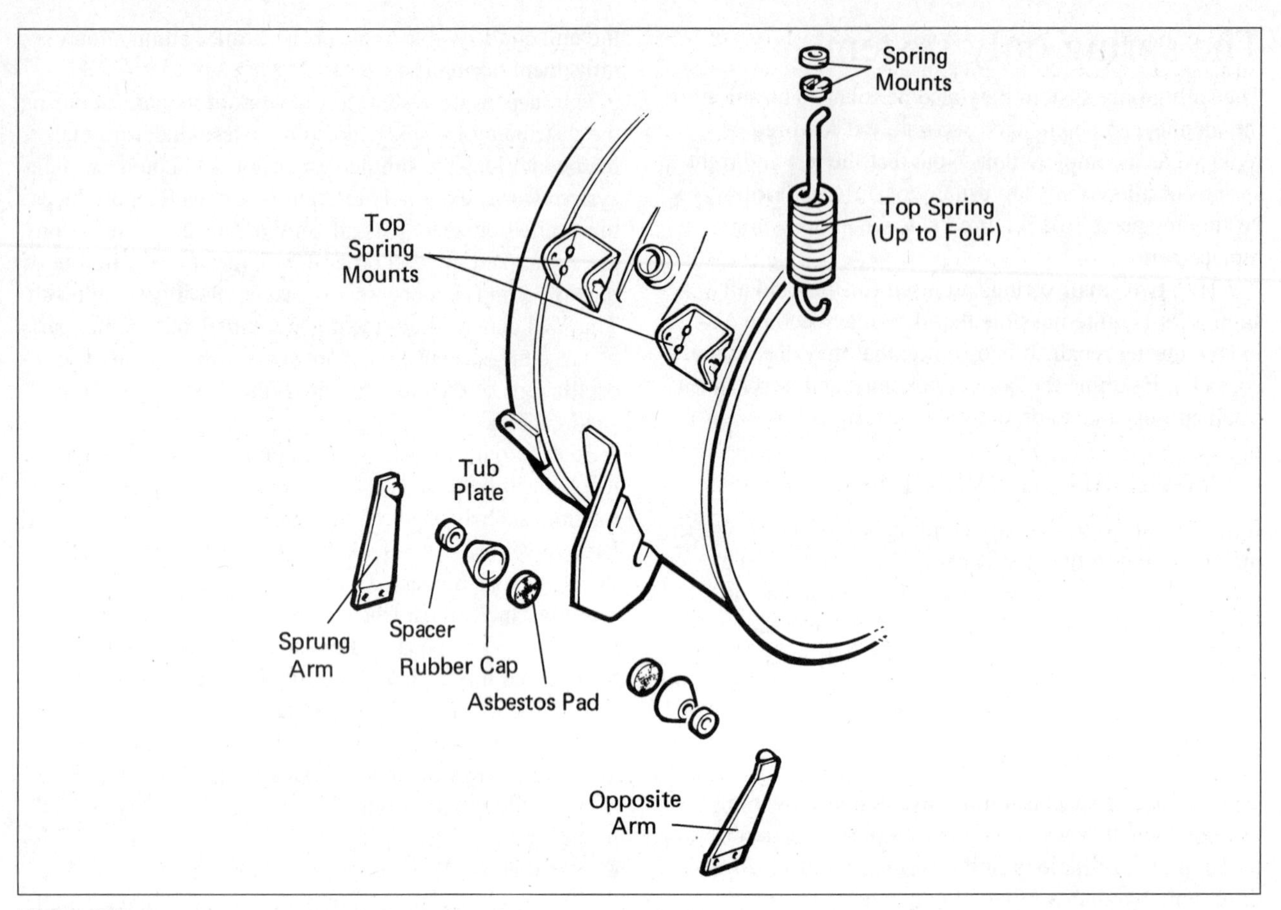

Friction damper.

Stopping the machine and removing and replacing the items individually is usually all that is required for this problem. It is wise to reset the machine to a rinse position before the spin to allow correct distribution to take place.

- Under-loading. This occurs when there are not enough clothes in the drum to distribute evenly around the entire surface. Half the drum surface is covered but the other half is not, creating a flywheel effect when rotated. This is a common fault on some machines and aggravated by the user removing items in the hope of improving the matter when extra items are required.
- Overloading. This results in little or no free movement of the wash load, so no distribute action is practically possible. This also results in a generally poor wash.
- An unsuitable wash load. Items such as trainers and sleeping bags create severe 'out of balance' situations and subsequently damage to the machine and/or its surroundings. Try to load the machine correctly.

The best way to avoid out of balance problems is to load and use the machine correctly. However, even after doing this, you may find an 'out of balance' situation still develops. Many modern machines, especially later computer-controlled models, now incorporate a means of detecting an 'out of balance' situation and will take steps to rectify it by re-balancing the load, by extending the distribute phase, or by terminating the acceleration up to full spin speed thus limiting or avoiding further vibration or damage.

Out of balance detection (OOB)

OOB detection works in one of two ways: electro-mechanically and electronically. The first method uses light action micro-switches strategically mounted on the edges of the outer tub or on the suspension legs. They may be actuated by a weight on the arm of the micro-switch or by a contact bar, either of which will be adjustable for calibration of OOB movement. The micro-switch is linked into the motor speed module control circuit *(see Motors, pages 110-122)*. Excessive movement or inertia resulting from an 'out of balance' situation actuates the micro-switch at a predetermined level. The impulse caused by the micro-switch's operation terminates the build-up to the spin speed selected and normally allows only the distribute speed to operate.

After a period of time governed by the circuitry of the module, a spin sequence is reinstated in the expectation that a second distribute has cleared the OOB problem. If it has not, the process is repeated. With mechanical programme timers, this process may continue, depending on the make and model of the machine, until the time allotted for spin has elapsed. Computer-controlled machines are often programmed to accept only three OOB impulses before terminating the spin or remainder of the programme completely.

The setting of the mechanical OOB detection micro-switch differs greatly between makes and even between models produced by the same manufacturer. Therefore, no specific adjustment details can be given here.

The electronic OOB detection system is at present limited to computer-controlled machines. It is only the detection process that differs from that already described. In this instance, detection is by monitoring the motor speed reference voltage from the tacho coil *(see Motors, pages 110-122)*. During an 'out of balance' situation the motor speed varies, that is, it rises and falls on each revolution of the drum proportional to the degree of the imbalance. If this reference exceeds the pre-programmed tolerances for that particular programme setting or spin, a fault sequence similar to that described above is implemented and distribution occurs. As before, this may be limited to three such cycles before termination of the set programme and, in the case of computer-controlled machines, the display of a corresponding fault code.

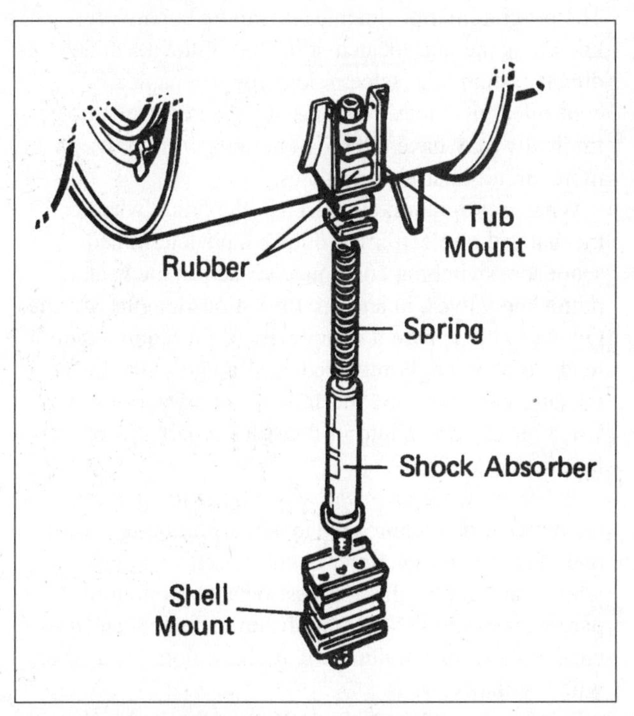

Large damper mounts

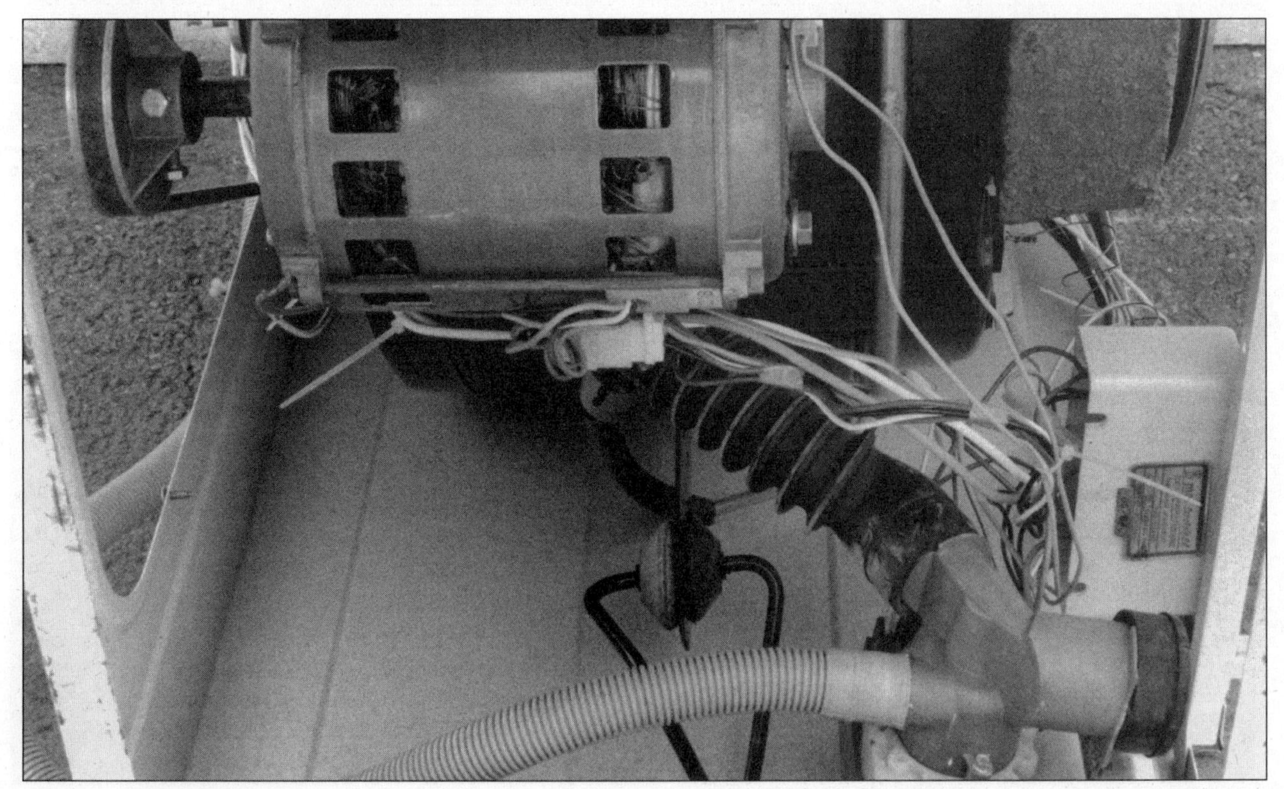

A friction damper system from a Fagor machine. This is viewed in situ, because the complete front panel can be removed. The front panel removal is easy: remove the lower panel and two securing screws. Do not lubricate this type of suspension.

Timers

The programmer or timer, as it is more commonly known, is the unit located at the top of the machine directly behind the selector knob on mechanically controlled machines. Machines that are controlled electronically may have the programmer split into two or more circuit boards or modules.

When a programme is selected, the timer, whether mechanical or electronic, follows a predetermined sequence, switching components, such as the heater, pump and valves, in and out for various lengths of time. Owing to the apparent complexity of this component, it tends to be wrongly regarded as a 'no go' area. In fact, the automatic washing machine is not mysterious and when broken down into its constituent parts, its operation is revealed as simple.

However, to describe the way a timer in a particular machine works requires the make, model number, date of manufacture and the timer number itself to ascertain which variation of timer and associated programmes a particular machine has. Manufacturers have seen fit to change their timers, numbers, wiring colours and so on with regularity.

Timers cannot be repaired. Without detailed information of the switching sequences of the faulty timer, internal faults are difficult to trace. Complete unit changes are needed for internal timer faults. However, on some timers, drive motor coils can be renewed if a simple open circuit has occurred *(see Electrical circuit testing, pages 23-25)*.

Units can often be difficult to fit, unless a logical approach is used. However, modern timers are very reliable and new units are often relatively inexpensive, although the actual cost differs from make to make.

ETN-style timer with two internally mounted motors, one on each side.

However, the price variations for similar parts, such as the cam barrel, can be extremely wide.

When a fault is suspected, it is not always the most complicated component that causes the most trouble. If a process of elimination is used, and all other parts of the machine are found to be working correctly, only then should the timer be suspected, except in the case of obvious failure, such as a burn out or damage to the timer.

Ensure that the power is turned off and that the plug is removed from its socket at all times. Do not remove the timer from the machine at this point. The removal and replacement of the timer can be a long and tedious task

Two timers are normally used on combined washerdrier machines – one for programme selection (knob on the right) and one for selecting the length of the drying cycle (knob on the left).

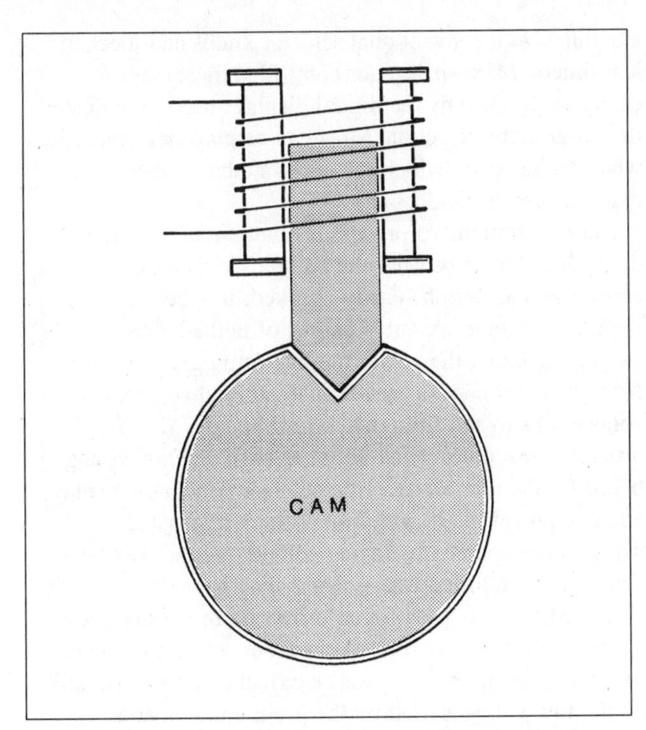

Fig. 1 Thermostop coil in de-energized mode preventing the timer cam advancing until the temperature required closes the thermostat and supplies the coil with power.

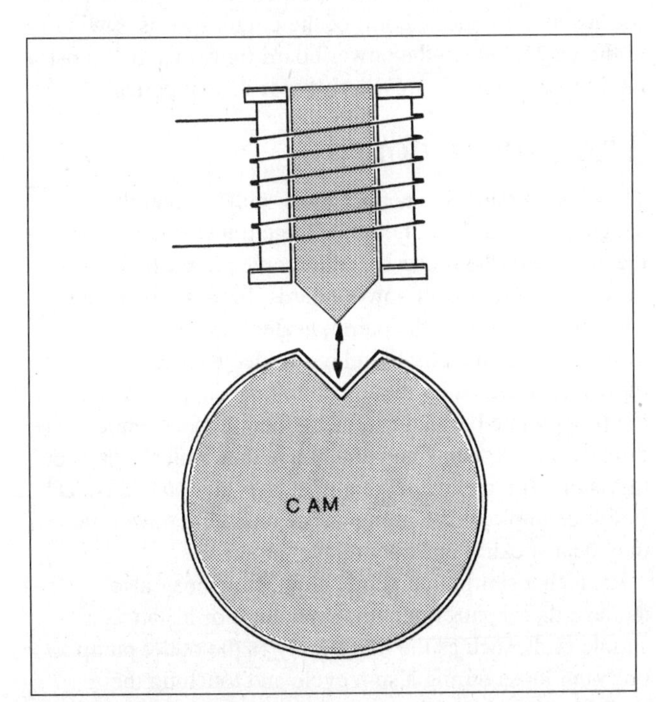

Fig. 2 Thermostop coil in energized mode allowing the advancement of main cam barrel, that is, thermostat closed and supplying power to coil.
Figs. 1 and 2 depict a thermostop system operating with normally open stats (NO). Versions with normally closed stats (NC) work in the reverse manner: when energized, main cam advance prevented; when de-energized, main cam advance allowed. Both variations are found – be aware of the way your particular system functions.

on some machines and should not be undertaken lightly. However, several more-modern machines have an improved way in which the wiring harness is fitted to the timer. Multi-block connectors are now used on many machines making timer renewal much easier. Do not fall into the trap of replacing the timer because of the ease of the job. Correct diagnosis of the fault, and a methodical approach to both fault finding and perhaps subsequent replacement of the timer, is essential.

Do not remove any wiring, but thoroughly check for any overheating of the connections to and from the timer spades (connections). For example, if a fault is suspected in the heater switch, trace the wire from the heater to the timer. This gives the location of the heater switch, which should be examined for any signs of burning or being loose, to confirm your suspicions.

Replacing a timer

Having decided that the timer is at fault, you should note all the numbers that are on it, together with the make, model, serial number and age of your machine. With this information you can obtain an exact replacement. Once you have the replacement, check that it looks identical to the old one, as manufacturers will not exchange timers once they have been fitted.

Having confirmed that it is the correct replacement, and read any accompanying documents, you can proceed to swap the wiring. Place the new timer in the same plane as the original and swap the wires or block connectors one at a time. Although very time consuming, this is by far the safest method.

An alternative method is to tag each wire with its corresponding timer connection number – 2H, 3B etc – before removal. However, a mistake at this point would be almost impossible to rectify without a wiring and timer diagram. It is advisable to ask a colleague to supervise operations.

Some timers have small metal clips that join or link terminals together. These generally are not supplied with the new timer. Ensure that they are swapped from the original.

When all connections have been exchanged, the timer can be fitted into position. Ensure that any parts that are connected to door interlocks etc, are positioned correctly, double-check the work carried out and check the earth continuity. Once the timer has been fitted into position, and the covers have been refitted, the power can be turned on and a functional test programme carried out.

Electronic timers

The functional parts of computer-controlled machines, such as the motor, drum and pump, differ little from

Internal view of thermostop type timer: main cam barrel right, narrower timing and motor reversal cam centre and thermostop coil and mechanism on the left.

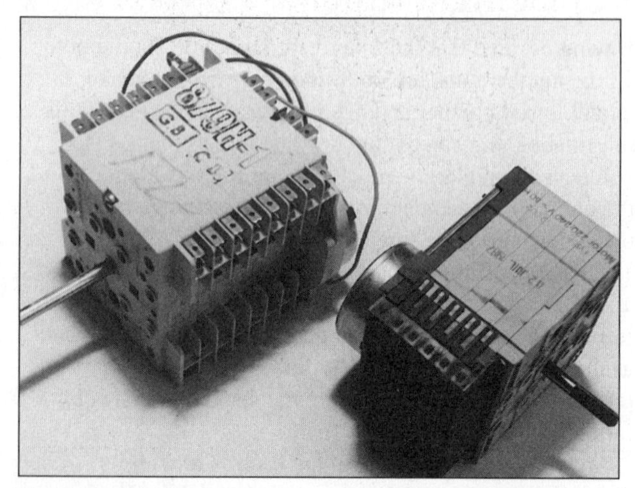

Early and late Crouzet type timers. The timer on the left has individual push on connectors, whereas the timer on the right has round pin block connectors to one side only and is, therefore, much easier to fit.

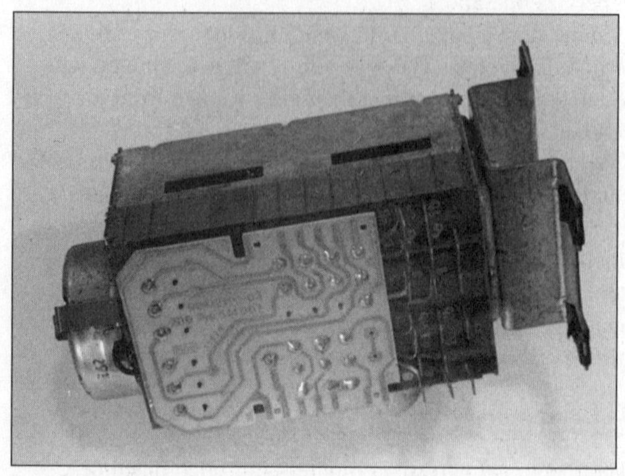

This timer from a Hotpoint machine has both edge connectors and individual amp tag connections to the front switches.

machines with conventional selector knobs and mechanical timers. Micro-processor controlled machines are easily recognized by the digital displays used to indicate the programme selection. Most can display an error code when faults occur, which relates to a table in the manufacturer's handbook.

Faults within micro-processor timer circuitry may be difficult to locate because the complex circuit board components cannot be easily checked. It is best to eliminate all other possible causes of faults before suspecting either the power module or programme unit. If all the other checks prove satisfactory, then check all connections to and from the control boards. Micro-processor machines generally have two: one low voltage board for the micro-processor, selector panel and display, and one power board with transformer, relays and thyristors to operate the mains voltage switching. Computer-controlled washerdriers may have three boards. The third board is used for auto sensing of the dry cycle.

The connections to printed circuit boards are prone to oxidation, giving rise to poor electrical contact, especially on the low voltages used by the programme boards. Check closely for poor connections. If the fault remains after all other components have been checked and found to be alright, the only option left is to change one or other (or in some instances, both) of the circuit boards. Owing to the way in which the power board functions, it is most likely to be a fault in this circuitry or its components.

The power board

The power board is generally much bulkier than the programme board and houses a large transformer to drop the voltage to the processor. Electronic circuit faults occur more often with power boards, because the mains switching, operating the pump, heater and so on, is switched mechanically by relays or electronically by thyristors operated by the lower voltage supplied from the programme board. Processors themselves cannot directly switch mains power and use mechanical relays or thyristors. If a mechanical fault is suspected, try to isolate it. For example, if the heater is not receiving power, there may be a sticking or faulty relay.

Items that short circuit and 'blow' fuses may also damage their control relay or thyristor. For instance, a simple fault, such as the live supply to the outlet pump breaking loose during a spin cycle and touching the earthed metal shell of the machine, would result in a direct short circuit. This fault on a machine with a mechanical timer would 'blow' the appliance fuse in the socket and rectification would be straightforward. However, the same fault on a micro-processor controlled machine is likely to damage the components of the circuit board used to switch the pump supply, unfortunately resulting in a much more expensive repair.

The programme board

The programme board may also be referred to as the display board or module. It differs from the power module in that it is much slimmer, although often wider, than the power board and lacks the larger components, such as relays, transformers and heat sinks. It is usually mounted behind the front fascia of the machine although the exact position varies between manufacturers and also between models within each range.

Obtaining individual components from the manufacturer is not possible because only complete boards/modules are supplied as spares. In any case, faults within the components or circuitry of the board are extremely difficult to trace.

Before contemplating a board fault, ensure all other components within the machine are alright and that all connections to and from the printed circuit boards are in good condition and firmly pushed into place. Check all connections, wiring and protective covers. Isolate the machine and look closely at all connections that carry mains voltage when in use, because loose connections can cause overheating and interference, which may affect the processor chips. Include all earth path connections in the wiring checks and renew any that are loose, have cracked covers or show signs of damage. Do not forget to include the plug connections.

With persistent or unusual faults, check the condition

Block connectors like the one shown are becoming more popular. Five individual connectors are used on this timer and must be fitted correctly. Ensure that connections within the block do not push out when fitted.

This AKO timer has an exposed thermostop coil and slide operating a plastic arm.

Typical timer with thermostop coil and plate to lower right-hand side. The single timer motor and thermostop mechanism is clearly visible with the rear cover removed. Timer coil renewal is also possible on this type of timer.

Temperature plus time programmer in situ in this condenser washerdrier machine.

of the supply socket. Such faults include intermittent operation, random displays or the machine works for short periods and then blanks memory. If a poor connection exists between any pin of the plug and its supply connection, interference will be present, which may corrupt the processor *(see Electrical basics, pages 10-13)*. Renew if suspect.

A faulty suppression unit may also be the cause of such random and difficult-to-trace faults. Ensure the unit is securely earthed *(see Suppressors, pages 127-128)*.

Some items, such as the main wash motor, may have small suppression units called chokes *(see Suppressors, pages 127-128)* to prevent any interference being transmitted along the wiring of the machine. Ensure that any chokes fitted within the machine are secure and in good condition *(check continuity, see Electrical circuit testing, pages 23-25)*.

If all the checks mentioned prove satisfactory, then it is quite likely that the processor chip is corrupted or the board has a fault within its circuitry. Both faults require a replacement unit. However, there are one or two other things you can do before taking this step.

Note all connections to and from the unit and remove it from the machine. (Note the advice regarding handling of the unit given below.) Inspect it closely for dirt or debris which may be affecting the circuit or its components. If any dirt is present, blow it free; do not use metal items, such as screwdrivers, for cleaning. Check the board for cracks or moisture damage because of faulty covers.

Finally, inspect the printed circuit board's soldered connections to verify that they are sound. Loose or poor connections, called dry joints, can often be easily rectified.

If all these checks prove negative, a new unit will be necessary. Take care to fit the unit correctly on all its mounts and ensure all covers and connections are sound. Take particular care to avoid direct contact with the components of the unit. Do not touch the processor board's components for any reason, because they are sensitive to static electricity and are easily damaged by careless handling. Power boards are more robust but care must still be exercised when handling them.

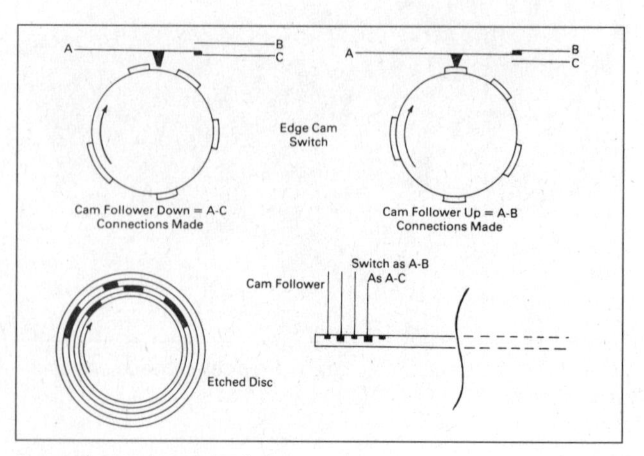

Types of timer internal switches.

FAULT FINDING WATCHPOINTS

1 **Carry out all checks with the machine isolated:** taps off and the socket switched off and the plug out.

2 **Do not try to test the processor board with even a low 9 volt or similar tester** because the micro-processor chip can easily be damaged.

3 **Use the 1.5 volt tester for continuity testing of the wiring** between the module units. Try to leave the block connectors in place and, using the probes of the tester, check for continuity between exposed printed circuit points close to the connector blocks. This tests both the wiring between the modules (usually ribbon cable) and the connection to the printed circuit board.

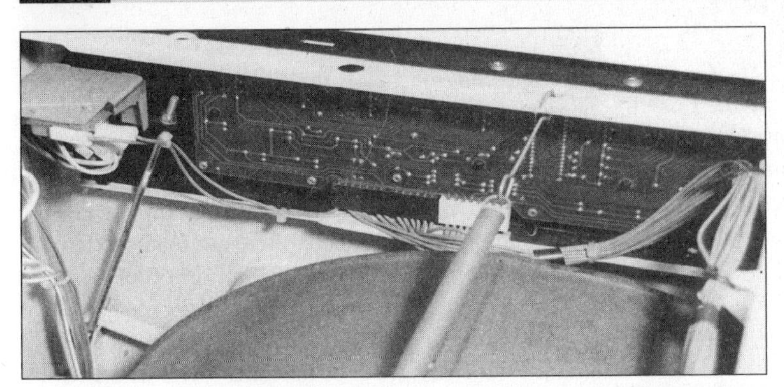

This internal view of a computer-controlled washing machine (left) shows the programme module in place behind the front fascia of the machine. The printed circuit board is clearly seen, as are the PCB edge connectors. In this machine, the wiring leading to the power module is ordinary cable not ribbon cable.

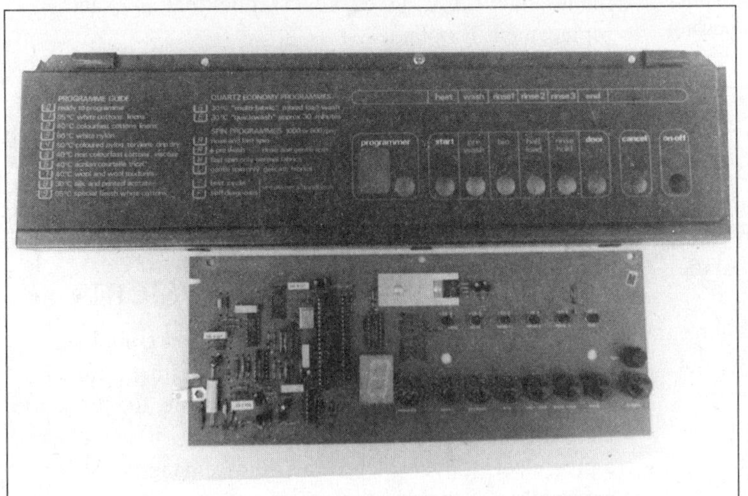

A typical computer-controlled machine fascia with selector buttons and LED display. The same fascia panel removed to expose the main programme circuit board housed behind it.

The photograph below shows the matching power module from the same machine.

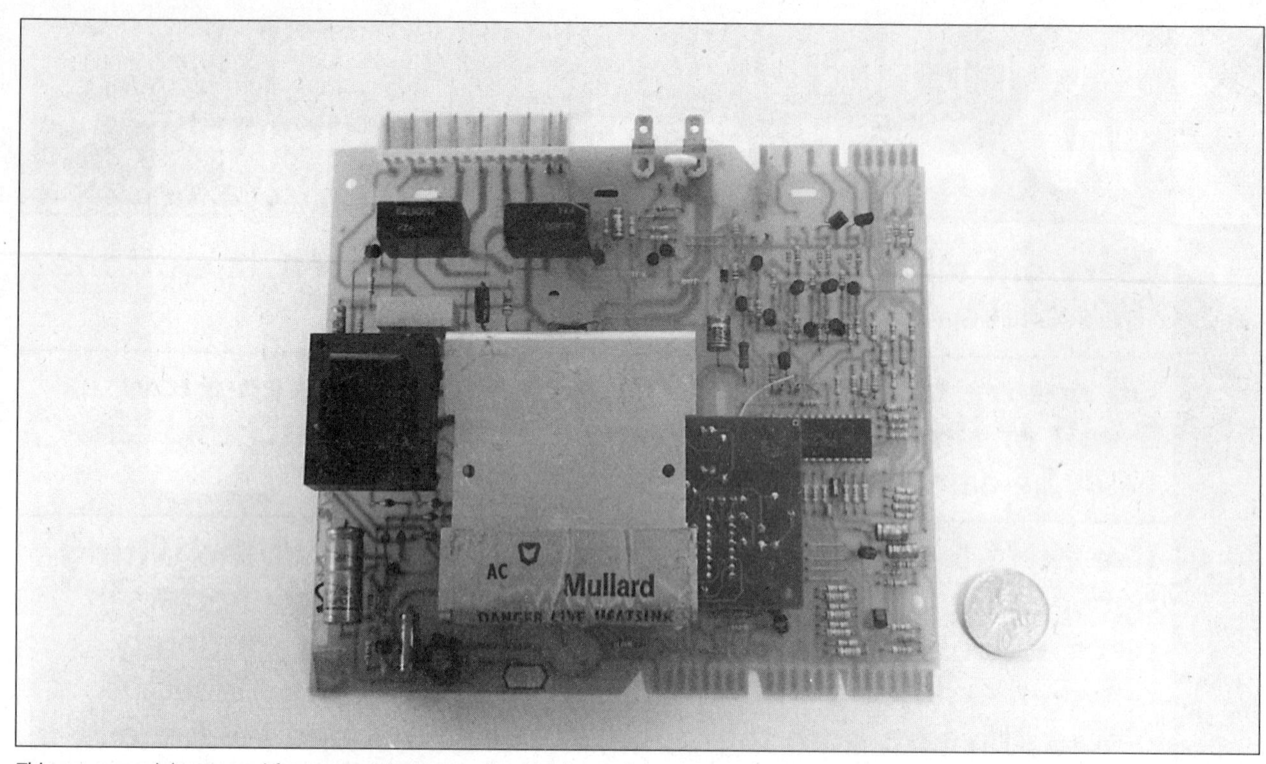

This power module removed from a machine shows the complex circuitry of this unit. This particular unit was severely damaged by a simple short circuit of the pump wiring. The fault damaged several circuit board components and the circuitry of the board itself. The simple fault resulted in a complete new unit being required.

Tumble dry only machine

Both mechanical and electronic (micro-processor) controls are used for controlling drying programmes and their operation is similar to that described above. Mechanical timers are much simpler on dry only machines, as fewer combinations of switches are required. Computer-controlled machines, although very similar in basic construction, usually contain some means of sensing the dryness of the load and match this to a setting selected by the user – iron dry, fully dry and so on. The programme terminates when the selected setting is reached, thus preventing over-drying.

Mechanical timers

Mechanical timers are the most popular type used in the lower and mid-price range of tumble dry only machines. They are smaller than the similar timers used in washing machines, but the cam action of switching is almost the same.

Failure of the clockwork drive and pitting of switches are the commonest faults. Individual spare parts are not available, so repair consists of a complete replacement unit. As there are several shaft sizes, hole fixings and timers available, make sure that you obtain the correct replacement. Fit it on a one-for-one basis.

On more recent machines, the clockwork-driven timer has been replaced by an electrically powered motor to drive the timer . Again, no individual parts are available from manufacturers or spares suppliers for these timers, so a faulty timer must be replaced with a complete new unit. In common with all timers, there are many different combinations possible although they all may look much the same externally. Fixing points, terminal positions, shaft sizes and length of selectable time differ with each make and even each model of machine from the same manufacturer. It is, therefore, essential that an exact replacement is obtained.

The clockwork timer is restricted to tumble dry only machines, whereas the motor driven timer may be used for controlling drying times in tumble dry only machines, vented washerdriers and condenser washerdriers. In both types of washerdrier, motor reversal for the main wash motor will be controlled by the main wash timer.

Auto sensing drying systems

Many modern machines, both dry only and combined types, now have the option of drying on either a timed basis or by auto sensing. With auto-sensing, the dry cycle ends when the clothes are completely dry or have reached a pre-selected level of dryness. Auto-sensing is an extremely useful and economical option.

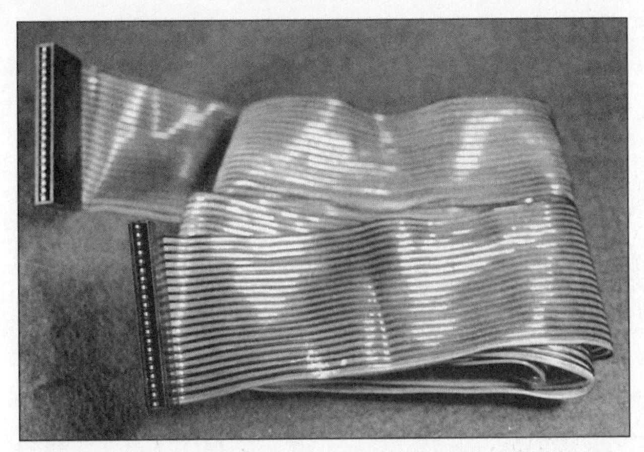

Ribbon cable is often used to connect programme board with the power board.

Several different systems are used in auto-sensing machines. A general description of three very common systems follows. The first is found in both combined vented machines and some dry only machines; the second and third versions are confined to the more expensive range of computer-controlled models.

The simplest system monitors the temperature of the air within the outlet duct by means of a bi-metal thermostat. A double thermostat is often used to allow for two levels of dryness: completely dry or iron dry (slightly damp to aid ironing). The normal thermostats on the inlet ducting control the temperature of the air used to dry the clothes and to prevent overheating.

A theoretical operation of a combined vented washer-drier is as follows. The machine is set for a dry cycle and starts to blow heated air through the wash load while rotating the drum. The length of the dry cycle, main drum motor control and supply to the drying group are all controlled by the one main timer unit. The main timer is of the thermostop variety (see Drying components, pages 144-153), and is prevented from advancing the main cam barrel by the intermittent action of the thermostop facility. Supply for the thermostop coil is via either of the exhaust thermostats, which are normally closed. Which stat is in circuit depends on the level of dryness required and is set by user switch selection – completely dry or iron dry.

Air passing into the drum from the heater unit is controlled, in this instance by a 100°F thermostat. As it passes through the wet clothing, its temperature drops. The temperature of the air leaving the drum via the exhaust duct is appreciably lower but will increase as the

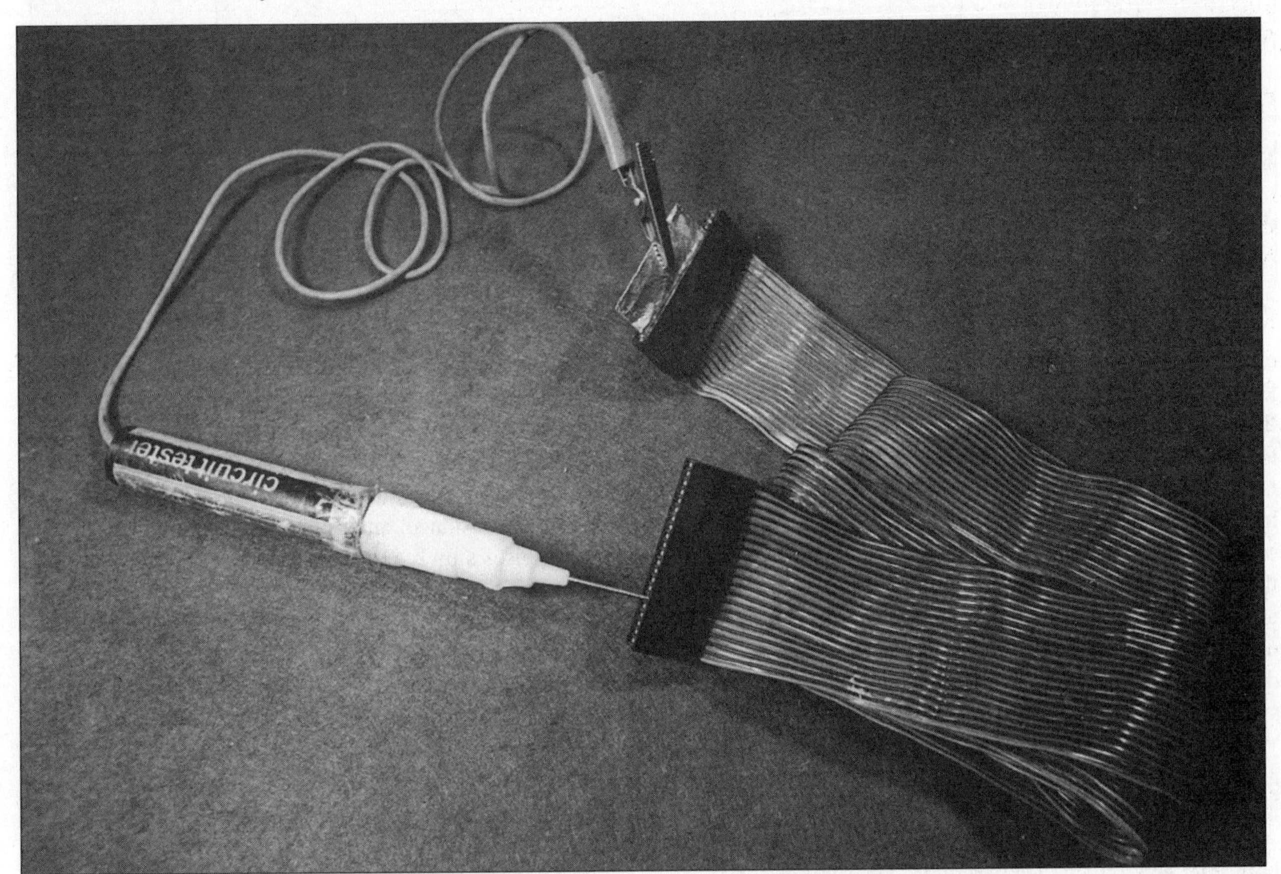

Test ribbon cable continuity by inserting a metal plate in one connection and testing individual wires at the other. A low-voltage test meter is essential for this type of test. Move the whole length of the cable on each test to check for intermittent open circuit.

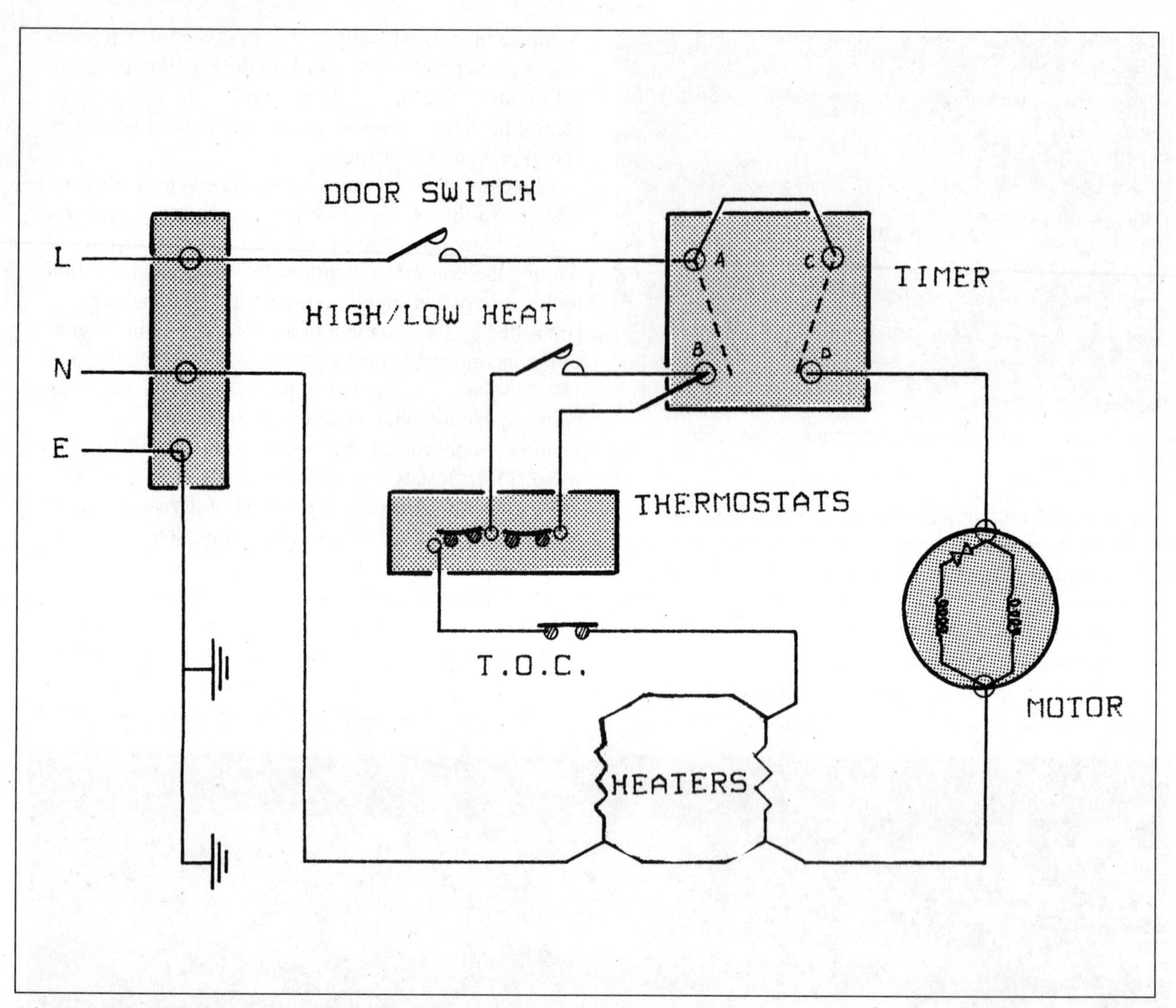

Schematic wiring diagram of a simple tumble dry only machine.

clothing dries. The exhaust thermostats monitoring the temperature will have lower settings than the inlet thermostat, in this instance 50° for iron dry and 55° for totally dry. Let us assume that the iron dry thermostat has been selected to control the thermostop supply. Once 50° has been detected in the exhaust duct, the thermostat goes open circuit and severs the thermostop supply. With the thermostop now disengaged, the timer is free to move on to the next cam position which removes power to the drier heaters. A cool tumble takes place for a pre-determined time before the machine automatically switches off.

This system is simple yet effective and does not require a separate drying time to be set by the user. Only the level of dryness required is selected.

The second system uses a micro-processor to monitor both inlet and exhaust temperatures. Measurement is by means of thermistors mounted at both inlet and outlet points. The resistances of both thermistors are monitored by the processor, which calculates the temperature differential between them. At the start of the drying cycle both thermistors have much the same resistance. The inlet temperature then rises quickly to its constant working temperature/resistance, but the exhaust temperature/resistance rises slowly in proportion to the dryness of the load until a pre-selected temperature/resistance differential value is attained.

The programme of the processor has resistance variations built into it. These correspond to the required levels of dryness of the load depending on which level has been selected by the user. When the corresponding resistance is achieved, the processor proceeds to the next step in sequence, that is, heater off and cool tumble for set time.

Both this system and the following one offer a greater range of different dryness settings. These may be user selectable or, in the case of combined machines, matched to the wash selected, such as delicates or cottons.

The third system is also part of a micro-processor – or module – controlled circuit, but sensing is somewhat different. Detecting the required level of dryness is not governed by temperature. This system monitors the moisture content of the load as it rotates in the warm airflow.

The electrical resistance of the load is proportional to how wet it is. When the load is very wet, the resistance is low because water is a good conductor of electricity; when it is dry, resistance is high. The sensing system may sound complex but, in fact, is quite simple. A metal probe, usually a simple domed bolt, is mounted on the drum surface at a point where it will make contact with the load as the drum rotates. The probe is insulated from the metal drum by plastic washers and is electrically connected to the module or micro-processor timer. Because the drum has to rotate, a moving contact system is employed at some point in the circuit. This usually consists of a bus bar and phosphor bronze bush. Within the circuit is a bleed resistor, designed to dissipate any static build-up and prevent it from being transmitted to the electronic control circuitry where it could cause damage.

The monitoring probe is part of the electronic circuitry and so operates at 5 volts. The variations in resistance are used in a similar way as that described in the second system.

When testing systems that contain micro-processor controls, ensure that all continuity testing is carried out with a low-voltage tester. Do not exceed 9 volts when continuity testing.

Before suspecting a fault in the control module, check the following:

- The filter is clean and free from lint/fluff.
- The vent/hose is not blocked or obstructed.
- The inlet is not blocked/obstructed.
- TOCs and fixed thermostats, connections and wiring are sound.
- Continuity of elements.
- Firm contact on movable connection on probe circuit.
- The insulation on the probe head is intact, that is, no cracks, etc.

Three versions of mechanical timers. The top timer is clockwork driven and restricted to tumble dry only machines. The lower two are electrically driven and may be found in both combined machines (vented and condenser types) and tumble dry only machines. Many variations exist but the basic function remains the same.

- All control board edge connectors.
- If pressure system *(see Dry only machines, pages 154-167)*, ensure all seals and panels are fitted correctly to prevent air leaks.
- Ensure that the circulation fan is not jammed. The circulation fan on some machines may be driven by a belt. Make sure that it is intact and not slack (some belts are elasticated and have a tendency to stretch). Renew if suspect. Some machines may have the circulation fan fitted to the motor shaft. Ensure that it has not worked loose from the shaft.

Many of the faults shown may be indicated by a code displayed on the front of computer-controlled machines. Codes differ from machine to machine so refer to the manufacturer's handbook.

Renewing the motor coil

In some instances it is possible to renew the timer motor coil if it is found to be open circuit. Do not mix the motors because they normally rotate in different directions. This procedure can be carried out with the timer in situ. It was removed in this instance for photographic purposes only.

TOOLS AND MATERIALS

- ☐ Small flat bladed screwdriver

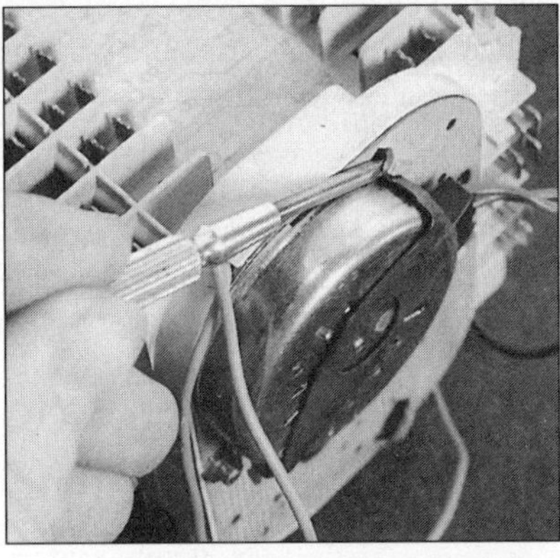

1 Insert a small flat-bladed screwdriver under the retaining spring clip and lift it slightly.

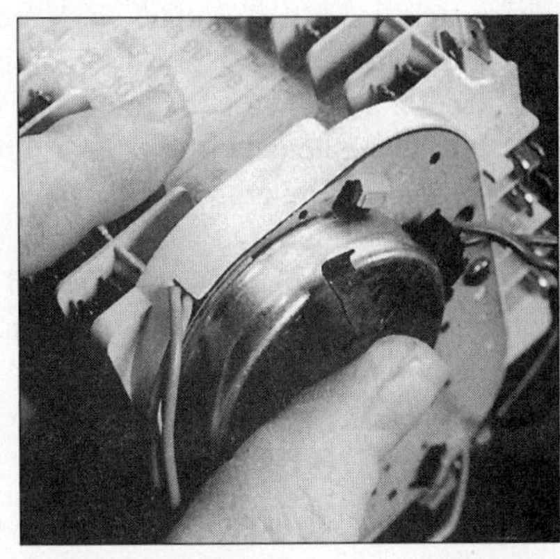

2 Mark the correct motor position and lift the motor free.

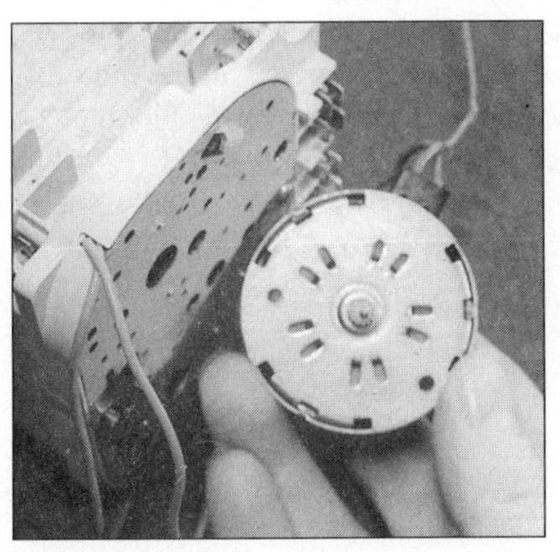

3 With the clip eased, slide it sideways to free it from its position.

4 Insert a small flat-bladed screwdriver between the two halves of the motor casing and ease them apart.

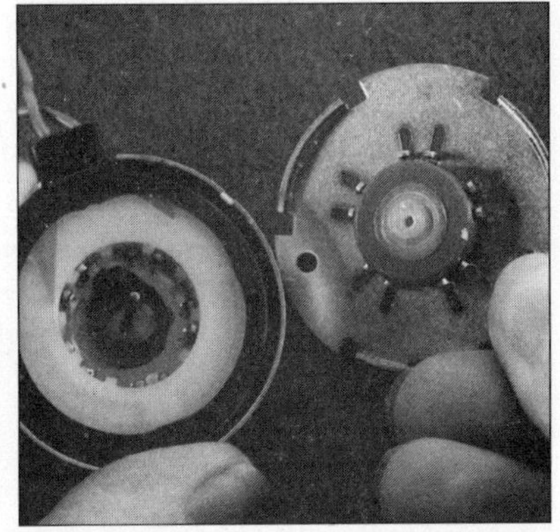

5 The two halves of the casing will part to expose the coil and permanent magnet rotor.

6 Remove the coil from its position. Ensure that the small plastic anti-reverse mechanism is in its correct position in the base of the casing.

7 Fit the new coil making sure that it is the right way round.

8 Re-assemble both halves and press firmly together to ensure that they locate correctly. Check that the motor rotates correctly (in only one direction) as original. Refit the motor to the timer.

Motors

There are three types of main wash motor used in today's machines: universal AC brush motor, permanent magnet (PM) motor and induction motor. The motor is usually bolted to the underside of the outer tub on washing machines and combined washerdriers. The Hotpoint front loader is an exception; the motor is bolted to the top left-hand side of the outer tub (viewed from the rear). Tumble-dry-only machines normally have the main motor mounted on the base of the machine.

Brush motor

This normally consists of two sets of electromagnets: an outer fixed set, called the field coil, and an inner set free to rotate, called the armature. The armature is made up of many separate windings configured in such a way that power is supplied to only one set of windings at a time. The corresponding movement induced in the armature continuously brings a new set of windings into circuit, while the previous winding circuit is broken. The windings are continuously out of synchronisation, inducing continuous rotation of the armature whenever power is supplied. Reversal of the motor is normally achieved by reversing the power flow through the armature windings via a set of reversing switches in the timer.

This type of motor can be used with alternating current (AC) from the mains or direct current (DC) from a battery. It is often used for the main drive motor in washing machines, especially on machines with a spin speed over 1000 rpm.

Induction motor – capacitor/relay start

Induction motors of this type are found in both washing machines and tumble dry only machines. The motor has a capacitor/relay to 'kick' the rotor into action by putting a delay into the motor's start windings. The resulting imbalance creates rotation in the direction of the run winding current flow. A reversal of power in the run winding reverses the motor. Speed is governed by the number of windings supplied with power.

Because all the work is done by a complicated set of windings in the stator, this motor cannot usually be

This type of unit normally requires renewal if faulty, although some brushes and patterned armatures are available separately.

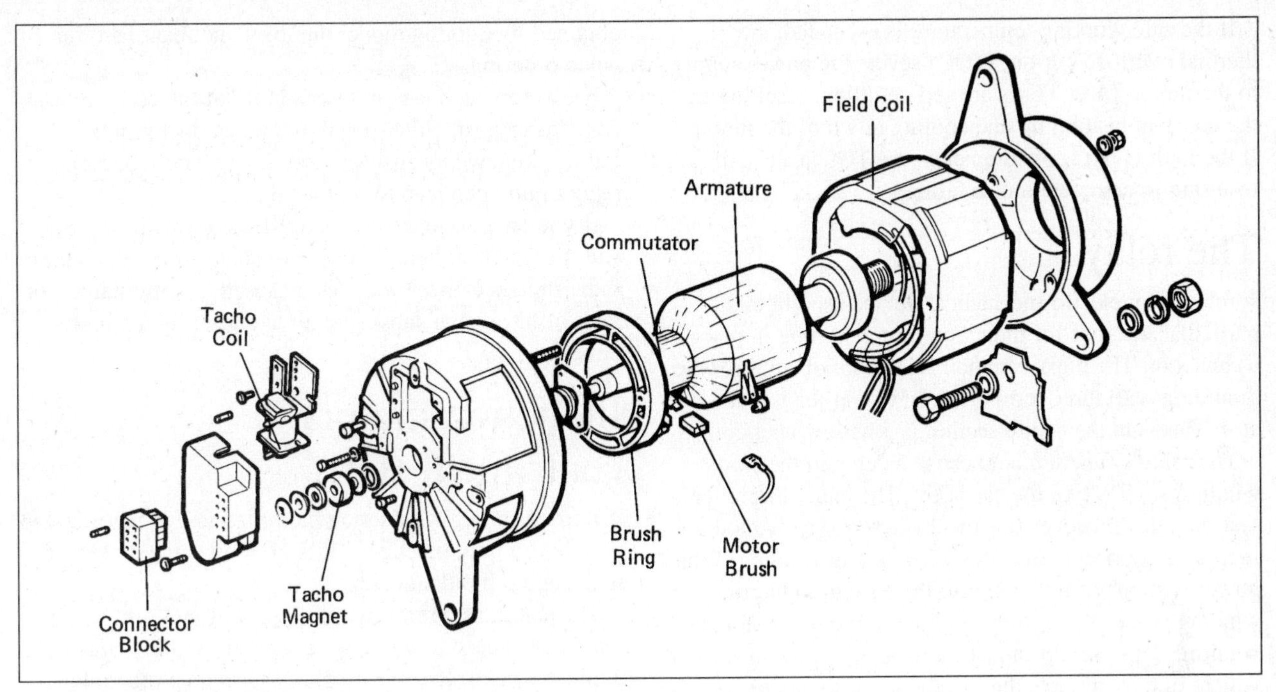

Exploded view of a brush motor.

repaired and must be replaced by a new unit if it fails. Capacitor failure often results in the motor's failing to run. This can result in burn-out as the rest of the motor windings are receiving power but no rotation is possible. Overheat is inevitable, even when TOC (thermal overload cut-out) protected.

When checking for faults, the machine must always be isolated from the mains. Turn off at the wall socket and remove the plug. However, the capacitor(s) will still contain a charge, which must be discharged using an electrically isolated screwdriver. Do this by 'shorting' the terminals of the capacitor with the shaft of the screwdriver, ensuring that you are in contact only with the insulated handle. It is not safe to proceed further until this has been done.

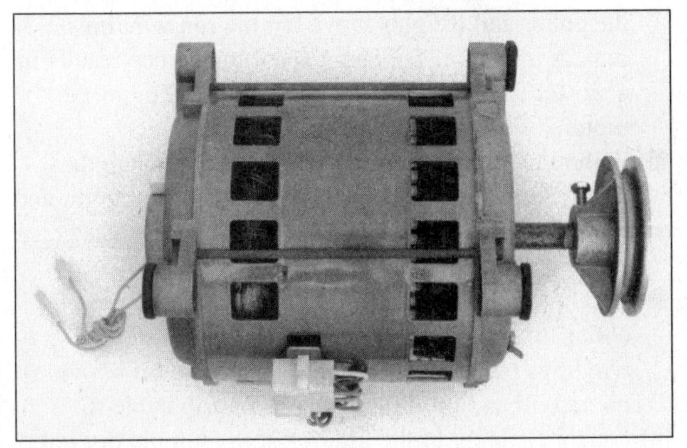

A Fagor motor, typical of the new-style induction motors, is capable of variable speed build-up via a module. Note the tacho connector at the rear of the motor.

If the stator windings of an induction motor are faulty, it may continue to run slowly and sluggishly, getting extremely hot when used even for a short time. If you have been running the machine to determine the fault, proceed with care because the motor will remain hot for some time. If the motor is very hot, the motor winding may be faulty and the unit should be replaced.

The capacitor

Capacitors used for motor starting have either metal or plastic outer casings with an insulated top with two terminals.

The storage capacity of a capacitor is measured in microfarads (μF) and is displayed on the casing of the capacitor. Any replacement must be of the same μF rating.

If the motor fails to run on wash but runs on spin, and there are two capacitors fitted, it is possible that one of them is faulty. Change the capacitor with the lower μF rating and re-test. If only one capacitor is fitted, then check the motor.

If the motor runs on wash but fails on spin, and there are two capacitors fitted, there are two possible faults. Change the capacitor with the higher μF rating and re-test. **(Remember the warning about an isolated capacitor retaining an electrical charge.)** If the door interlock is connected directly to the motor spin circuit, and the door is not closed properly, the spin will be prevented from operating. A fault within the interlock also prevents the spin *(see Door locks, pages 75-78).*

If the safe working temperature is exceeded, the thermal overload cut-out (TOC) severs the power supply to the motor. Most TOCs are self-resetting, resulting in the constant heating up and cooling down of the motor. If the fault is not spotted quickly, the TOC itself will fail, resulting in complete motor failure.

The relay

A relay is an electro-mechanical device used in this particular instance for induction motor starting in place of a capacitor. The most common relay consists of a plastic moulding with three terminal tags, two at the top and one at its base. On the centre section is a wire wound coil.

The relay's function is to cause a delay in the start winding supply, like the capacitor. The main difference is that the relay achieves this mechanically. The wound coil section is connected in series with the run winding. When power is supplied to the motor, the current to the run winding passes through the coil and on to the motor run winding. This current induces a magnetic force in the coil which, in turn, attracts the metal core of the relay. The metal core is linked to an internal contact switch and when 'made', allows current to pass to the start winding. This operation provides the required time delay to induce starting of the induction motor.

When power is switched off, gravity resets the relay core. It is essential that the relay is in its correct position and the machine upright; otherwise the relay will not function correctly.

The relay may also be matched to the run winding of the motor. As initial power draw is high, the magnetic attraction of the relay coil is great enough to attract the core, but when the motor is running, the initial high power draw drops and weakens the magnetic pull of the coil. The core drops and open circuits the start winding allowing the motor to continue running more efficiently. Always make sure that the correct replacement is obtained by quoting model numbers and manufacturer when ordering.

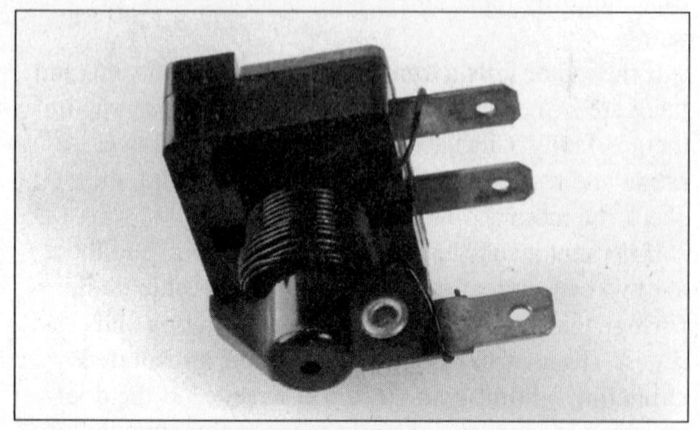

A relay may be placed in circuit to cause the phase displacement necessary to start the induction motor. This is a mechanical delay and it is essential that the relay is upright when energized.

Faults to watch for are open circuit of the coil, metal core sticking (in either position) and contact points failing. Renew any suspect relay immediately because relay failure can lead to motor failure.

If you have to renew a damaged stator coil or motor, and it is relay started, it is wise to change the relay at the same time because it may have caused the original motor fault or have been subsequently damaged by the motor failure.

Induction motors – centrifugal start

A third system for induction motor starting is restricted to tumble dry only machines. It is used on some makes of machine both old and current.

The stator, as before, is split into start and run windings. The start windings are connected in series with a micro-switch. Its switching arm rests on a movable circular collar forming part of a spring and weight system fixed to the rotor shaft. When power is supplied to both windings of the motor, rotation is induced by the phase displacement of the two windings. This may be increased further by the addition of a capacitor within the start windings supply. As the rotor increases in speed, centrifugal force lifts the weights, and the movement produced pulls the collar away from the micro-switch. This switches off the start windings. The motor continues to run with only the run windings in circuit.

There are three possible mechanical faults in addition to the normal motor faults.

- The micro-switch may fail in either the closed or open position. If permanently open circuit, the motor over-heats when first turned on as only the run windings receive power. Rotation does not occur. If the micro-switch fails in the closed position, the motor starts, the collar and weights move but the run windings remain in circuit. This permanent imbalance results in a slower speed than normal and overheating of the motor.
- Failure of the centrifugal system to reset when the motor stops, leaves the start windings open circuit and the fault will arise when the machine is next used *(see symptoms above).*
- The failure of the centrifugal system to move the collar from its rest position, results in the start windings being left in circuit *(see symptoms above).*

The second and third faults may be attributable to infrequent cleaning of the interior of the tumble dry only machine. This may lead to fluff jamming the centrifugal mechanism.

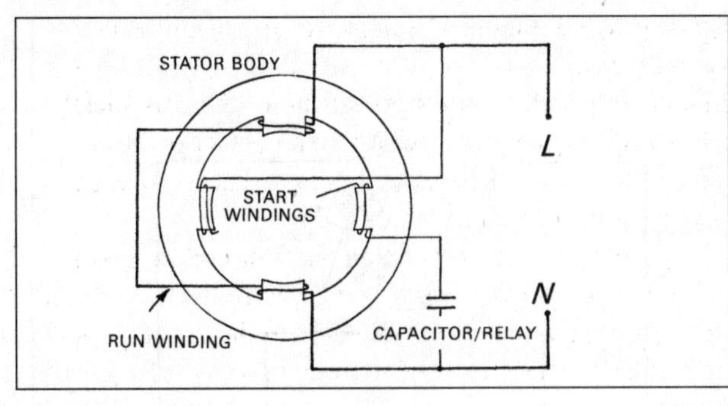

Basic induction motor.

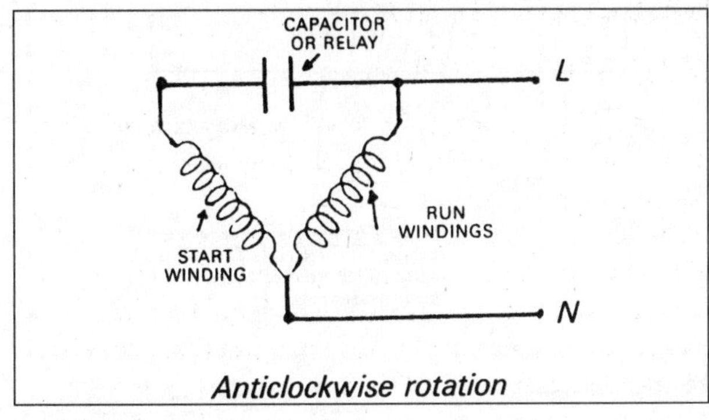

Anti-clockwise rotation.

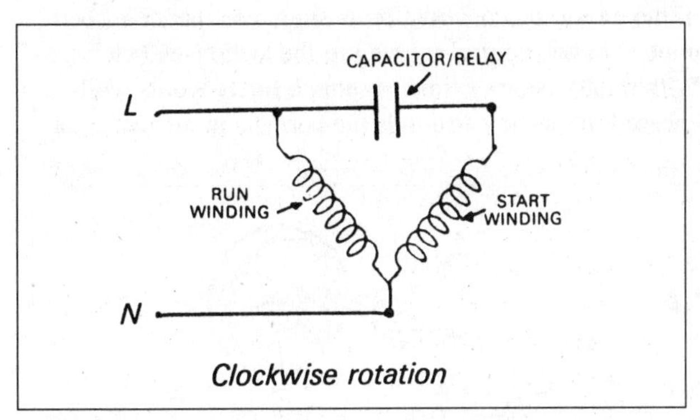

Clockwise rotation.

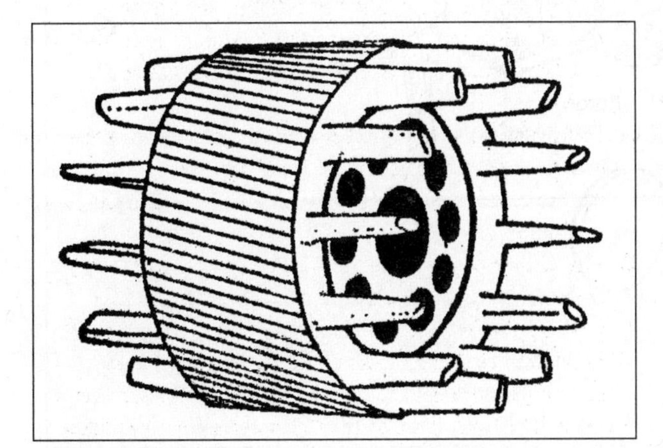

Induction motor rotor.

Machines with a centrifugal switch system can be identified by the distinct noise that the motor makes shortly after initial start up and just before stopping rotation when the motor is switched off. The noise (a faint whir and click) is caused by the movement of the weights and collar and switch, both on initial run up to speed – switching out the start windings – and run down when the motor is turned off – resetting of weights, collar and switch.

Ensure all moving parts are securely fitted, clean and free to move. Do not over-lubricate because this can itself cause a fluff build-up. Apply a very small amount of light machine oil only to the moving parts of the mechanism.

Induction motor – shaded pole

This type of motor is associated with pumps and low-power air circulation fans, because it has low starting torque. This means that the motor is impeded from starting easily, because the initial rotation is only from copper segments bound into the stator. When power is applied to the stator coil, the copper segments create a permanent imbalance in the magnetic field produced. This induces rotational movement.

The shaded-pole motor is one of the simplest of all induction motors. It is similar to other types in basic format of rotor and stator, but only one stator coil is used to create the magnetic field. Obviously this alone would not induce rotation of the rotor, only a constant magnetic field. To start rotation, an imbalance in the magnetic field is required. This is achieved quite simply by copper band inserts at the pole ends of the stator laminations. The copper bands within the mild steel stator laminations (dissimilar metals) distort the magnetic field in a given direction, thus inducing rotation in the stator. Reversing the supply to such motors does not change the motor direction because this is governed by the direction of the fixed shaded poles. These motors do not have a high starting torque and because of the magnetic imbalance

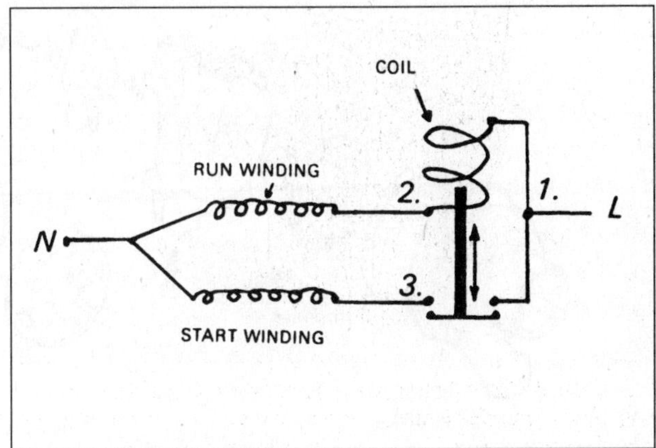

Permanent magnet motor

being fixed, heating of the stator occurs. This does not normally create any problems, but most stator coils are protected by TOCs for safety.

Induction motor – permanent magnet rotor

This extremely simple motor is used to drive all versions of mechanical timers both for timing and cam advance. Consisting only of a wound circular coil fixed around a permanent magnet rotor and supported at both ends by simple sleeve bearings, it can be extremely small and manufactured at low cost. As with all induction motors, its simplicity of construction limits it to use with AC supply only.

A larger version of this principle is now being used to power outlet pumps in a wide of variety of modern machines. The PM rotor is housed within a sealed plastic chamber, but is still free to rotate by the alternating current (AC) supplied to two externally mounted stator poles. The rotor drives the impeller of the pump in the normal way, only the motor of the pump differs.

This type of motor has been used for many years to power the timing and advance mechanism of washing machine programmers. It is the simplest of electric motors, quiet to run and cheap to produce. The larger version now used to drive the outlet pump of some modern washing machines avoids some common pump problems.

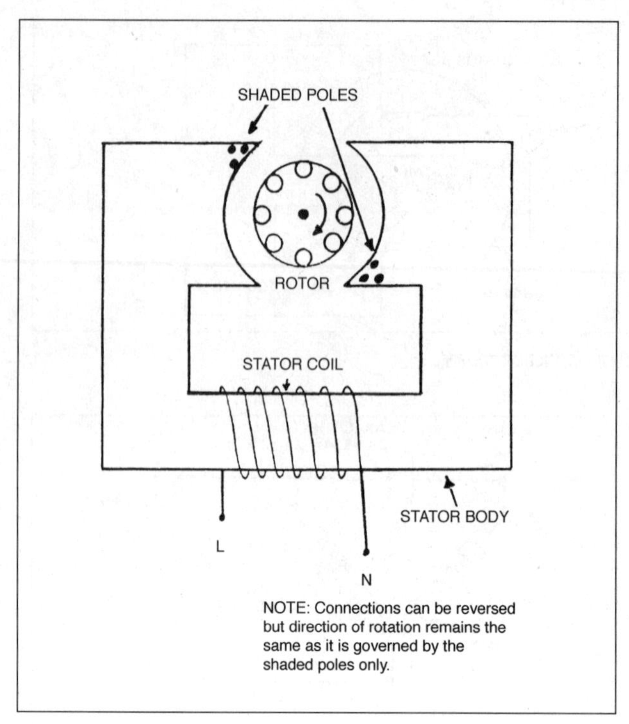

Shaded-pole induction motor.

A circular multi-pole permanent magnet forms the rotor of the motor. Its construction is similar to that described under Module control relating to the tacho (see below). Within the casing of timer motors a finely wound coil encased in plastic surrounds the permanent magnet rotor.

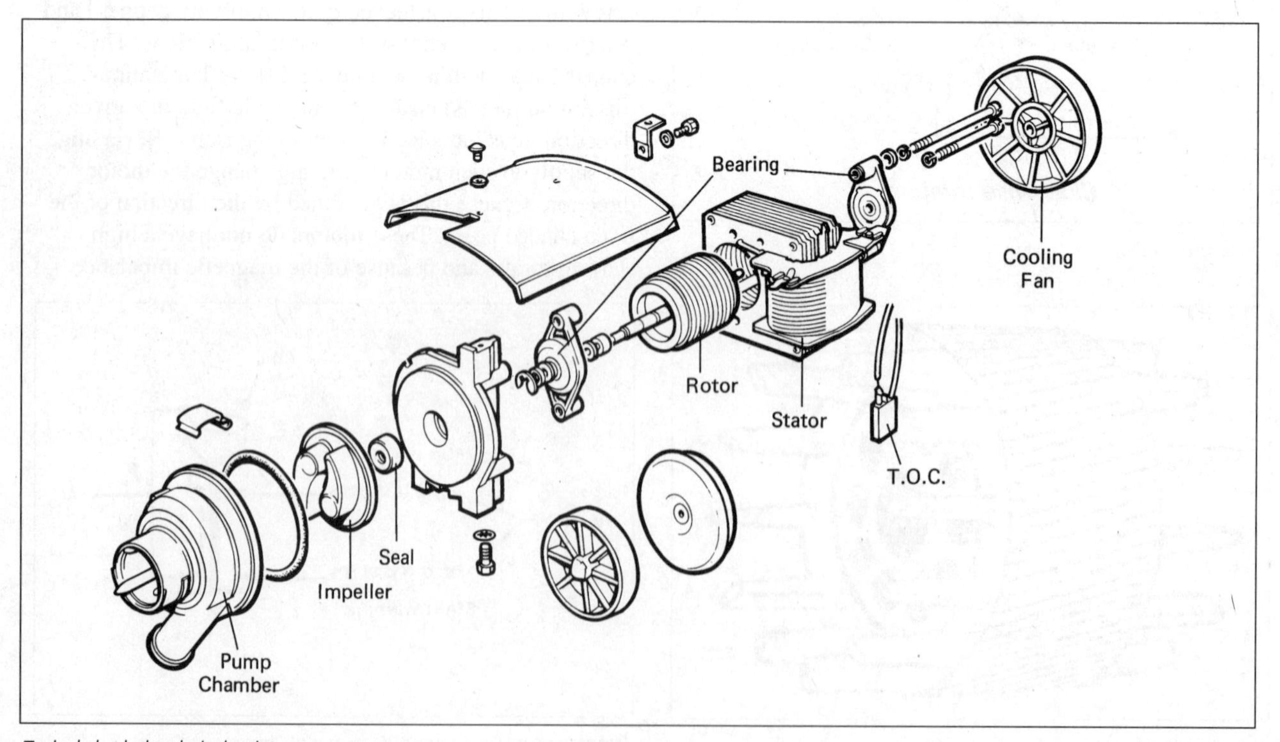

Typical shaded-pole induction motor pump.

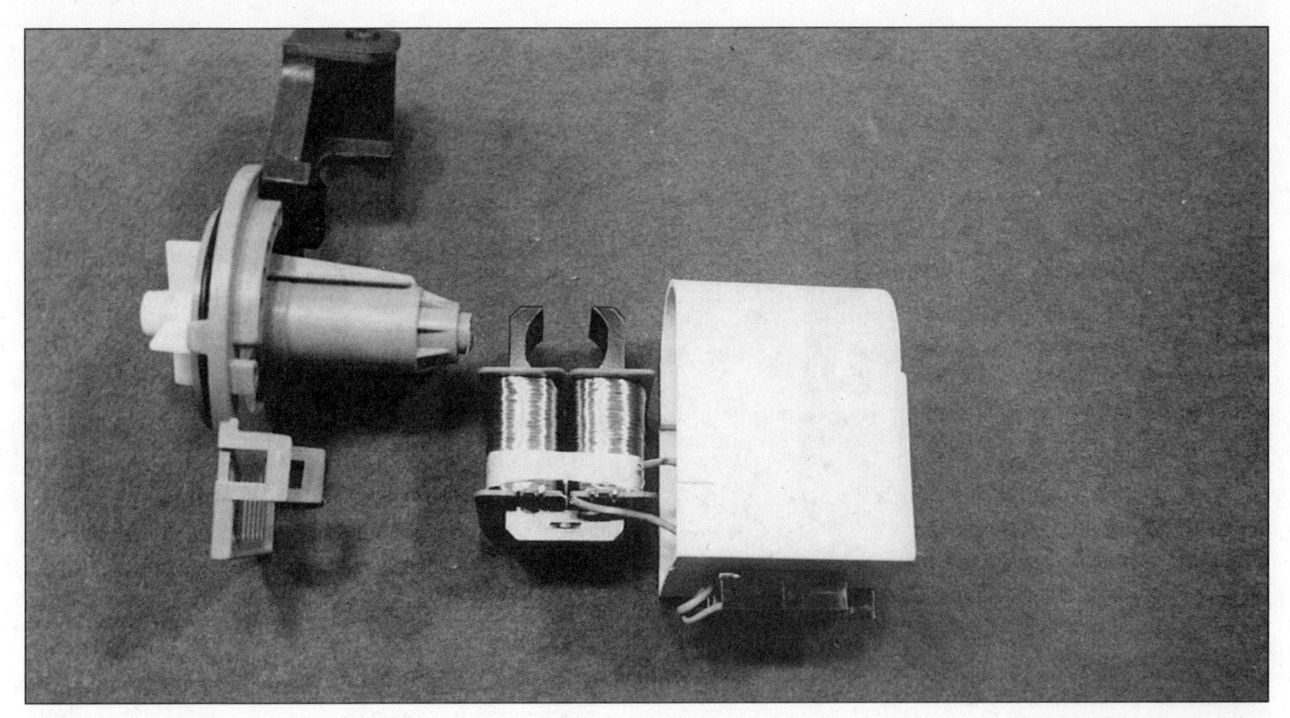

Component parts of a permanent magnet rotor pump. Only the complete pump is available as a replacement. It is shown here dismantled to illustrate its simplicity.

Permanent magnet pumps have two coils wound on to a laminated steel stator. In both instances, supplying AC power to the coil(s) induces rotation of the magnetic rotor. However, rotation could start in either direction and this, in the case of timer motors, would be most unwelcome. To ensure that rotation occurs in the required direction – clockwise or anti-clockwise – a small plastic cam is positioned within the casing (seen as a small plastic pip on the rear of the motor casing), which allows rotation in one direction only. Should the motor try to start in the wrong direction, it hits the plastic cam which flicks it back, thus inducing correct rotation.

This ability to run in both directions is utilised when this style of motor is used to drive an outlet pump impeller. If the impeller of the pump comes into contact with an item such as a button, which would usually jam or completely stall a normal shaded-pole motor, the motor may be nudged into revolving in the opposite direction and clear the blockage or, alternatively, continue to pump while running in the opposite direction.

The construction of the PM pump helps alleviate the problem of shaft seal leaks and bearing failure, which are common to shaded-pole versions.

Failure other than brush wear requires complete motor renewal – make sure only the correct replacement is obtained.

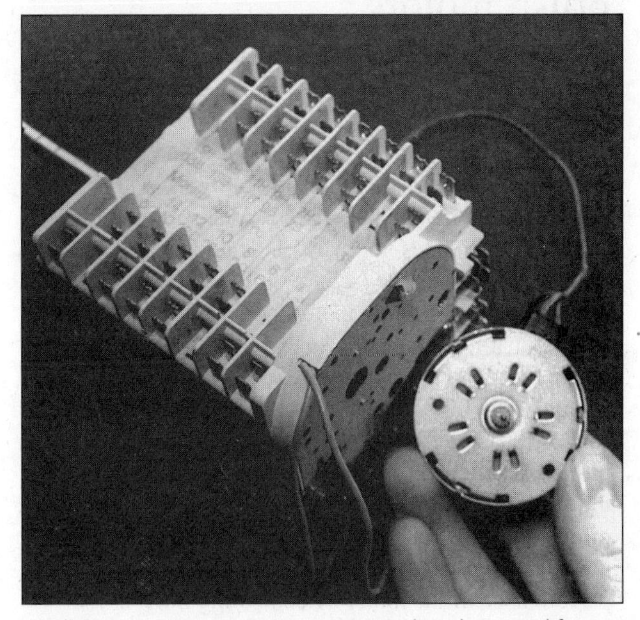

Small permanent magnet rotor type motors have been used for many years to drive mechanical timers

The induction stator of a pump. The shaded poles are clearly visible by the two bands of copper inserted in opposing poles. Note the orientation of the copper bands before removing the stator from the pump. If the stator is refitted back to front, the pump will run in the opposite direction and not pump at all.

Module control for brush motors

Not all motor speed faults are directly attributable to the motor. The fault may be caused by the module. This controls the speed of the motor and is connected between the timer and the motor. There is no standard location for the module, but it is easily identified by its distinct printed circuit board and large heatsink. Computer-controlled machines normally integrate the speed control into the power module *(see Timers pages 98-109)*.

On the rear end of the motor's armature is a circular magnet that revolves in unison with it. Close to this magnet is a coil of copper wire (this may be encased in plastic), which is called a tacho generator. If a magnet is rotated next to or inside a coil of wire, a current is produced which is proportional to the rotational speed of the magnet. Therefore, the faster the motor is running the more current is produced. This current is fed to the module as a reference voltage and is used to monitor the performance of the motor by comparing the relative speed of the motor with a known voltage via a comparator circuit. If the reference voltage is lower than the comparator voltage, the module increases the pulse rate, so increasing the speed of the motor. If the voltage is higher, the pulses are slowed, therefore decreasing the motor's speed. This happens undetectably many times a second.

To check whether the module is at fault, isolate the machine from the mains. Turn off at the wall socket and remove the plug. If any of the internal components of the module have burned out, that is they are charred or burned looking, check the motor for any shorting, loose wires or low resistance, because these may be the cause.

Checks on the tacho magnet and tacho coil

- If the magnet is loose or broken, this results in incorrect speeds at lower motor speeds.
- Severe damage or complete loss of the magnet causes the motor to spin on all positions.
- A break in the coil results in a spin on all positions. This is because a 'good' coil is usually about 200 to 1600 ohms resistance. If there is a break in the coil, the resistance is 0 ohms. The tacho generator is not returning any current, so the module speeds up the motor. The increased speed is not transmitted back to the module, so the process is repeated ad infinitum. Note: Most modules fitted to modern machines have an in-built tacho test circuit and will not operate if the tacho circuit is open.
- Breaks and/or poor connections of the wires leading to and from the tacho can have the same effect as a break in the coil. This is especially true at the connection block with the motor and at the connection with the module.

Module faults

Any loose connection will be aggravated by the movement of the tub on the suspension; take this into account when testing for such faults. Do not attempt to adjust the tachometer other than as shown in the armature section.

The modules shown are a small selection of the very many different types fitted to today's machines. Their appearance and function differ very little from one another, but they are strictly non-interchangeable. Always ensure that the correct replacement unit is obtained by quoting the make, model and serial number of your machine when ordering spare parts.

If the fault persists, the module is probably at fault. This should be replaced with a new unit, ensuring that the correct type is purchased. To fit, make a note of the connections, remove them and replace them on the new unit. It is important that the 'Duotine' (edge) connector fits tightly on the module. The connections can be closed slightly by inserting a small screwdriver between the back of the tag and the plastic duotine. Take care not to close it too far, however, as this may result in the tag's not making proper contact by being pushed back into the connector.

The reverse side of the printed circuit board with the heatsink removed to show the components of the module clearly. The faulty components can be seen directly behind the area of discolouration. The component (a resistor in this case) has been overheating and subsequently failed. Repairs to modules are not merely a simple replacement of obvious components, because micro-chips within the circuit may have been damaged. It is, therefore, advisable that the module be removed and replaced with a complete new unit.

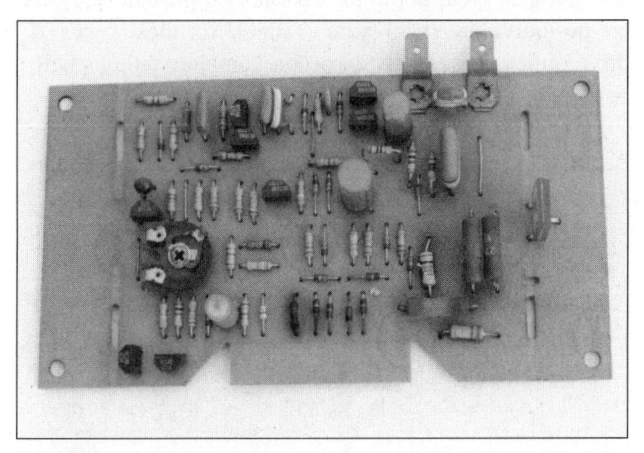

The module from a Hotpoint Automatic: note the discolouration that has occurred on the centre of the printed circuit board (front). This is a sure sign that the module is at fault or will fault soon.

The burnt connection on this Ariston washerdrier motor module was caused by a loose connection creating arcing and localised overheating

The washing machine must be isolated from the mains. Turn off at the wall socket and remove the plug. The large metal back of the module is used as a heatsink. This means that it is live when in use, and therefore should be fitted correctly and securely to its plastic mounts. Even when testing, any contact with the earthed shell of the machine will render the unit completely useless.

Module control induction motors

Many modern machines with induction motors now use electronics to help control the selected speed more precisely, and smooth the transition from one speed to another. They differ little from their predecessors and still require a complex set of windings within the stator to give a series of fixed speeds. However, they do have the addition of further speed control via a speed control module and tacho coil and magnet arrangement which is similar to the system used to control brush motors.

Faults in either motor or module require a complete change because, except for the tacho coil, no internal components are available.

Centrifugal pulleys

The centrifugal or variomatic pulley is the large pulley that can be seen on some induction motors, such as in the Candy, early Ariston, Indesit and Philips machines. It is fitted to help increase the drum speed when the machine spins.

The reason for having adjustable pulley drives relates mainly to the need for faster spinning at the end of the wash cycle. Adding a large pulley to the motor, to create an increase in drive ratio to the drum pulley for the spin, causes problems. The low-speed wash action would require extra windings to slow the rotation, while high spin speeds would be affected by lack of torque because the number of poles are reduced to increase the motor speed. In the past, when spin speeds were much slower (500-800 rpm), the induction motor was the ideal choice. However, the need for ever-faster spin speeds has outstripped the capabilities of the normal induction motor. To compensate for this inability to reach higher drive speeds comfortably, further mechanical additions have been made to the drive pulley and, in some instances (Candy), to both drive and drum pulley. Some Zanussi machines use a gear and clutch arrangement to help increase the drive speed of their induction motor. Both these systems are still limited to a drive ratio that produces a maximum

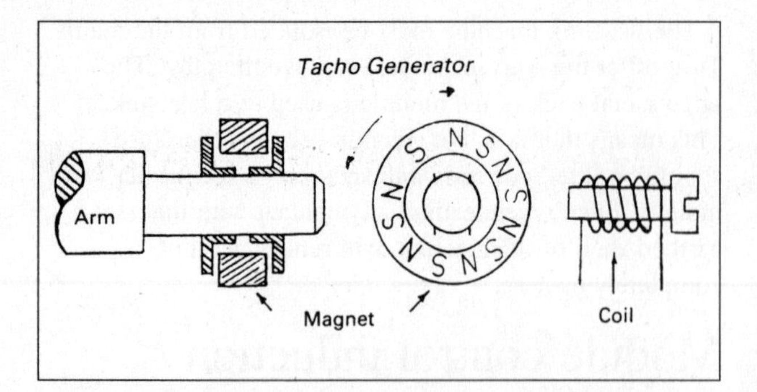

spin speed of 1000 rpm (of the drum). Universal brush motors are required for spin speeds that exceed 1000 rpm. Many modern machines now attain 1400 rpm.

Weights within the pulley are pushed outwards by centrifugal force when a fast motor speed is selected. Because the pulley is constructed in two halves, the movement narrows the gap between the front and back plate of the pulley, therefore increasing its diameter. This, in turn, increases the drive ratio between the drum pulley and the motor pulley. When the motor speed allows, the reverse occurs: the back plate moves away from the front plate and the belt rides on the smaller diameter of the pulley.

Do not overtighten the drive belt on machines with a centrifugal pulley. Some are self-tensioning by the weight of the motor, that is, the motor is only supported by rubber mounts on the rear end frame, allowing the weight of the motor to tension the belt.

The addition of a mechanical pulley system to an otherwise simple and reliable motor increases the risk of faults occurring. In addition to the normal and expected induction motor faults, problems occur with the mechanical action of both centrifugal motor pulleys and drum pulleys. Because the belt is constantly squeezed along its edges, belt wear is accelerated and regular checks are advised. Renew if suspect because spin efficiency will decrease and/or excessive noise will occur.

Pulleys are made of plastic or cast aluminium and wear ridges form on the belt contact faces. This may lead to restricted movement of the belt, resulting in poor drive, poor wash, spin or both, noise, excessive belt wear or jamming and, finally, possible motor failure.

The Zanussi geared systems are similar, with noise and internal gear wear being the commonest problems. There are no individual spare parts available for these types of drive pulley, so if faults do occur, complete pulleys will be required.

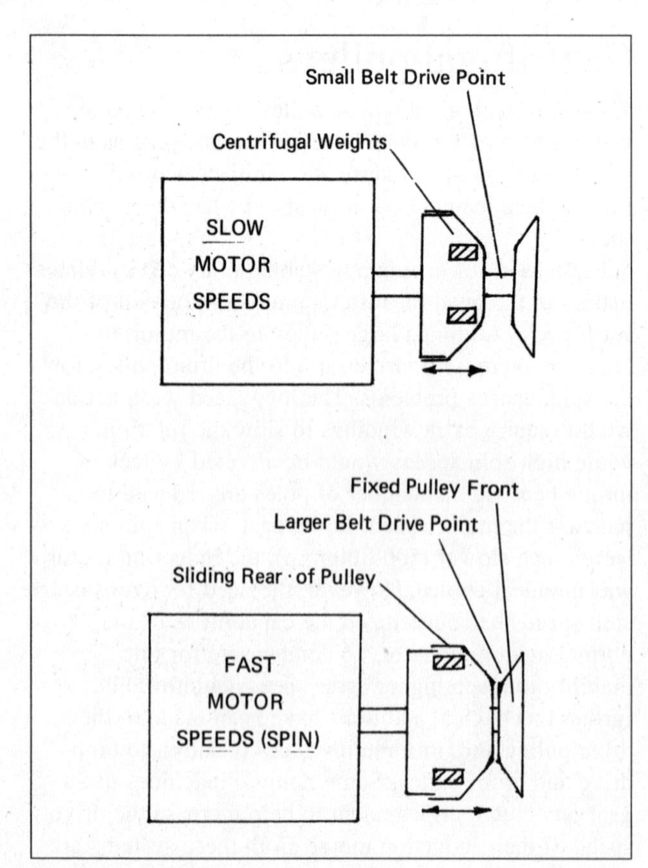

Operation of a centrifugal pulley system.

This large induction motor with centrifugal pulley is supported only by the large rubber mounts to the rear of the motor. This allows the weight of the motor to self-tension the drive belt and to rise and fall as the pulley increases and decreases in diameter with speed. Increased belt wear is common with this system.

Armature change - Hoover type

These step-by-step photographs show the removal of a motor from a machine, and the detailed removal and refitting of a new armature and brush ring. The motor shown is of a universal type found in many early wash only and vented washerdrier machines in the Hoover range. Later machines and condenser washerdriers are more likely to have motors where only the motor brushes are available as replacement items. This practice is now common to a large proportion of machines from all manufacturers.

Remember to note any connections before removal to ensure correct refitting at the end of the repair.

TOOLS AND MATERIALS

- ☐ Screwdriver(s)
- ☐ Test meter
- ☐ Drill/chisel
- ☐ Spanner
- ☐ Hide mallet
- ☐ Armature
- ☐ Brushes

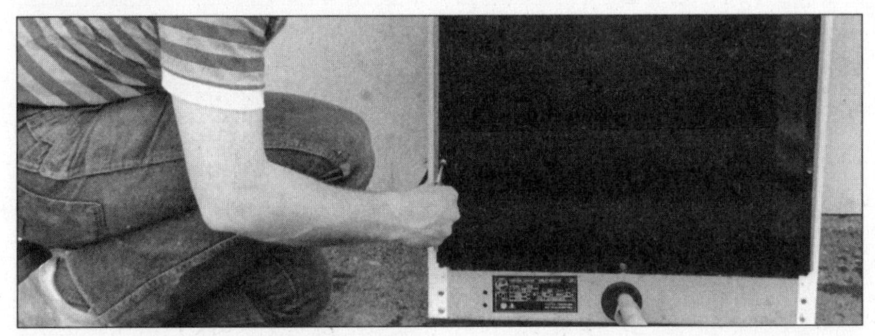

1 Isolate the machine and remove the rear panel.

2 A 'low-insulation test' *(see pages 172-175)* discloses the motor fault. Remove the motor bolts and withdraw from the machine.

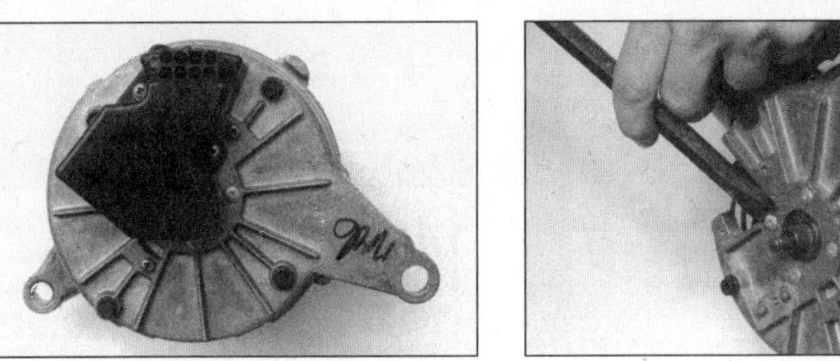

3 After noting the motor block colours, positions and connections, remove the plastic cover to reveal the tacho coil. Carefully remove the tacho coil and magnet. (The clip on the shaft can be lifted with a small screwdriver).

4 Drill out the four end rivets or remove with a sharp chisel.

5 Mark the position of the end frames with a pencil. Then remove the four securing bolts.

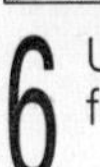

6 Use a hide mallet or similar tool to free and remove the front end frame.

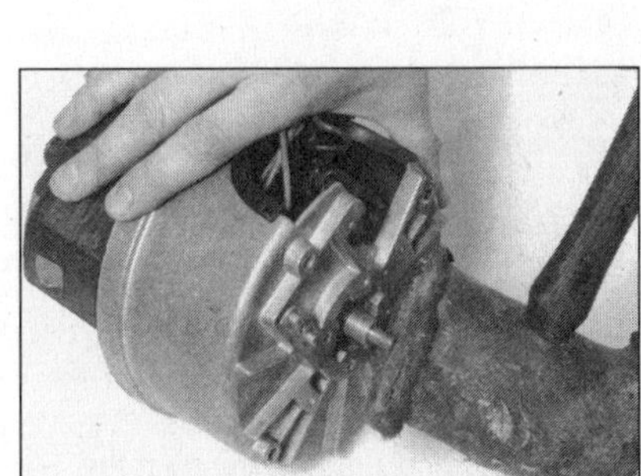

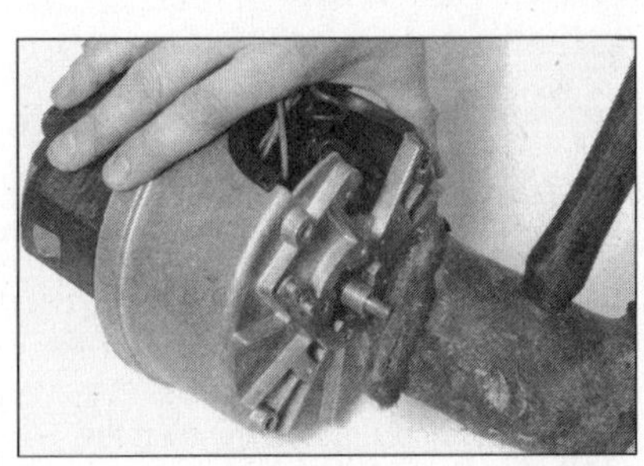

7 Knock the armature tacho end shaft free. Remove the armature and inspect for faults.

8 Check the copper segments on the armature for damage, such as burned or loose/raised segments, and for carbon build-up. This one is badly damaged.

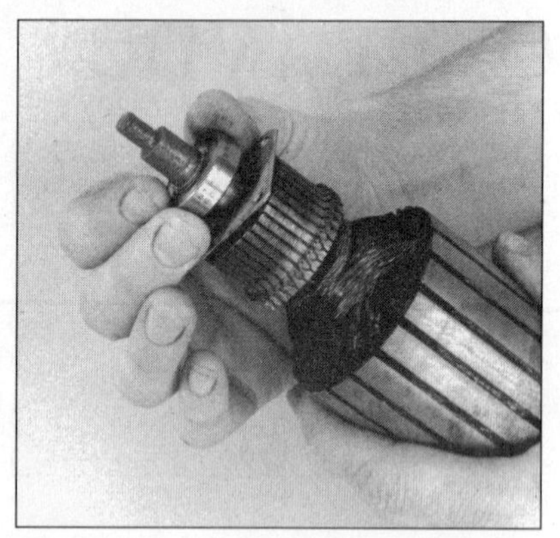

9 Check the bearings for free and quiet running by spinning them on the shaft. Also check for tight fit to shaft. This one has damaged the shaft.

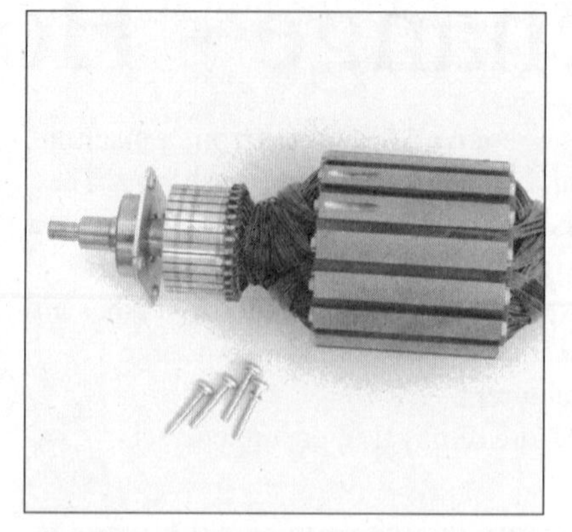

10 The new armature is ready to fit. Note screw plates and screws instead of rivets to aid fitting. Fit a replacement unit if you are in any doubt about the condition of the old unit.

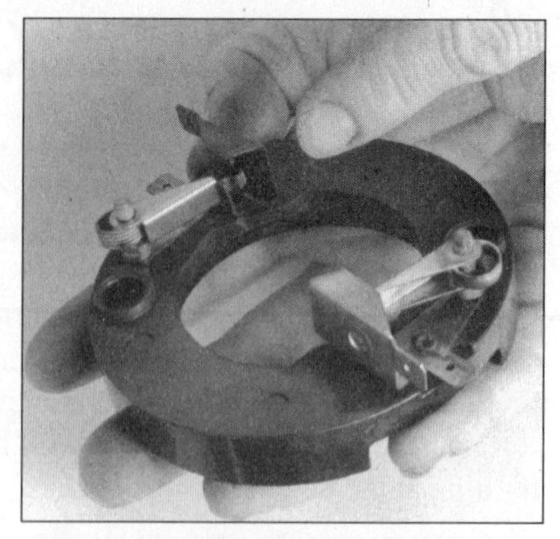

11 Inspect the old brush ring for damage. Carefully check for smooth brush slides. Also check that no carbon deposits have caused low insulation. Change if in any doubt. This one has burned slides.

12 New and old brushes. The top rows show tagged and non-tagged types of brushes. The lower rows show split and worn brushes.

13 Fit the new armature and brush ring to rear end frame.

14 Refit tacho magnet and clip. Ensure that when fitted the magnet will not turn on the shaft: it should be locked to the armature.

15 Adjust tacho setting if necessary. Screw centre up to the magnet and turn back 1½ turns only.

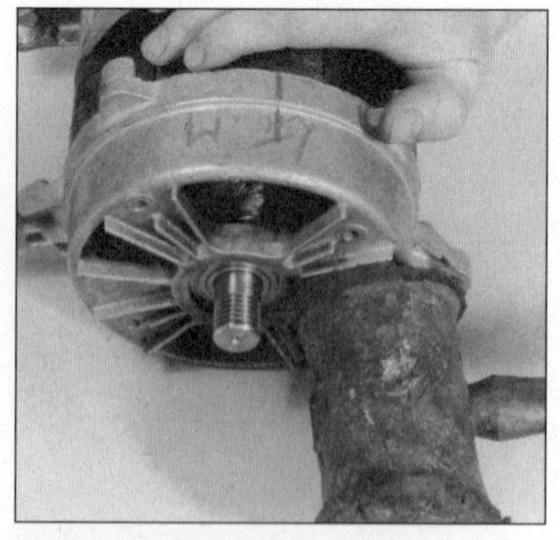

16 Refit front end frame and reassemble motor, lining up the marks made in step 5.

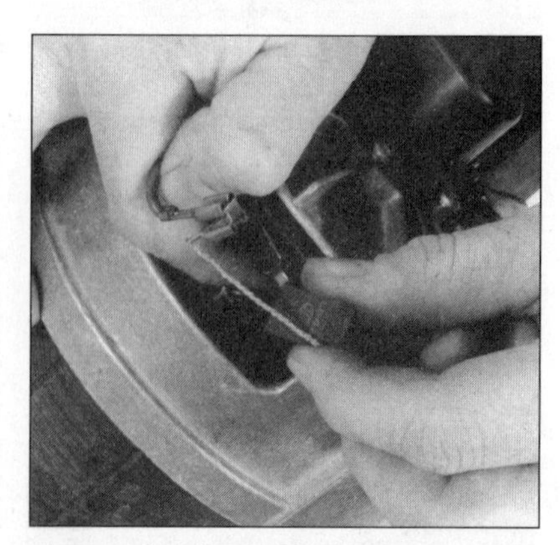

17 When fitting new brushes, ensure free movement of the brush in the slide.

18 Make sure all connections are tight and do not foul the metal body of the motor. Ensure the insulation strip is fitted to the brush opening. It will fit easily if warmed first. The motor is now ready to fit to the machine for functional testing after all panels have been refitted.

GEC-type motor

Shown here is the brush replacement for the GEC-type motor, variations of which can be found in Hotpoint, Indesit and Creda machines. Any other fault with this type of motor requires a complete change of unit.

TOOLS AND MATERIALS

- ☐ Screwdriver
- ☐ Brush
- ☐ Pipe cleaner

1 GEC-type motor – the early type can have a similar armature change as the Hoover type *(see pages 119-120)*, but otherwise only brushes can be fitted.

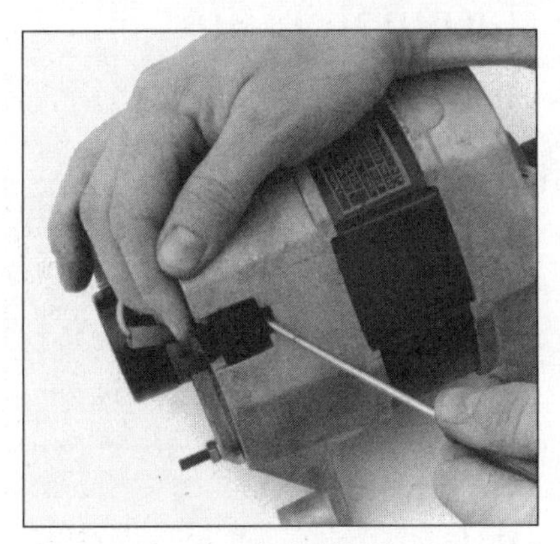

2 To remove brush and holder, insert a screwdriver and lift the tongue of plastic at the base of the holder.

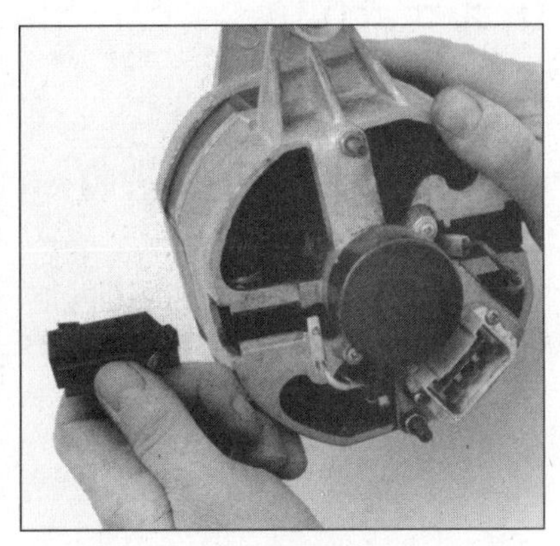

3 Slide out the brush holder complete with brush. Note the length of the new brush and check for good movement of brush in slide. This brush has worn very short.

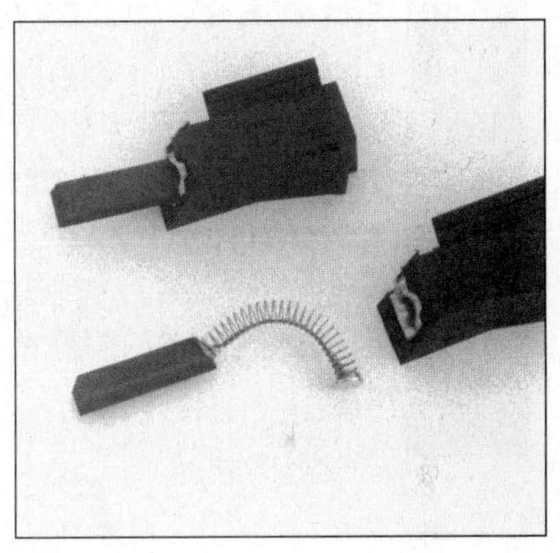

4 View of new brush and holder complete. Early screw-on-type brush holders have separate brushes as shown above.

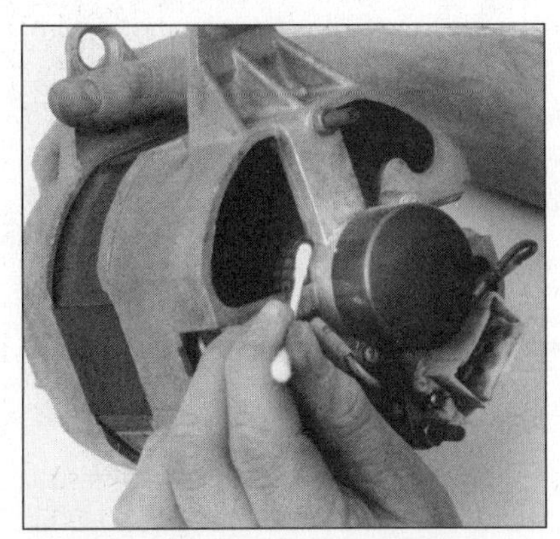

5 Ensure that any carbon dust deposits inside the motor casing and armature are removed. Blow out dust and clean with a pipe cleaner or similar object. Take care – do not inhale the dust

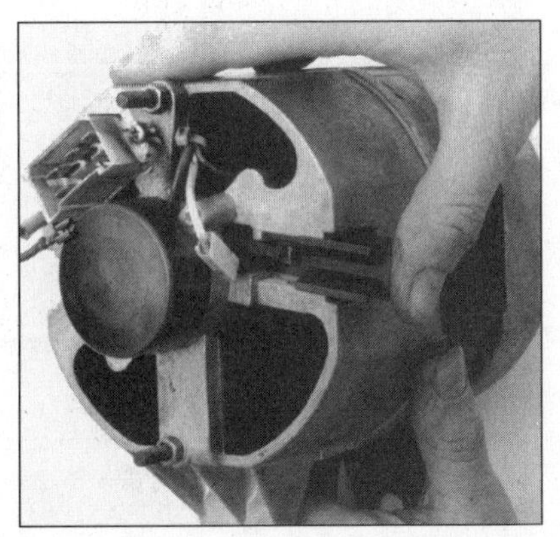

6 To refit the brush holder and complete brush assembly, carefully slide back into position, ensuring that the tongue of the holder engages into position.

MOTOR REPAIR WATCHPOINTS

1. **Always isolate the machine** before inspecting, cleaning or repairing the motor.
2. **'Short' the terminals of the capacitor** with an electrically isolated screwdriver before repairing an induction motor.
3. **Take care when checking a sluggish motor** – it may still be very hot.
4. **Check regularly** that machines with relay start motors are standing upright.
5. **Apply lubrication** lightly to tumbledrier induction motors to avoid fluff build-up.
6. **Always ensure that you buy the correct replacement** for your machine.

Belts

Two different types of main drive belt are used in automatic washing machines (wash only and combined washerdrier versions) and dry only machines. Both kinds are found in many sizes, but each machine must be correctly fitted with its exact size and type and no other. The two types of belt are the 'V', so called because of its location on V-shaped pulleys, and the multi-'V' belt, which is much flatter and has a series of 'V' formations on the drive face. The use of a multi-'V' formation gives a greater contact surface area in relation to the belt width. This is necessary because the belt is designed to be driven by a much smaller and, consequently, much faster-rotating drive pulley than the larger single-'V' drive pulleys.

In general, 'V' belts are found on washing machines with induction motors and multi-'V' belts on washing machines with brush gear motors of all types. Tumble dry only machines mainly use multi-'V' belts (of a smaller width and fewer Vs) to rotate the main drum via an induction motor. However, a combination of multi-'V' and 'V' belts are found on some early tumbledriers, such as Burco and Hotpoint. Also some tumble dry only machines use an elasticated belt to drive the independent fan unit. Machines, such as Candy, Hoover and Burco appliances, can be found with this configuration *(see Dry only machines pages 154-167)*.

Both 'V' and multi-'V' belts consist of woven nylon cords on which a synthetic rubber is moulded. The single 'V' belt has sides of approximately 40° and terminates in a flat base. The multi-'V' belt has a series of peaks and troughs, the number of which varies with the work load requirement.

Always ensure that a replacement belt is the correct size and width. Most belts have sizes or size codes printed on the outer face. However, these marks have often become virtually illegible on old belts because of general wear. Take a note of the make, model and serial number of your machine along with any legible belt code when obtaining a replacement belt.

Removal and renewal

The single-'V' belt drive system requires both drive pulley (the one on the motor) and drum pulley (the larger one on the drum shaft) to have a recessed groove the same dimensions as the belt. There are two variations of multi-'V' systems. Both drive pulley and drum pulley

Typical simple 'V' belt drive. Belt tension is by motor adjustment.

may be grooved to accept the multi-'V' configuration of the belt. Alternatively, only the drive pulley is grooved (to aid grip on its much smaller surface area) and the drum pulley is smooth and slightly convex in shape. Grip is created on the drum pulley purely by its having a greater contact area to the belt when in use, even though this is only on the peaks of the belt. The convex shape keeps the belt in place on the grooveless pulley, and reduces wear if misalignment occurs.

A system that is very similar is used in tumble dry only machines. The belt runs directly on the outer of the drum, the size of which gives a very much greater surface area, and because of its width, allows the belt position to line up with both drive pulley and tension wheel.

Belt care

Ensure pulleys are in good condition, that is, not chipped or buckled, and are aligned correctly. Misalignment shortens the working life of both types of belt. Poorly aligned belts shed the rubber compound coating from the cords, leaving tell-tale dust or flakes in the base and surrounding area of the machine. This may block the 'V' section of multi-'V' belts and cause the belt to fly off, usually on a spin cycle. Single-'V' belts may twist within the 'V' section when misaligned. Close inspection for damage to belts is essential, and reversing the belt and bending it is the best way to carry this out. Check the whole length in this way. If any defects are found, renew the belt.

When removing a belt for damage inspection or for repair take care to avoid inadvertently inflicting damage on the belt itself. Do not use screwdrivers or similar tools to prise belts on or off because this can easily damage the belt cords and the moulding of the soft aluminium pulleys used on washing machines. Slacken off the motor bolts to reduce tension (on washing machines) and pull the belt towards you midway between the pulleys, while carefully and slowly rotating the drum pulley clockwise. This allows the belt to ride out of position smoothly. Reverse this process for refitting. It is advisable to wear protective gloves because the pulleys can have very sharp edges.

Belt tension

It is normal for some degree of stretching and wear to occur during use, so re-tensioning is needed. Some machines are self-tensioning, that is, the weight of the motor keeps the belt under tension. This system was popular with some early washerdriers. On most tumble-dry only machines, a spring-tension jockey pulley (so called as it rides on the belt) is used to keep the belt under tension. Note that two may be found on reversing

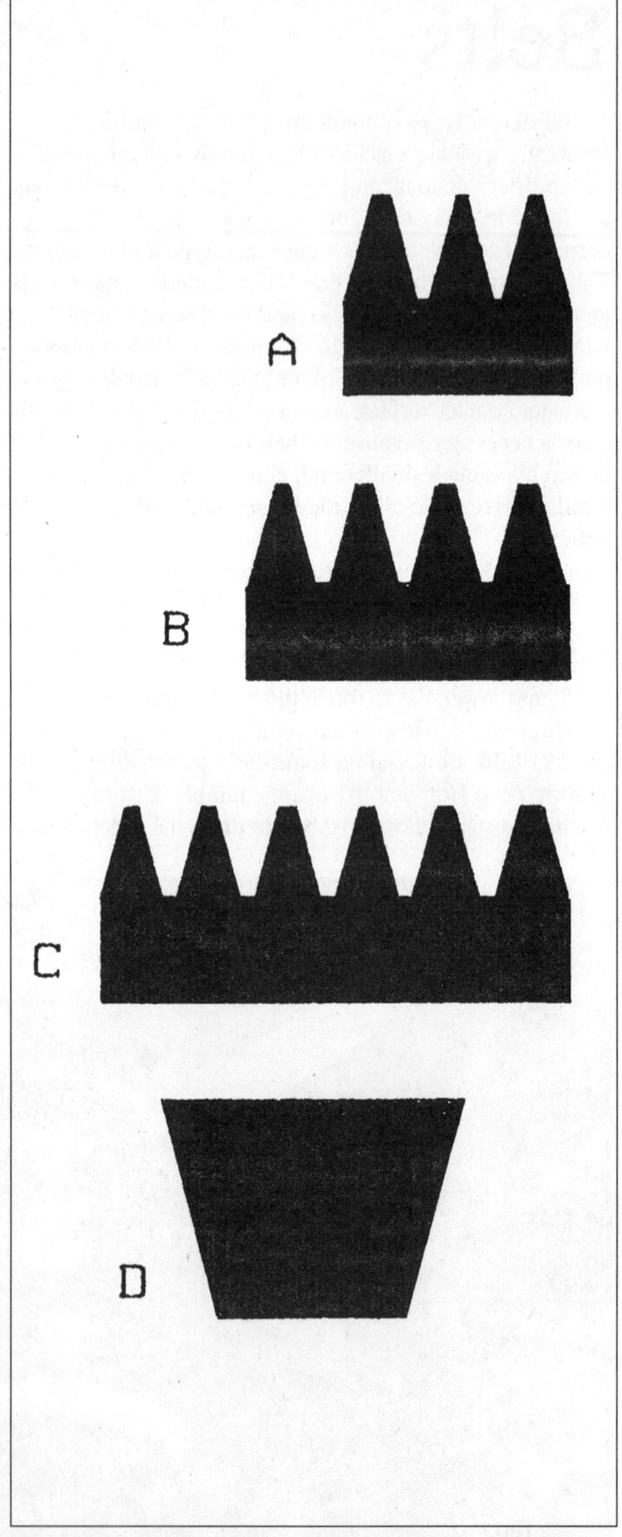

A, B and C are multi-'V' belt variations. A is predominantly used in tumble dry only machines, whereas A, B, and C are found in both wash-only and combined machines. D depicts the profile of a 'V' belt and can be found in wash only, combined and tumble dry only machines. E is a cross-section of an elasticated belt restricted to use in certain tumble dry only machines.

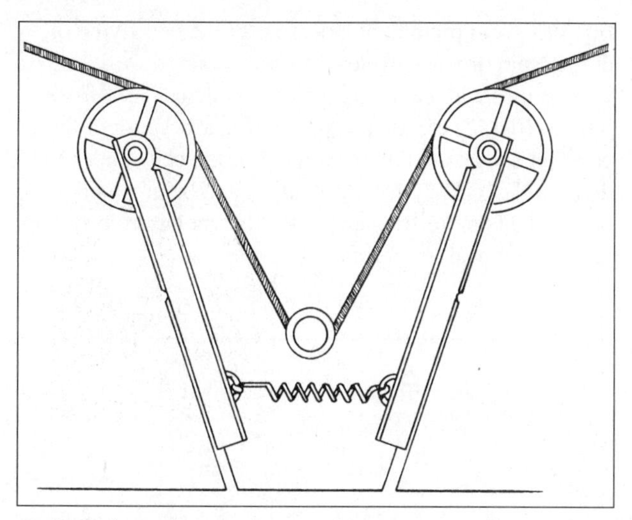

Double jockey pulley system required for reversing tumble dry only machines.

tumbledriers. However, the majority of washing machines and washerdriers rely on motor adjustment – two fixed bolts and one slotted – to tension the belt.

Setting the correct tension is essential. Too tight wears the belt quickly and, worse still, damages the drive pulley causing premature motor bearing failure. Too slack causes belt slip, resulting in poor wash, excessive vibration or heating of the belt, which finally causes belt damage or failure. On machines with aluminium pulleys, slipping can create ridges on the pulley grooves. If this happens, the damaged pulley or pulleys must be renewed because any new belt fitted to such a pulley will soon become damaged by the uneven surface.

The correct tension of a belt depends on its free distance between pulley contact points. As a rule, a 12mm (½in) deflection per 30cm (12in) of free belt is required. Most washing machines have about 30cm (12in) of free belt between pulleys and, therefore, a 12-13mm (½in) deflection is optimum. When fitted and tensioned correctly, the belt has a springy feel.

Some stretching will inevitably occur to new belts, but modern, good quality belts are much less affected. The belt will need to be checked at a later date and readjusted if necessary. Do not over-tighten a new belt in the misguided hope that this will overcome any initial stretching.

Common faults

It is not uncommon for belts to be warm after use, even when correctly tensioned. This is as a result of the energy absorbed as the belt flexes, and is proportional to the load. If the belt is hot or even very warm after use, this is an indication of incorrect tension or overloading causing belt slip and friction heating. Correct the problem, but check the belt for any cracking caused by overheating. If in doubt, renew the belt.

A squealing noise, most often heard on wash rotation and before spin (distribute), when the belt is under most

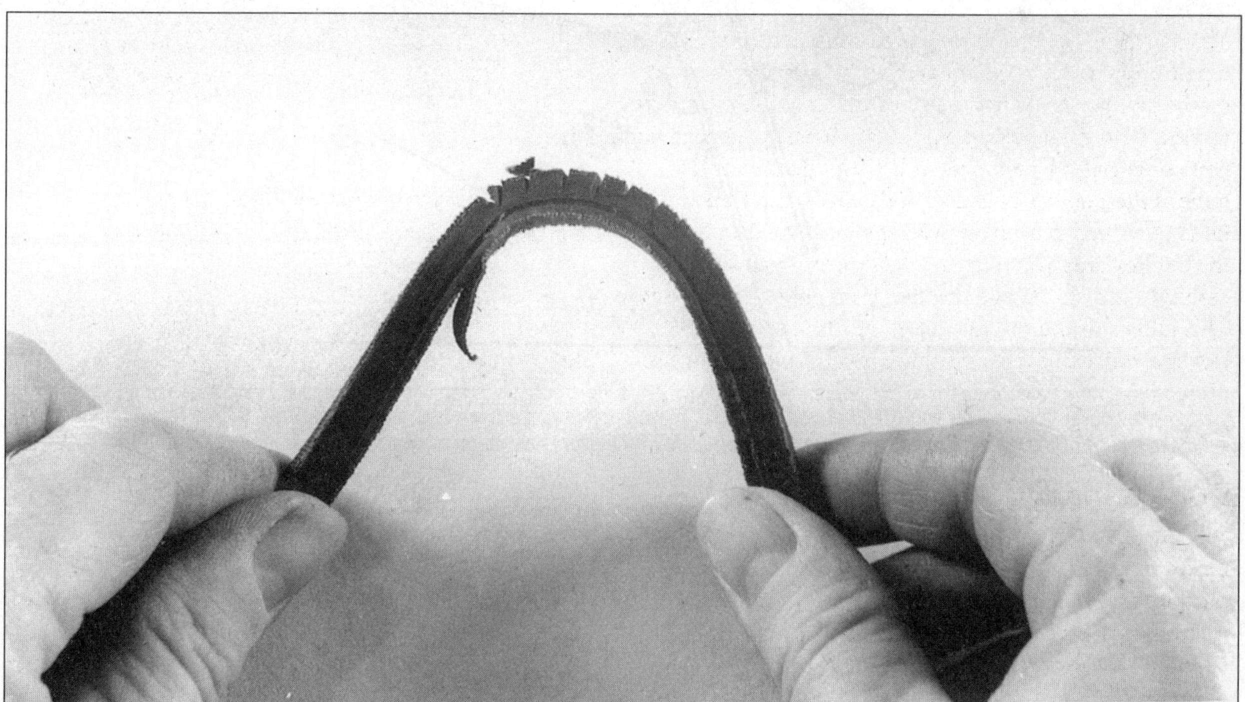

Turn belts inside out to check for problems. Do this over the whole length of the belt. This belt has a detached outer cover and cracked inner wedge. A fault such as this requires renewal of the belt.

load may simply be incorrect belt tension, slack, or worn pulleys, ridges on the pulleys, misaligned belt or too big a wash load. Isolate and correct the fault and inspect the belt closely for damage. If in doubt, renew it.

Ensure multi-'V' belts align correctly if both drive and drum pulley are grooved, note the position of the original belt, that is, first groove on drive pulley is used, then first groove on drum pulley is used. Misalignment of multi-'V' belts is easily done, so ensure that they are correctly fitted to avoid premature belt wear or the belt flying off during spin or wash cycles.

Machines with centrifugal clutch systems *(see Motors, pages 110-122)* create quicker belt wear because of the constant squeezing and movement of the belt. Check the drive pulley closely for wear ridges and be prepared to change belts more frequently on this type of drive system to maintain peak performance.

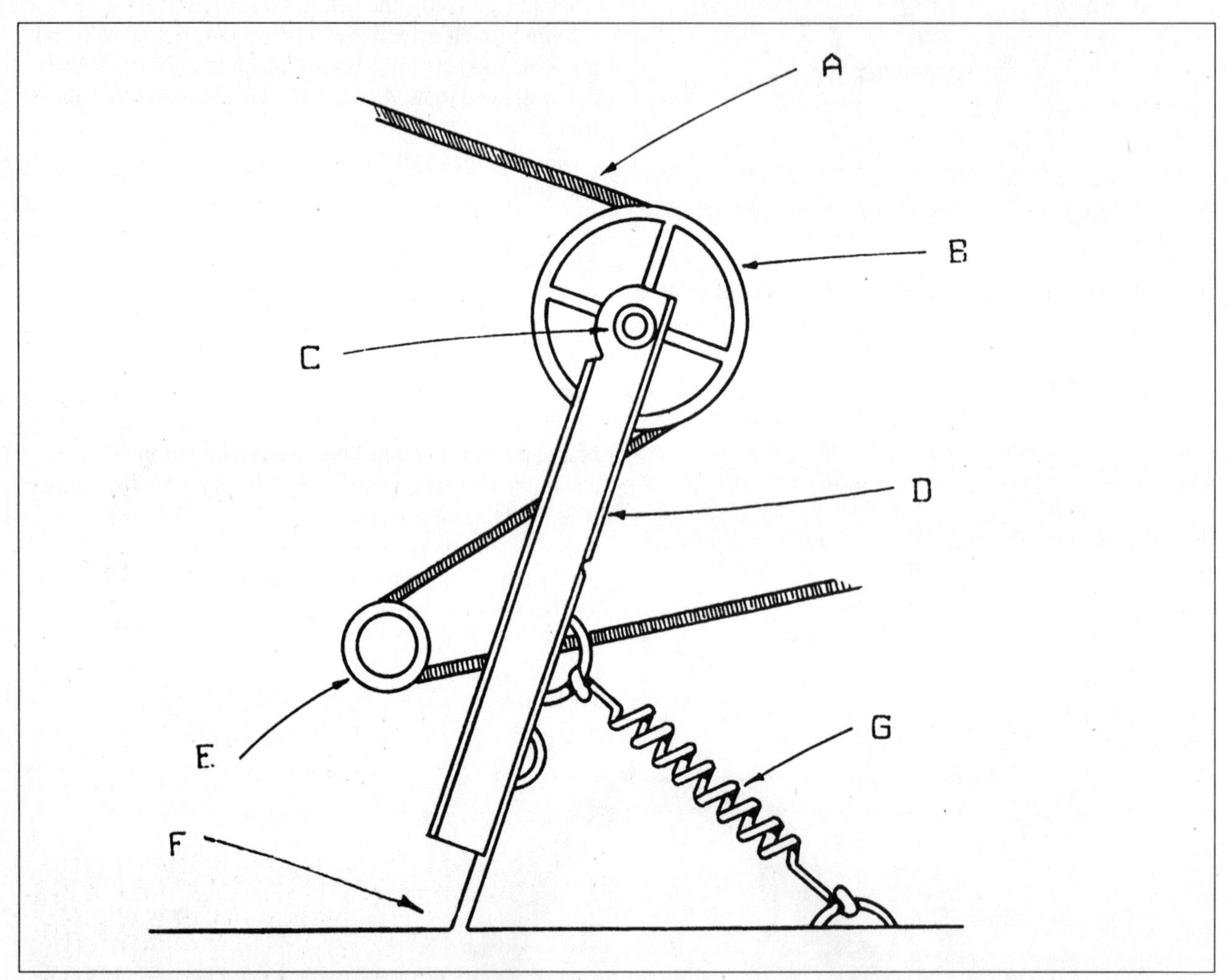

Typical jockey pulley components and configuration. Note that most, but not all, belts are tensioned as depicted. Before removing the old belt make a note of the correct position in your machine as some of the alternative methods of tensioning are not so obvious.

A Main drum multi-'V' drive belt
B Jockey pulley
C Pulley shaft
D Tensioning arm
E Multi-'V' drive pulley on motor shaft
F Pivot point
G Tension spring

Suppressors

A suppressor is a device designed to eliminate the formation and transmission of spurious radio waves that may be produced by the operation of the motor and switches within an appliance during normal operation. When switching occurs within the appliance, and it is not suppressed, small sparks at the contact points or brushes may produce interference on radio and television channels or audio equipment plugged into the same electrical circuit, that is, not only through air waves but also down the mains cable.

By law, all domestic appliances must be suppressed to conform to the regulations on radio interference. It is an offence to use an appliance that is not suppressed to these standards.

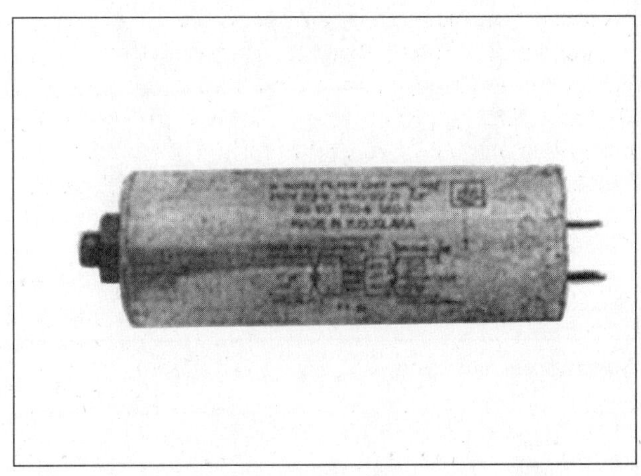

In-line suppressor 4TAG type (variations possible).

Location

Suppressors vary in style, shape, position, size and colour. Sometimes individual parts are suppressed, but more often the mains supply is suppressed at or just after the entry point into the appliance. This is called 'in-line' suppression, because both the live and neutral supply go through the suppressor and on to supply the whole of the appliance with power.

Do not confuse suppressors with capacitors which may be used for induction motor starting. They may look very similar but carry out distinctly different functions. Suppressors may also be called 'mains filters' because of their ability to remove spurious radio transmissions.

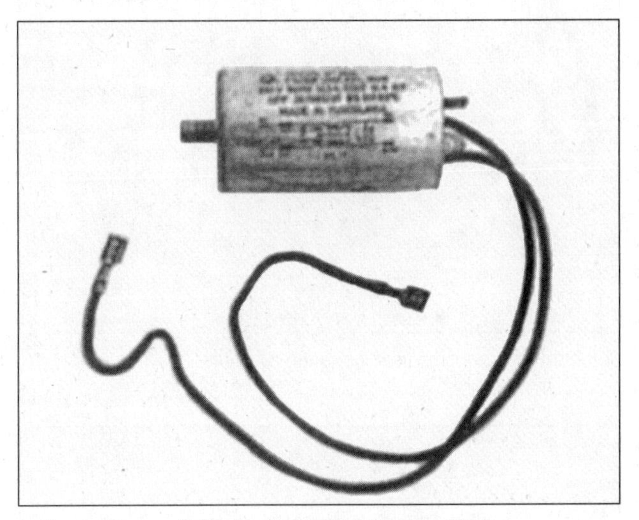

In-line suppressor flying-load type (variations possible).

In addition, a revised version of a suppression unit may be found. This is an induction coil fitted in series between the neutral position at the terminal block and the shell of the machine. Because an induction coil is of a far heavier gauge, it passes only suppression current, whereas the two versions described above carry the full voltage load. Some machines have a combination of both in-line and induction coil types of suppressor.
All versions require a good earth path of both plug and socket *(see Electrical basics, pages 10-13)*.

An additional means of suppression may sometimes be found, especially on micro-processor controlled machines. This is a choke-type suppressor and may be fitted in series between live and neutral positions or individually in-line on both live and neutral supplies to individual components. Such units consist of a ferrite core or ring around which the conductor is wound.

Some appliances use a combination of different types of suppressor. Suppressors of any type should not be by-passed or omitted, because to have an unsuppressed appliance is an offence, owing to the interference that it

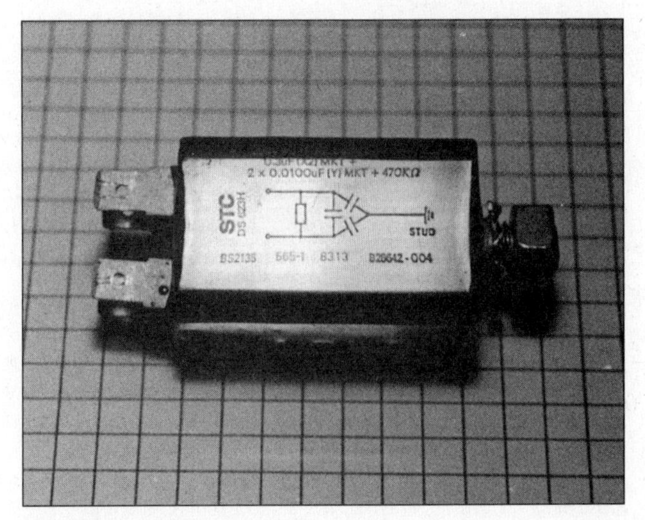

New-style square suppressor.

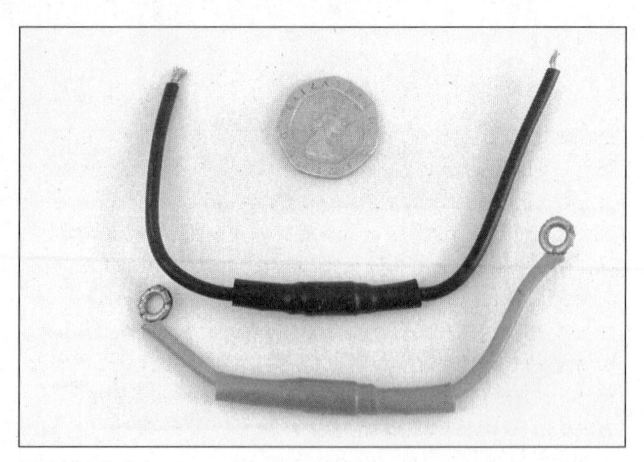
Typical 'choke' suppressors.

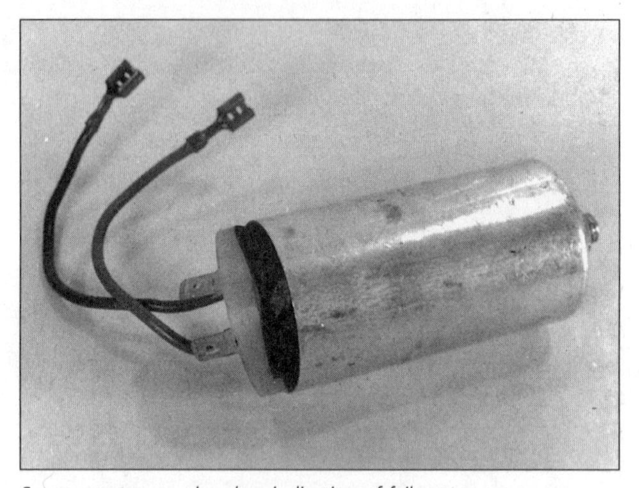
Some suppressors give clear indication of failure.

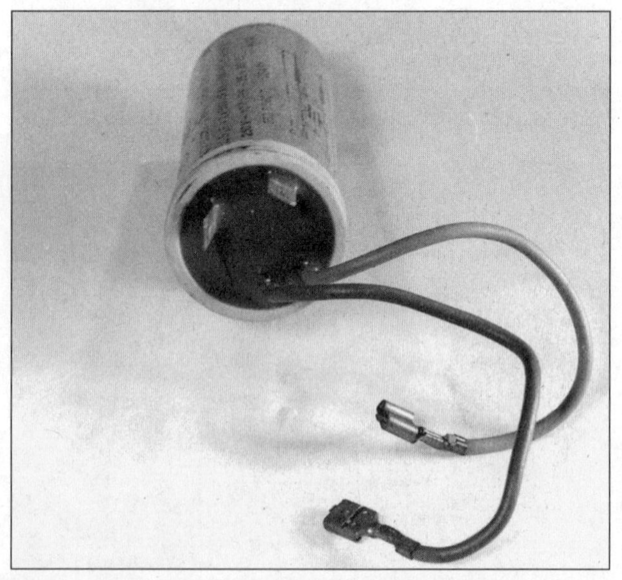
Check suppression units closely for cracks or expansion. If in doubt, renew. This unit has a small crack across the top but otherwise it looks alright. However, internally it is a direct short circuit and potentially dangerous.

may cause to others. In the case of 'chokes', failure or omission of this device may cause damage to or corrupt electronic circuits within the appliance *(see Timers, pages 98-109)*.

Common faults

The main and most commonly occurring fault is one of short circuit to earth, usually resulting in the unit's 'blowing' both the main fuse and itself. This is frequently accompanied by a horrible pungent burning smell. Repair is a matter of straightforward one for one replacement.

Open circuit problems can occur and the unit will then fail to allow current to pass through as normal. The suppressor can very easily be checked for continuity using a meter *(see Electrical circuit testing, pages 23-25)*. When checking, inspect the top insulation closely and, if it is cracked or at all suspect, renew the complete unit.

It is common for in-line suppressors to use the earth path as part of their filtering circuit although very little power passes through it. It is essential for all appliances to have a good earth path. If an appliance with an in-line suppressor or filter has a break in its earth path – as a result of cable, plug or socket fault, for example – small electric shocks may be experienced when the user touches the metal parts of the machine, especially if he or she is in touch with a good earth path, such as by holding a metal sink. It is absolutely essential that such faults are traced and corrected immediately, before the machine is used again *(see Safety guide, pages 6-8 and Electrical basics, pages 10-13)*.

Although the continuity of a suppressor (lead-through type) can be checked easily, its function of suppression cannot. If all other checks, that is, for good earth connection (check for loose/poor connection to shell of machine) and no cracks or loose/heated terminals, prove to be alright, and interference to other equipment persists, renew the suppressor.

Bearings

This chapter deals predominantly with the main drum support bearings, although many of the associated problems also relate to other areas, such as main motor bearings on both washing machines and tumble dry only machines, and also fan and pump motors on machines that use ball race or roller bearings. For information on sleeve bearings *(see Pumps, pages 69-74 and Motors, pages 110-122)*. For information on tumbledrier drum bearing variations, read the following and, in addition, see Dry only machines *(pages 158-171)*.

Types of bearing

There are three types of bearing used in washing machines and combined washerdriers.

- The sleeve bearing. This is simply a phosphor bronze bush in which the motor shaft is free to rotate. It is more commonly used in pump and fan motors.
- Ball race bearing. This type consists of an outer ring in which a small inner ring is supported by circular ball bearings and is free to rotate.
- Taper roller bearing. The taper roller bearing uses rollers in place of the ball bearing and, as its name implies, the design angles the rollers to give a tapered appearance. The outer ring (shell) is not fixed, as with the ball race type, and is fitted into position separately. It is essential that taper bearings are fitted as a matched pair, that is, inner and outer.

The type of bearings used on the main drum bearing assembly differs not only between manufacturers, but also between models. One range may use ball race bearings when a similar model from the same manufacturer has taper roller bearings. Although only two types of bearing are used, both outer and inner vary greatly in size, so it is essential to fit identical bearings when replacing old or damaged ones and to renew all shaft seals and spacers at the same time. If possible, buy a bearing kit to make sure that all relevant parts are renewed. Do not cut corners by replacing only one bearing in a set of two or fitting a new taper roller inner to an existing outer shell because the old outer shell is difficult to remove. These short cuts will lead to premature failure and further damage.

Common faults

There are several reasons why bearings fail, apart from normal wear and tear over a long period of time. The commonest causes and points to watch out for are described here.

The commonest failure of both ball and taper bearings is the ingress of water and detergent into the bearing housing. This usually results from wear or premature failure of the shaft seal.

Two types of seal are used. The first is a simple shaft seal which is pressed into the bearing housing on top of the front bearing. When assembled, the lower section of the drum shaft locates within the seal (usually a raised metal shoulder or collar of mild steel or, more commonly phosphor bronze) and is fixed to the base of the drum shaft. The seal has either one or two spring-loaded lips which press firmly around the collar to create a watertight seal. Normal wear, fluff or scale may break this seal down and allow water and detergent into the

Typical ball race bearings open on the left and shielded on the right.

Top view of this Zanussi washerdrier shows that for drum renewal, it is necessary to remove the outer tub unit from the machine. This is because only the front section of the outer tub is removable. However, bearing renewal can be carried out without outer tub removal.

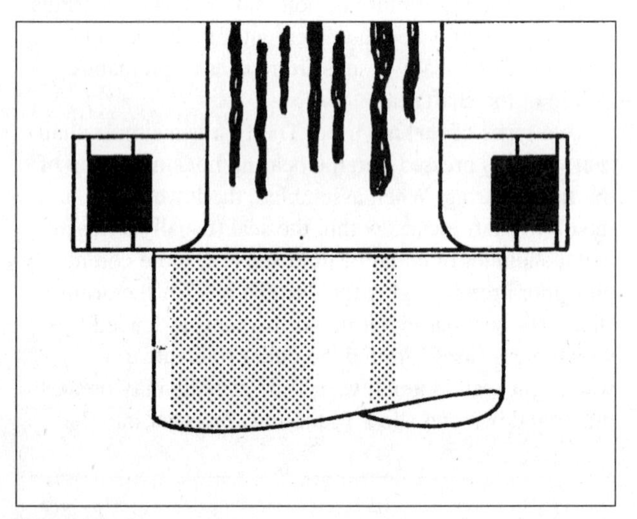

Scoured shaft.

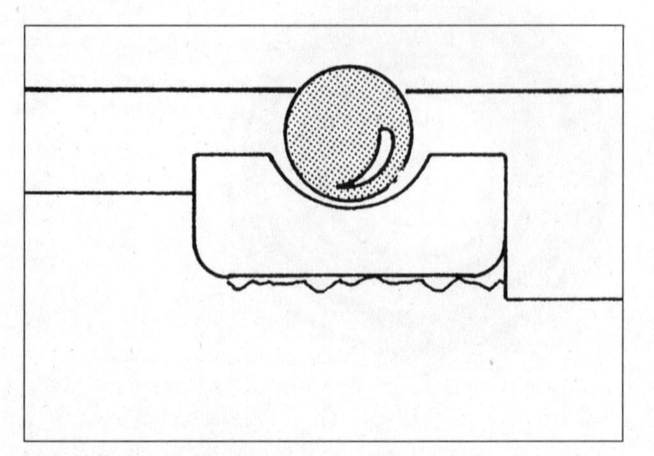

This shows how the bearing rests on a scoured shaft.

bearings, quickly resulting in failure. Ensure that the collar is clean and secure on the shaft and that both seal and bearings are renewed at the same time.

The second type of seal is a carbon-face seal. This system relies on two smooth faces of carbon (or ceramic), which are pressed firmly together when assembled. One face is free to rotate, being fixed to the base of the drum shaft, whereas the other, as before, is fixed on top of the front bearing and is spring loaded. The two smooth surfaces held under the pressure of the spring within the seal creates the movable watertight seal. Again, fluff, scale and normal wear will lead to water ingress. Ensure that the fixed carbon ring is smooth, not cracked and securely fixed to the base of the shaft. Check the new spring-loaded face seal in a similar way and make sure that it is seated correctly into the bearing housing. Application of a little sealant is recommended to make certain of a watertight fit to the housing. Do not allow any sealant or dirt to get on either of the faces of the seals when assembling them. Bearings will also fail if they are incorrectly fitted. Do not hammer in the bearing – or shell, if taper roller type – directly with a hammer because this may crack or chip it. Do not force the bearing or shell into a distorted or dirty housings because this will distort the bearing and create overheating resulting in failure. Inspect the drum shaft closely for ridges or rust which, again, will distort the inner race of the bearing, resulting in overheating or, if worn, create a loose fit. Take care to clean all areas and fit only the correct size of bearings carefully.

Inspect failed bearings closely for they will have sizes stamped on them which can be used to ensure the new bearing is the same size. Failed bearings may also give an indication of why they failed. Rust indicates that the seal has failed and prompts closer inspection, cleaning or renewal of the shaft, collar or carbon face. Bearings that have become coloured dark blue or black are usually the result of overheating. Check the shape of housing and shaft and do not over-grease. Flakes of metal from the bearing also indicate some form of distortion or ingress of dirt during assembly.

Do not overtighten bearings, especially taper roller types, hoping that it will help seating. The result is usually overheating and premature failure.

Noise can be a good guide to early recognition of bearing problems. Simply removing the drive belt should help ascertain if the faulty bearing is in the motor or the drum shaft support bearings. A selection of noise faults is listed on page 131 along with possible causes.

Fault finder

Symptom: Noises

POSSIBLE CAUSES	ACTION
Loud rumbling especially on spin.	Collapsed front drum bearing or seized front bearing. This often results in shaft damage if not attended to quickly.
Rattling with intermittent knocking.	Ball or roller of bearing defective.
Rattling/knocking proportional to speed.	Inner or outer of bearing faulty.
High-pitched metallic noise.	Common on worn motor bearings at high speed and on new bearings if they have been forced on to a damaged or over-sized shaft.
High-pitched ringing noise.	Bearing has been fitted to a damaged housing or fitted carelessly.
Grating and crunching noise.	Collapsed bearing cage or dirt between inner and outer race.
Squeaking.	On old bearing, usually due to lack of lubrication or ingress of water past shaft seal. May also occur when new bearings are fitted to a machine with carbon-face-type seals if care was not taken to keep both surfaces clean or if check was not carried out for cracks or scouring on carbon faces.

Tips on fitting

To gain maximum life from any bearing, take considerable care when handling and fitting it. Before stripdown and removal of the old bearings, make sure that you note or draw the position of seals, clips and washers. Inspect both the shaft and housing closely for defects. Ensure the shaft, housing, work area and hands are clean. Remove the new bearing from its protective packaging only when you are ready to fit it.

Ideally, bearings should be pressed into position but this is not always possible and some means of drifting the bearing into place may be required. Great care must be taken if this method is used because the bearing can easily be damaged. Endeavour to use a tube when fitting the bearing to enable an equal force to be applied. The tight fitting part of the bearing should take the force only. If fitting to a shaft, contact should be only with the inner race and when fitting a bearing into a housing, only the outer race should take the force. Do not, under any circumstances, apply any force to the part of the bearing that is free to rotate during fitting.

To assist in fitting bearings, expansion and contraction with heat can be used, although excessive heat must be avoided. Fitting a bearing or taper roller bearing outer shell into its housing can be assisted by simply putting the bearing or shell into the freezer for a while and warming the housing with the aid of a light bulb for an hour. This simple technique can help enormously. When fitting a bearing to a shaft, a reversal is required. Cool the shaft and warm the bearing, but be careful not to overheat the bearing (especially sealed bearings) because this may create problems. Do not exceed 100°C.

Many ball-type bearings are greased during manufacture and sealed on both sides. These do not require any extra lubrication before assembly. Roller bearings and some open-cage ball bearings do require packing with grease. This should be done sparingly because over-greasing causes churning of the grease and heating takes place, which results in loss of lubrication. Pack the bearing with grease and rotate both inner and outer with the fingers to allow any excess to be pushed out from the moving parts.

Typical bearing change

The step-by-step photographs on pages 140-141 show the renewal of a set of drum bearings. In this particular machine, the main drum support bearings are worn and water damaged. This was caused by the carbon seal failing and allowing water and detergent to enter the

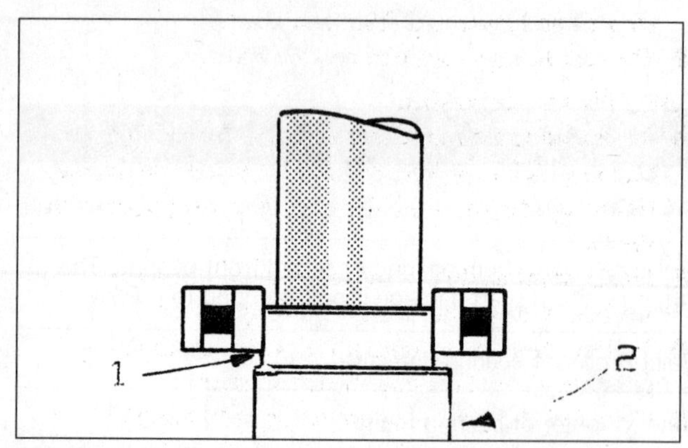

Position 1 shows a bearing that is incorrectly positioned and is proud of its correct location point. Position 2 indicates the collar on which the shaft seal will ride. Ensure it is both clean and secure.

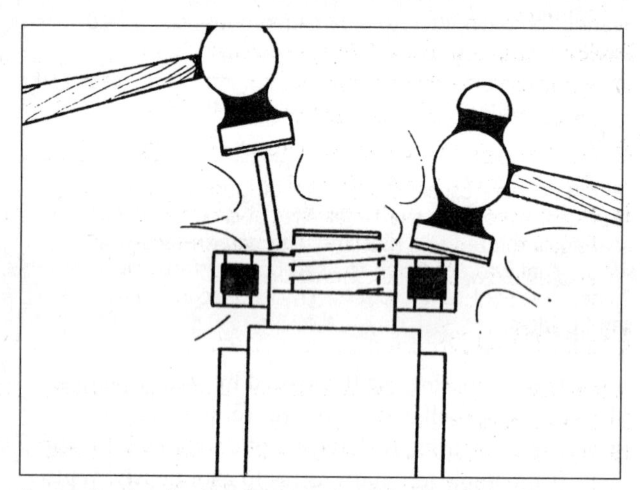

Do not fit a bearing in this manner because it can easily be damaged.

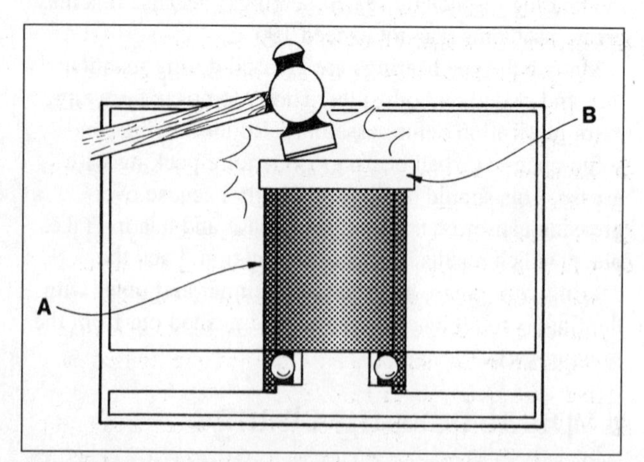

Press the bearing home, ideally by using a bearing press tube and metal plate. However, if access to a press is not possible, carefully use a hammer in its place after first applying heat as described on page 133. An alternative method is to use two large metal plates/washers and a long bolt and nut to press the bearings into position by tightening the nut and bolt.

bearing and housing. This fault was suspected because of the noise of the machine, especially on spin. It was confirmed by removing the drive belt from the motor to the drum pulley, spinning the drum slightly by hand and listening for any grating noise. If it had been quiet, then the motor would have been spun in the same way to test its bearings.

Another way of confirming the drum bearing fault is to open the door of the machine and move the bottom of the door seal in order to see the inner drum and outer tub gap clearly. Then try to lift the top lip only of the drum. The gap between the outer tub and inner drum should neither increase nor decrease and there should be no movement other than that of the outer tub on its suspension. This applies to all machines irrespective of bearing types.

Step 12 shows that the drum and spider mount is breaking loose from its position on the drum as a result of a form of metal fatigue fracturing the drum. The only possible remedy is fitting a complete new drum assembly. The initial fault is also rectified – renewal of the bearings. The user is unfortunate in needing the drum unit, especially when the support shaft itself is in such good condition.

Removing the drum and back half assembly on this style of machine is quite straightforward. This style of drum and back half is fitted to many leading makes of machine, including Hoover, early Hotpoint, Creda and Servis, although in each case the size of bearing and type of seal differ slightly between machines. As the bearing kits are complete matched sets, no problems should arise.

After completely isolating the machine and laying it face down on a suitable surface, removing the back panel in this instance reveals the back half and outer securing nuts. Remove all the nuts and bolts securing the back half and tub. If necessary, the top of the machine may be removed to gain access to the top bolts.

Mark all the connections to the heater and thermostat and disconnect them. The back half assembly can now be manoeuvred from its position and removed from the back of the machine. The back half is made watertight by a rubber seal located on its outer edge. If the back half sticks to the tub, remove all the necessary nuts and bolts and gently prise it from the tub without using excessive force. The drum and back half can now be manoeuvred free of the machine. Removing the pulley and bearings can now proceed.

If the bearing set for your machine is of the ordinary ball bearing type, the job is fairly easy because the old bearing should knock out in one piece. However, if the bearing 'collapses' or leaves its outer shell, the shell can be removed as shown for the taper bearing shell in step 18. Care must be taken not to go too deep into the soft aluminium housing with the drill.

Take care not to over-grease it, as this will not help lubricate it and, in fact, will considerably reduce the bearing life. Greasing is obviously not necessary on the sealed bearing type.

The new taper bearing kit comes complete with an odd-shaped aluminium washer that fits between the rear bearing spacer and pulley. This is known as a torque washer and is essential for the correct operation of the taper roller bearings. Always fit a new torque washer to this type of bearing system if the pulley is removed for any reason. Do not re-use the old one. The torque washer is a simple way of putting the taper bearings under a known pressure without the use of a torque wrench. When fitted together, the torque washer collapses at a given pressure and dispenses with the need for a torque wrench for tightening the pulley bolt. A full set of instructions for the torque washer comes with the new bearing kit.

If your machine has ball bearings, disregard the paragraph concerning torque washers. Renewal of this type of bearing is a simple reversal of the stripdown procedure.

Instructions for bearing change

Dismantle the old pulley and bearings.

- Remove the bolt (1) and washer (2 and 3) securing the tub pulley and remove the pulley (4).
- When removing the back half gasket, you will notice that it is compressed. It is advisable to replace this item to ensure a true watertight seal between the back half housing and outer tub.

Extract the old bearings.

- The existing bearing sleeves (9 and 11) are found inside the tub back plate (10). Extract the bearing sub assemblies and then apply heat to the area containing the sleeves. The sleeves can then be gently tapped out with a small chisel, drift or screwdriver.

Renew the bearings.

- Push the new sleeves into place, ensuring the inner surface of the tub back plate is thoroughly cleaned.
- Insert the first (larger) bearing and washer (12).
- Gently push the seal (13) into place. It is advisable to use a waterproof adhesive around the seal to aid fitting and prevent leaking past the outer edge.
- Clean the new carbon-face seal thoroughly to remove all traces of grease, oil, etc.
- It is advisable to fit new washers to the front and rear (12 and 7) because it is likely they have been scoured by the faulty bearings. Many bearing kits do not contain items 12 and 7 and these may have to be obtained separately.
- With the drum face down, fit the back half (with the new front bearing spacer and carbon seal fitted) to the cleaned and inspected drum and shaft.
- The rear bearing race can now be fitted.

Insert the torque washer.

- Re-assemble the remaining parts in reverse order to that in which they were removed. Do not use the old torque washer (6) – use the new spacer supplied with the kit.
- Fit the shim washer from the kit under the torque washer. In the position shown in (6A).
- Tighten the bolt (1) against the pulley as far as possible without locking the tab washer.
- Complete the tightening operation until the 'D' washer fits firmly against the shoulder of the spider unit (5).
- The spacer is now correctly pre-set. It is essential to do this to ensure the correct loading pressure on the bearings.

Discard shim and complete re-assembly.

- Remove the bolt (1), washers (2 and 3), pulley (4) and spacer (6). Discard the shim (6A).
- Re-assemble in reverse order again. This time lock the tab washer against the bolt.
- If this procedure on taper roller bearings is not followed correctly, the life of the new bearing set could be considerably shortened.

Useful tips

A few helpful hints on the refitting of the assembly back in the machine follow.

- It is advisable to remove the heater from the back half before refitting the back half and drum into the machine. The subsequent refitting of the heater in this fashion ensures the internal heater securing clip is correctly located *(see Heaters, pages 79-84)*.
- A smear of sealant helps slide the thermostat pod (if applicable) and the heater grommet into position. If the thermostat grommet looks perished, it should be changed.
- Remember to reseal any hoses on the pressure system if they have been disturbed.
- Remember to fit a new tub back half seal.
- Check the tension on the main drive belt and adjust the belt if necessary. This is done by moving the motor up or down to slacken or tighten the belt (like adjusting a fan belt on a car). Do not overtighten the belt *(see Belts, pages 123-126)*.
- Many condenser washerdriers have connections to the back half. There are numerous variations of fixings but correct and secure fixing is common to all. Ensure all connections are tight. Any damaged clips should be renewed along with any sealant used at connection points.

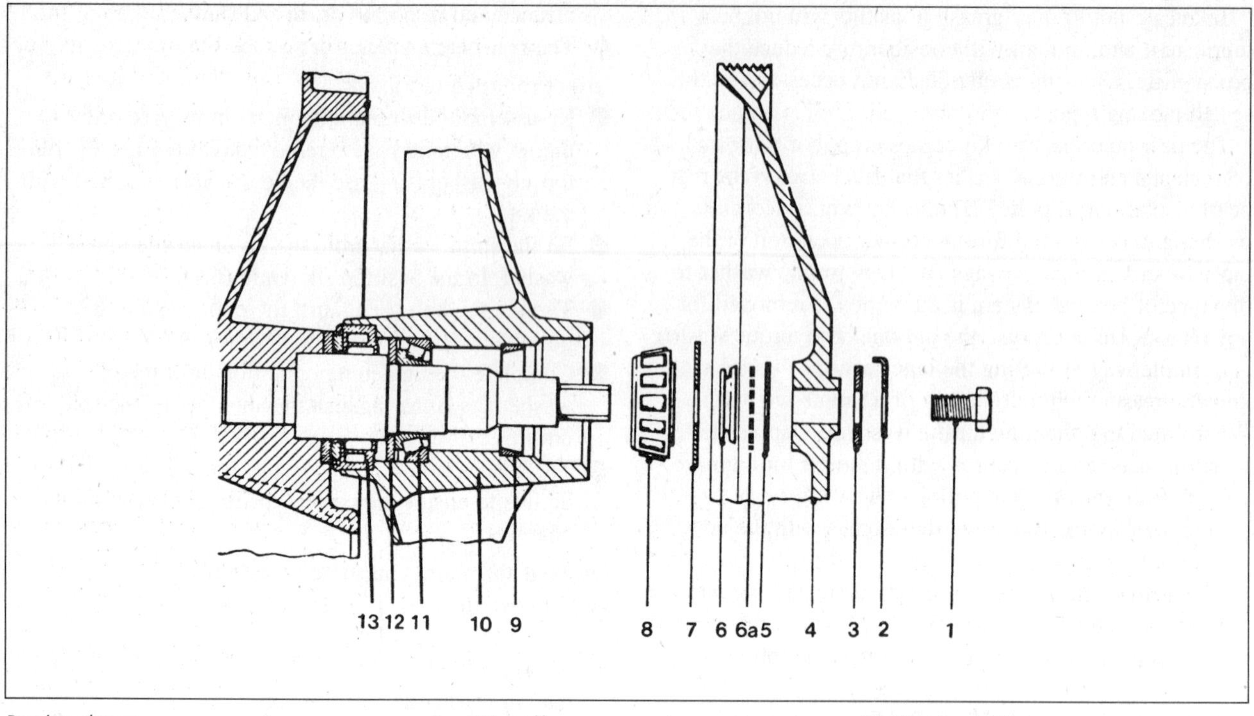

Bearing change

1 Bolt
2 Locking washer
3 Washer (small)
4 Pulley
5 'D' washer
6 Spacer
6a Shim (disposable)
7 Washer (thin)
8 Small bearing
9 Bearing sleeve
10 Tub back plate
11 Bearing sleeve
12 Washer (thick)
13 Carbon-face seal

Bearing changes needing tub removal

Often, the restricted rear access or reverse tub construction does not allow for the renewal of the bearings or drum in the manner previously described. In such instances, for repairs to, or renewal of, the drum bearings, tub seals or outer or inner drum, it may be necessary for the whole of the outer and inner drum unit to be removed from the shell of the machine before further stripdown can take place.

There are two main reasons why this course of action may be required. Access to the rear of the outer tub is restricted because of the very small access panel on the rear of the shell of the machine. However, on machines with external cast iron bearing holders, it is usually possible to manoeuvre the whole unit out through the opening, after first laying the machine face down and removing the securing bolts and pulley. Several Zanussi and Ariston models have bearings fitted in this way, but for inner drum removal/renewal the whole unit needs to be removed as described below. The second reason is that the construction of the outer tub is jointed at the front of the machine, not the rear as is usual in most machines. Consequently, the inner drum can only be removed by first removing the front of the outer tub. Unfortunately, in most instances, the front shell of the machine is not removable and so the whole unit needs to be removed as described.

Most models in the Candy range of washing machines require tub unit removal for both bearing and drum problems.

It is not uncommon for manufacturers to buy in products from another manufacturer and then 'badge' them as their own. This leads to a mix of model designs throughout the range. Until fairly recently, the current production machines were merely updated variations on a basic design and so some continuity and standard format existed. This is not always the case nowadays, which means that each machine has to be assessed before repairs are carried out. For example, are the bearings mounted in a detachable housing? If access requires removal of the front of the outer tub, can the front of the machine be removed to allow the main unit to remain in situ?

If no other option exists, the outer tub unit, complete with drum and bearing, has to be removed to allow for complete stripdown. The removal of this large unit is via the top of the machine shell. The following text describes the removal of the outer tub unit after removing such items as the top/bottom tub weights, pulley, all connections and hoses and electrical connections to the unit.

Remove all knobs from the front of the machine to reveal the fixing screws of the items behind them when servicing drum, bearings, outer tub and seals which entails the removal of the outer tub unit from the machine via the top of the machine's cabinet as a complete unit. These should be unscrewed and the components laid over the front fascia of the machine. If possible, do not disconnect any wiring, but make notes of all connections and fixings in the event of items becoming misplaced or dislodged.

Release the screws securing the dispenser unit to the cabinet and remove the hose from the dispenser unit. Lay the dispenser unit over the front of the machine. Remove the top tub weight (if fitted) and release the front fitting of the door seal. To help slide the tub unit out of the machine, two pieces of wood, 50mm x 25mm x 1380mm (2in x 1in x 4ft) can be inserted down the left-hand side of the machine between the tub and cabinet to support the tub during its removal. If your machine has the timer on the opposite side to that shown in Fig. 1, the wood should be inserted down the right-hand side and the machine laid over correspondingly.

Now lower the machine on to its left-hand side after making sure that the cabinet side and floor are protected. Release the shock absorber or friction damper mountings. Now disconnect the sump hose, pressure hoses, heater connections, thermostat connections and motor block connections. Remove the drive belt and drum pulley and check that all connections are free from the outer tub unit. At the top of the machine, release the suspension springs by pushing the tub unit towards the top of the machine. It is now possible to slide the tub assembly out of the cabinet. Help is useful as the unit will be quite heavy and needs manoeuvring out of position. Ensure that the lower friction plate is being supported by one of the wooden strips. Hold the door open during the tub withdrawal.

Once removed, the unit is easily accessible and the bearing renewal is similar to that shown in the step-by-step photographs. Make a note of all clamp positions, tub front and back positions and so on. It is advisable to re-seal and check all hoses and their fixing points before refitting the unit. It is important to do this when the tub is out of the machine because this may be difficult when the unit is replaced.

Refitting is a reversal of the removal procedure. After refitting, ensure that all electrical and earth connections are replaced correctly and securely and an RCD protected socket is used when the functional test is carried out.

Many manufacturers now produce machines with detachable front panels; Philips, Fagor and Hotpoint are three. With this type of machine, the drum and bearing assemblies can be removed and changed with the outer tub in situ, thus avoiding the extra work involved in tub removal. Plastic/nylon outer tubs are now used on some combined washerdriers.

With the Hotpoint machine, the pulley is threaded to the shaft and is secured by a locknut. To release the pulley, chock it with wood and rotate the drum anti clockwise from the front of the machine. When refitting, apply some locking compound to the shaft thread. This can be obtained from any good D-I-Y or motorists' shop.

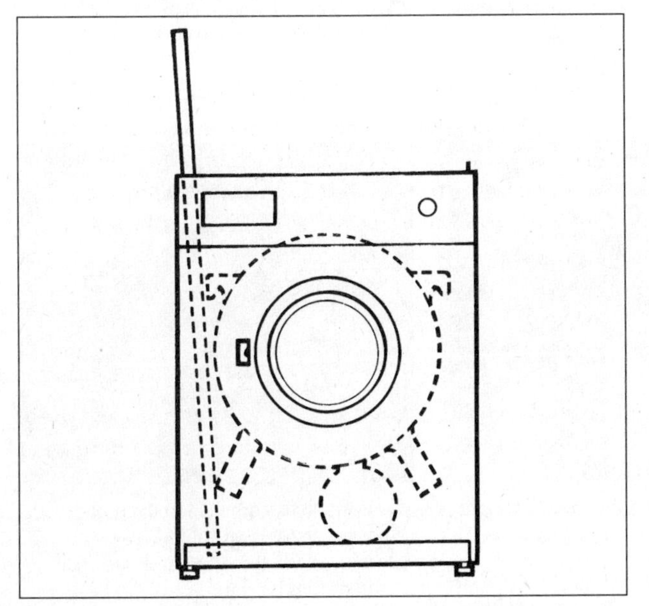

Fig. 1

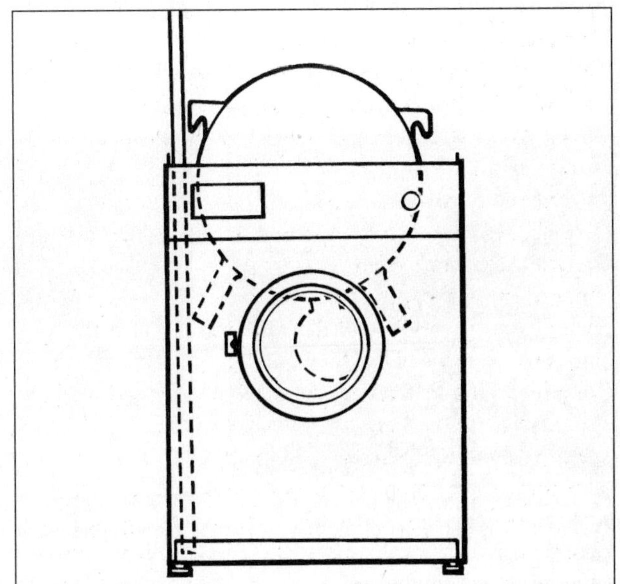

Fig. 2

Hoover taper roller bearing type replacement

Approach this lengthy task with organisation, following the instructions given on pages 132-133 and the steps shown here.

TOOLS AND MATERIALS

- ☐ Screwdrivers
- ☐ Spanners
- ☐ Pullers
- ☐ Soft metal drift
- ☐ Hammer
- ☐ Drill
- ☐ Sealant
- ☐ Bearings set
- ☐ Grease

1 Isolate the machine and remove the rear panel. Note all connections to the back plate.

2 Protect the face of the machine and gently lay the machine on its front.

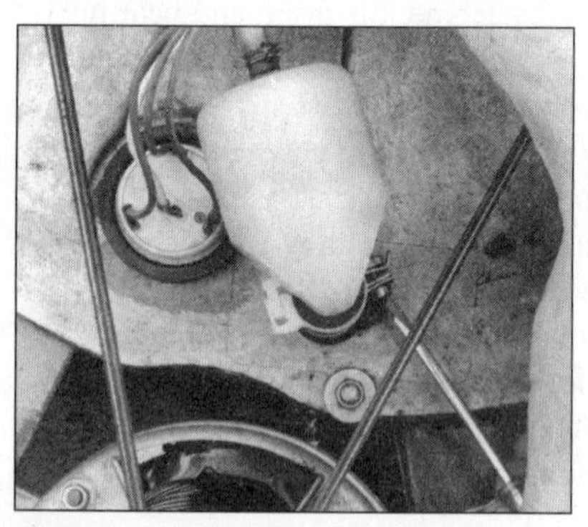

3 Note the position and angle of the pressure vessel and remove.

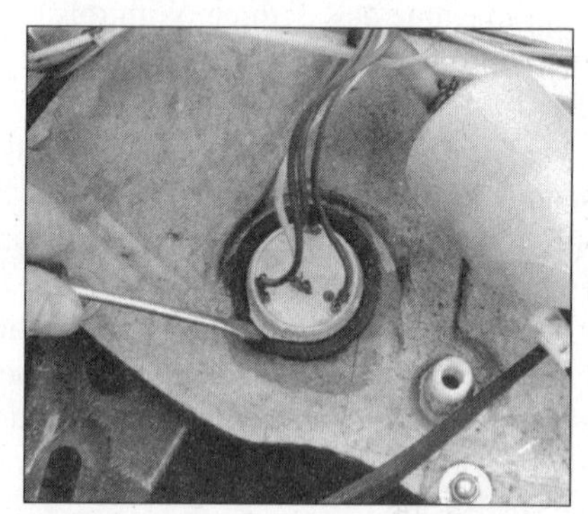

4 Note connections on the thermostat. Remove the thermostat with the wires still connected. Use a flat-bladed screwdriver to ease it from the seal.

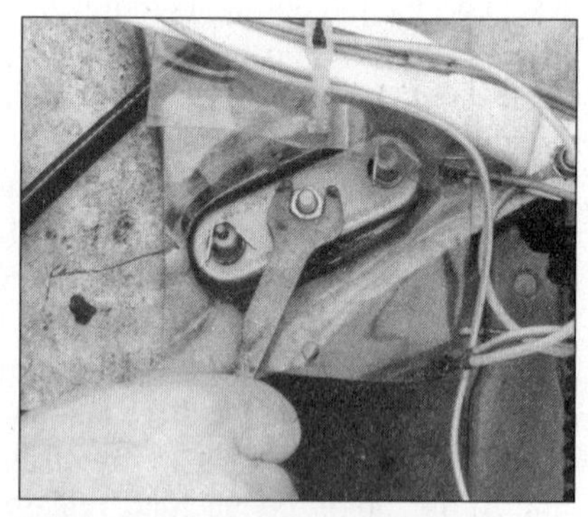

5 Remove the wiring connections. Slacken the heater clamp nut and gently prise the heater free.

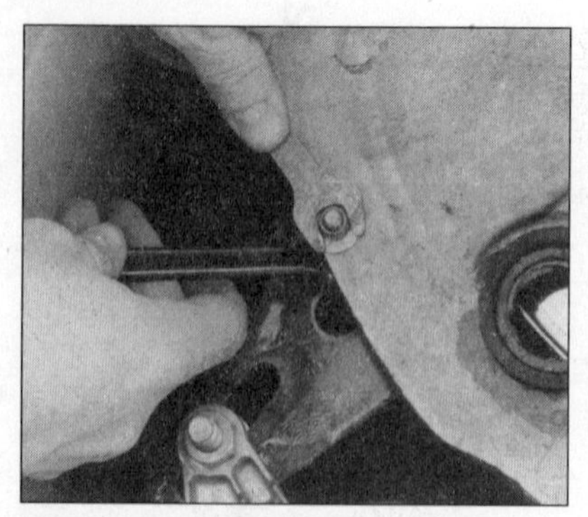

6 Remove the back half bolts from around the perimeter of the tub.

7 The discharge from the drain hole can be seen here and indicates water penetration of the bearings. The discharge is a mixture of grease, rust and water.

8 With all fixing bolts removed and wires secured out of the way, manoeuvre the drum and back half free from the tub.

9 With drum and bearing unit completely removed from the machine, it is much easier to work on.

10 Unlock the tab on the pulley bolt with the aid of a flat-bladed screwdriver.

11 Remove the pulley bolt, pulley and spacers, noting their correct order.

12 Back plate freed from the drum shaft. The front bearing may seize on the shaft and will have to be drawn off with pullers. If this happens, protect the shaft end by refitting the bolt on to the end of the shaft. This protects the shaft and aids the location of the puller centre.

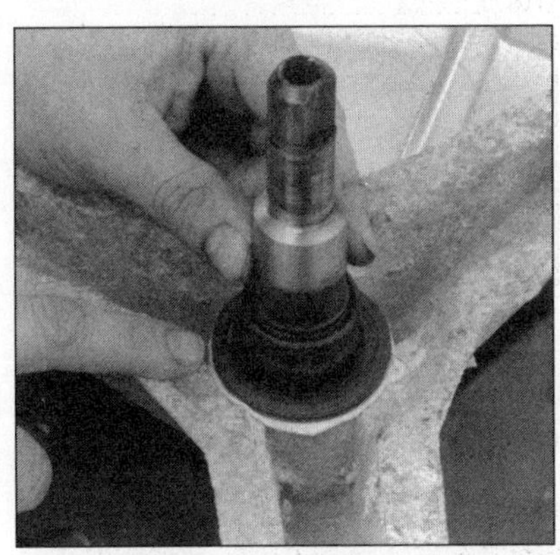

13 Check the shaft for wear and ridges at the bearing support points. Check carbon face for cracks or loose fit. Check the three mounting points for cracks.

14 Insert screwdriver and prise out bearing on carbon-face seal.

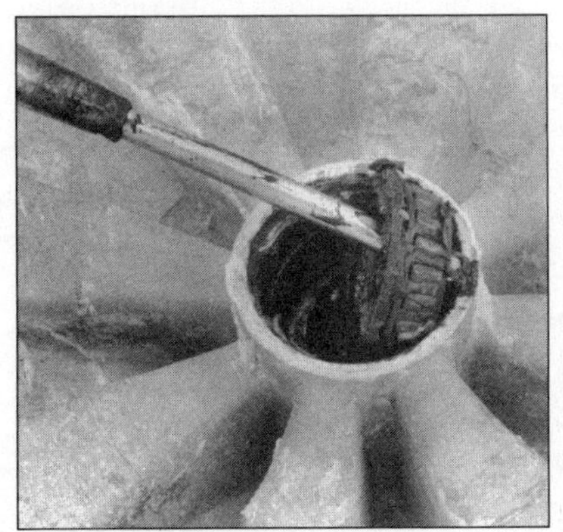

15 Remove old washer and front bearings. (Ball bearings will have to be knocked out with a metal drift.)

16 With rear bearing removed, knock our rear bearing shell/liner by the inner lip.

17 If there is no lip to the front bearing shell, drill two shallow slots opposite each other in the inner of the housing.

18 Position of drill mark on front inner.

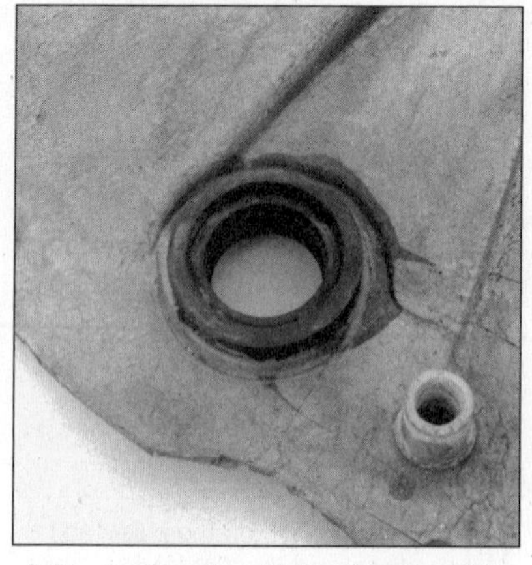

19 Remove the old thermostat seal if it is a poor fit or perished. Apply sealant to the new seal to ensure it is watertight.

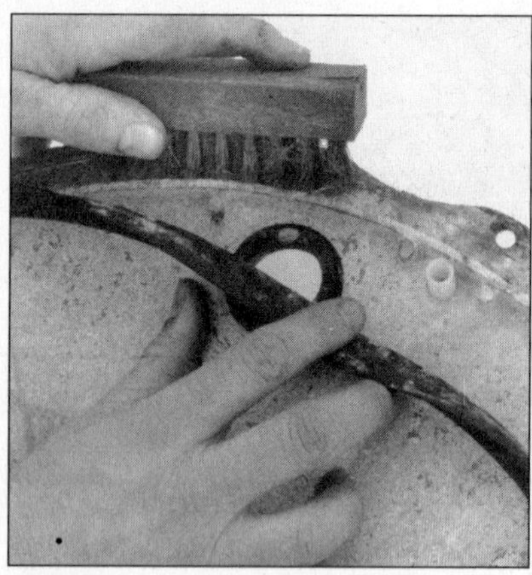

20 Clean all seal and bearing surfaces before fitting new parts.

21 New set of taper roller bearings, torque washer and carbon seal suitable for this machine.

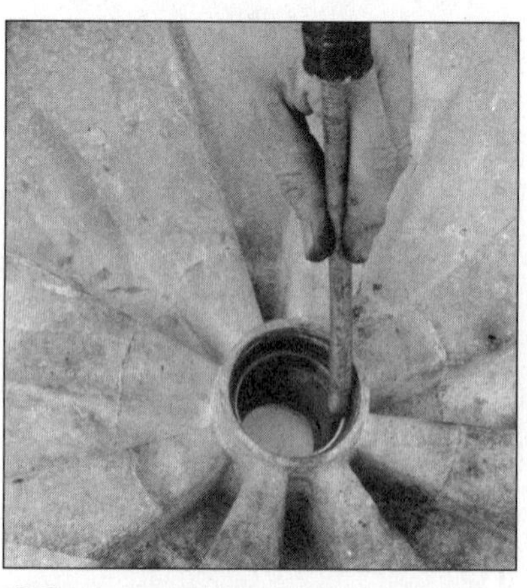

22 Insert rear bearing liner, and tap into position firmly, seating to its base using a soft metal drift.

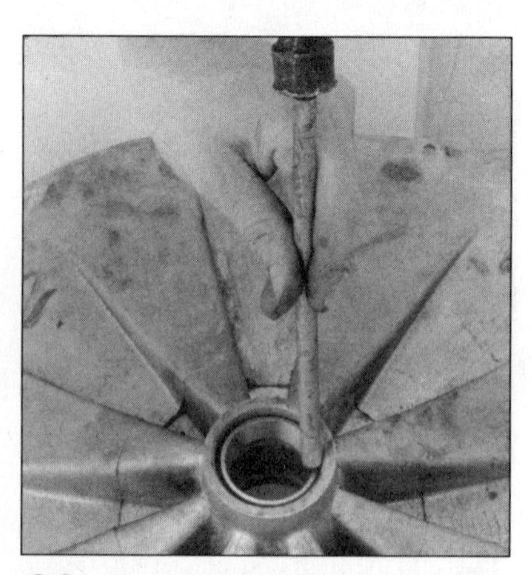

23 Now insert front liner, and tap into position firmly, seating to its base using a soft metal drift.

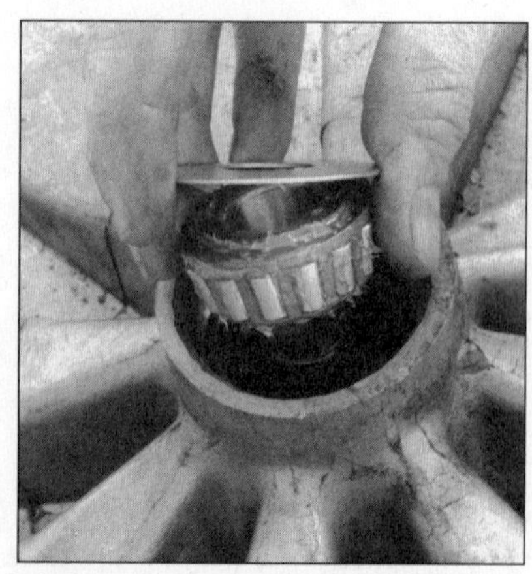

24 Grease bearings back and front and re-assemble as described on page 138.

25 Back half ready to be fitted to the drum shaft.

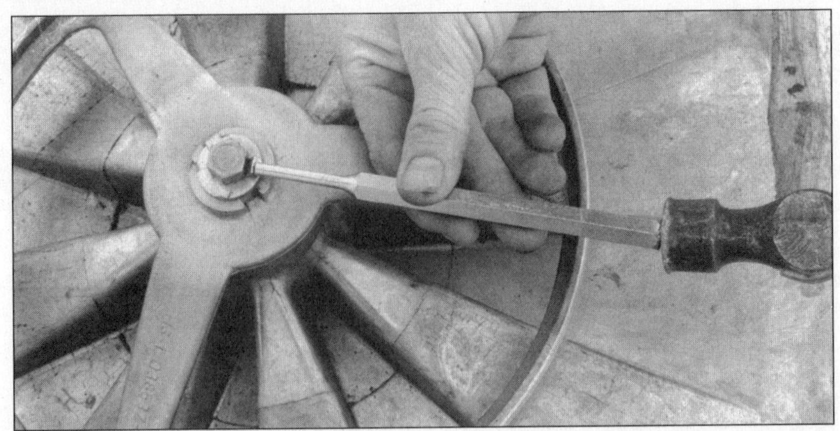

26 Follow the manufacturer's instructions for use of torque washer and shim to make the unit ready for fitting back into the machine. Make sure that you reset the lock tab on the pulley bolt.

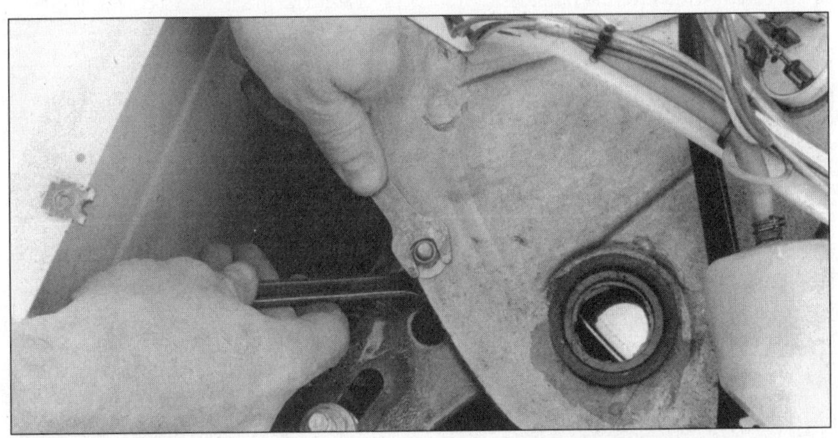

27 Tighten the bolts in sequence slowly using opposing bolts. Do not overtighten.

28 Seal and refit all hoses and grommets and secure all connections to the heater, etc. Adjust the belt tension before the functional test with all panels in position.

Hotpoint ball race type renewal

TOOLS AND MATERIALS

- ☐ Screwdriver(s)
- ☐ Spanner
- ☐ Wooden wedge
- ☐ Soft-headed mallet
- ☐ Soft metal drift
- ☐ Bearings

1 Isolate the machine and remove the securing screws for the plastic surround for the door seal.

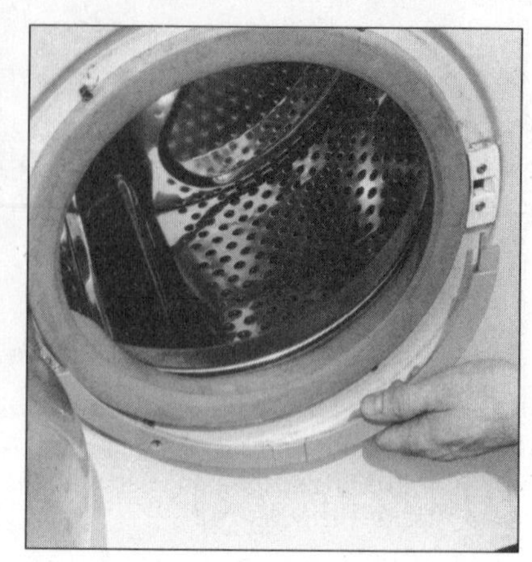

2 Remove the door seal surround carefully. This is in two halves (top and bottom).

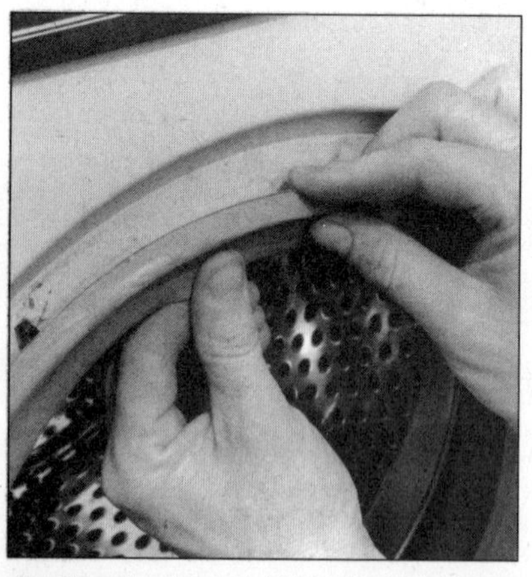

3 Free the door seal from the outer shell lip of the machine and allow to rest on inside of the front panel.

4 Remove the screws holding the timer knob in position. On early models, there is only one plastic screw.

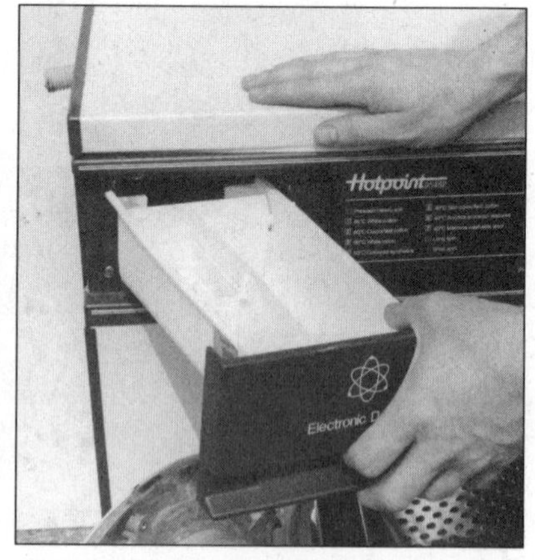

5 Pull to remove the soap drawer and remove the exposed screws. This allows the front fascia to be removed.

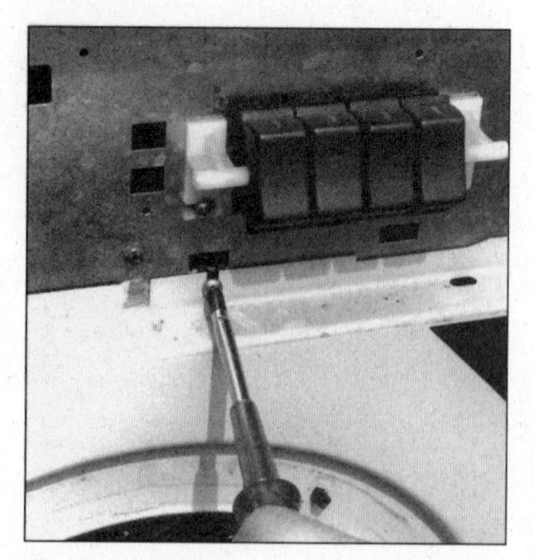

6 Remove the screws securing the top of the front panel.

7 Remove the screws securing the bottom of the front panel.

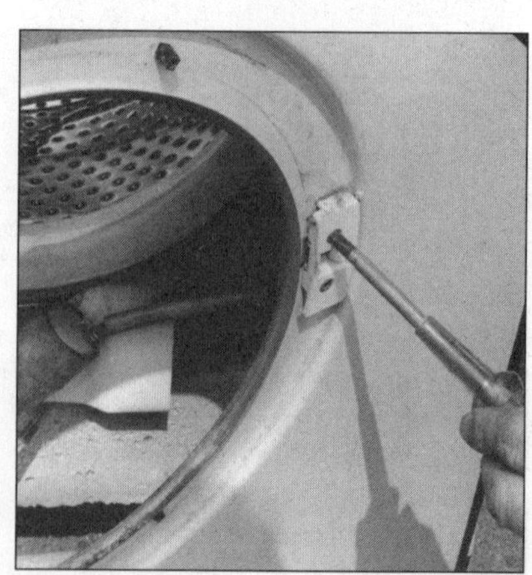

8 With the front panel off, remove the door catch and interlock (if fitted).

9 Note the position of the clips securing the front of the outer tub and remove carefully.

10 With the clips removed, the tub front can be removed completely. Take care not to damage the heater or connections.

11 Remove the screws securing the rear panel to expose the drum pulley.

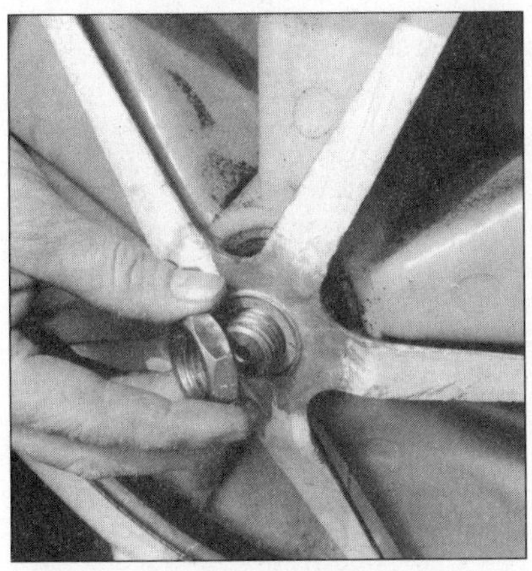

12 Remove the pulley lock nut (right-hand thread) and 'chock' the pulley against the tub with a wedge of wood.

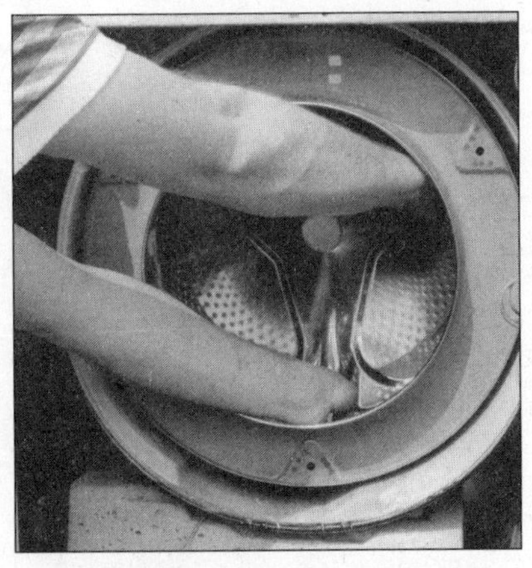

13 With the pulley securely wedged, grasp the inner paddles of the drum and turn anticlockwise. This will unscrew the pulley.

14 With the pulley removed, tap the drum shaft free from the bearings using a soft-headed mallet.

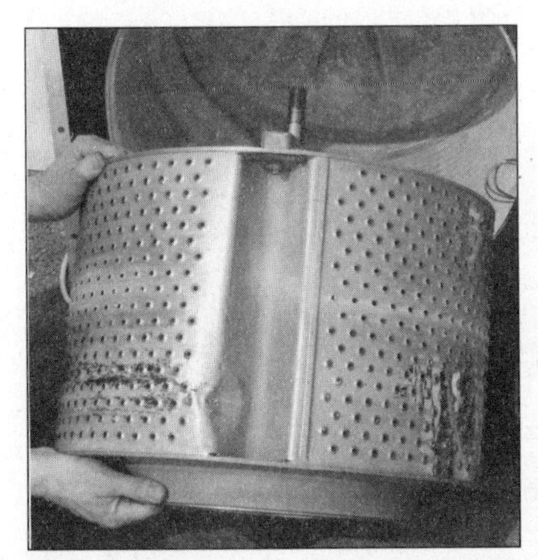

15 Withdraw the drum and shaft from the tub. Using a soft drift, knock the new bearing home, taking care not to damage the new seal. Re-assembly is a reversal of the previous procedure.

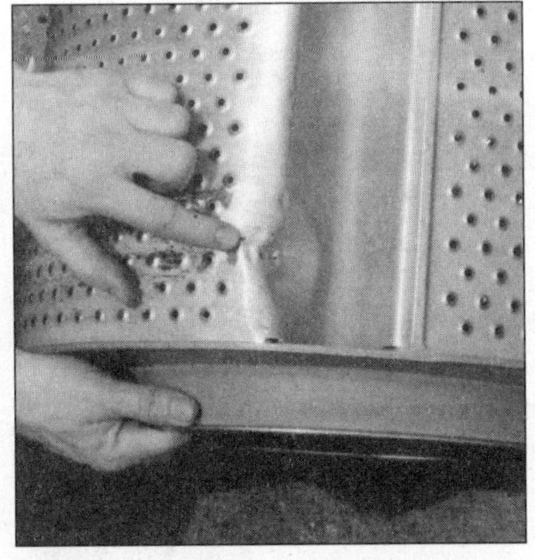

16 This machine also has severe drum damage.

17 This style of machine has a 'catch-pot' style filter in the sump hose. When removed, a large number of coins, metal screws, curtain hooks and various other household items are revealed.

The wiring harness

The term 'harness' is used for all of the wires that connect the various components within the appliance. On large appliances they are usually bound or fastened together in bunches to keep the wiring in the appliance neat and safely anchored. Smaller appliances, however, may sacrifice neatness for safety, and route the wiring to avoid contact with heat or sharp edges.

At first sight, the harness may look like a jumble of wires thrown together. This is not the case. If you take the time to inspect the harness, you will find that each wire is colour coded or numbered, either on the wire itself or on the connector at either end. This allows you to follow the wiring through the appliance easily. With practice, any wiring or coding can be followed.

As most of the wires in the machine either finish or start at the timer unit, it may be helpful to think of the timer as the base of a tree, with the main wiring harness as the trunk. As the trunk is followed, branches appear – wires to the valves, pressure switches, etc. Continuing upwards, the trunk gets slowly thinner as branching takes place to the motor, pump, module and so on.

Each item is therefore separate but linked to the timer. This, also, can be likened to a central command post, communicating with field outposts.

The connecting wires to and/or from a component are vital to that component and any others that rely on its correct functioning. Luckily, wiring faults are not very common. However, when they occur, they sometimes seem to result in big problems, although, in reality, only a small fault has occurred. For example, one poor connection can cause a motor to cease to function, rendering the appliance completely – and worryingly – unusable.

Do not fall into the trap of always suspecting the worst. Many people, including engineers, blindly fit parts such as a motor or a heater, only to find it does not cure the problem. Often the timer is blamed and changed and this is an expensive mistake to make. Stop, think and check all wires and connections that relate to the particular fault. Always inspect all connections and ensure that the wire and connector fit tightly. Loose or poor connections can overheat and cause a lot of trouble, especially on items such as the heater. Poor connections to items such as the main motor or pump will be aggravated by the movement of the machine when in use and may not be so apparent when a static test is carried out.

One of the most easily missed faults is where the metal core (conductor) of the wire has broken and the outer insulation has not. This wire appears perfect from the outside but will pass no electrical current. To test for this

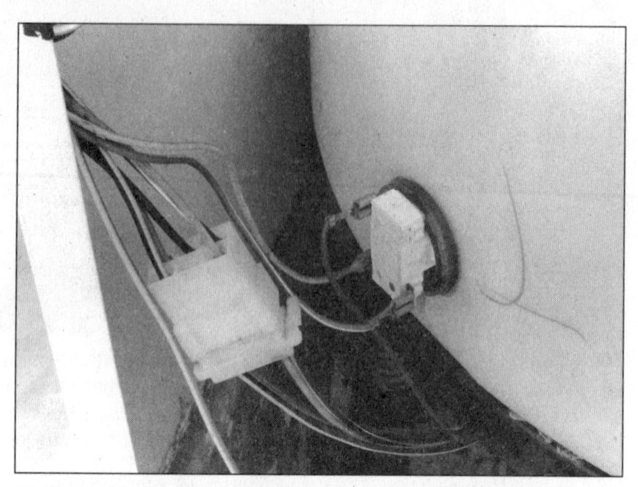

Harness connector block. Loose connectors will overheat and cause problems. Ensure a secure fit.

The terminal block is the first distribution point of the power into the machine. Ensure all connections are sound, otherwise heat will be generated.

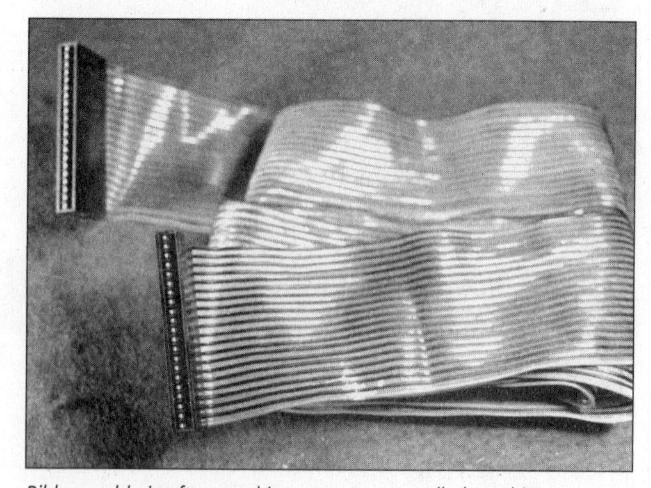

Ribbon cable is often used in computer-controlled machines.

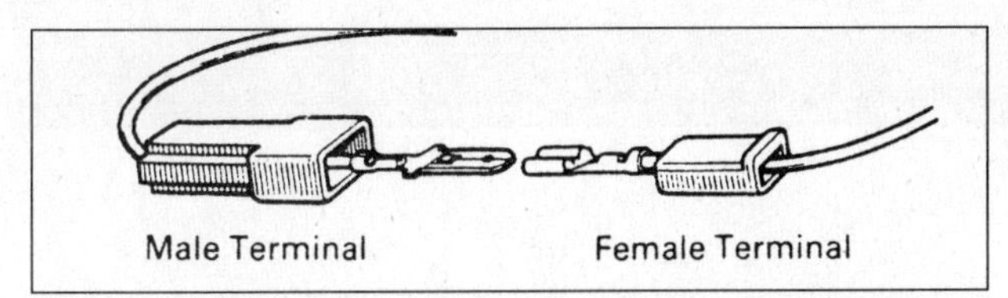

Harness connection. Can often be of a multi-block fitting of several wires in one moulded block.

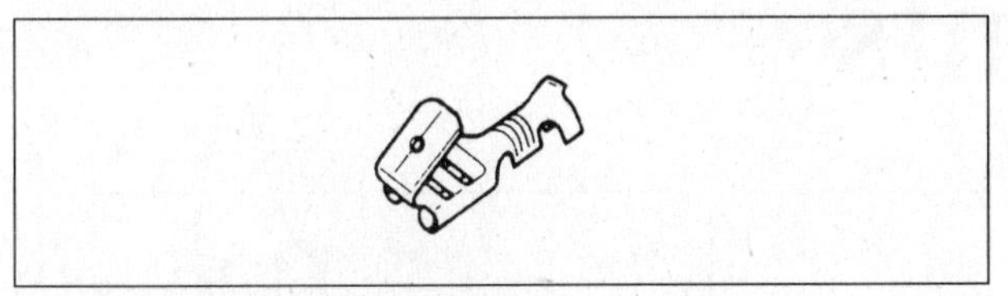

Piggy back terminal for two wires to one terminal.

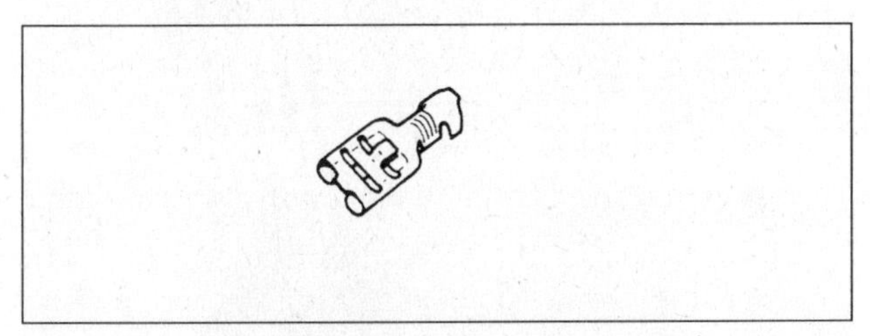

Female spade terminal.

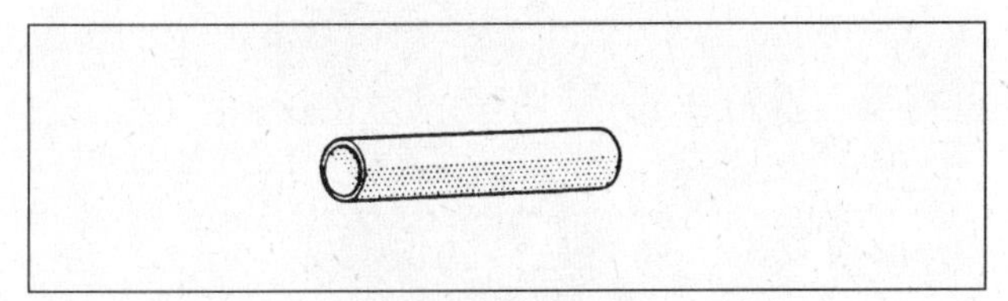

In-line connector used for low amperage.

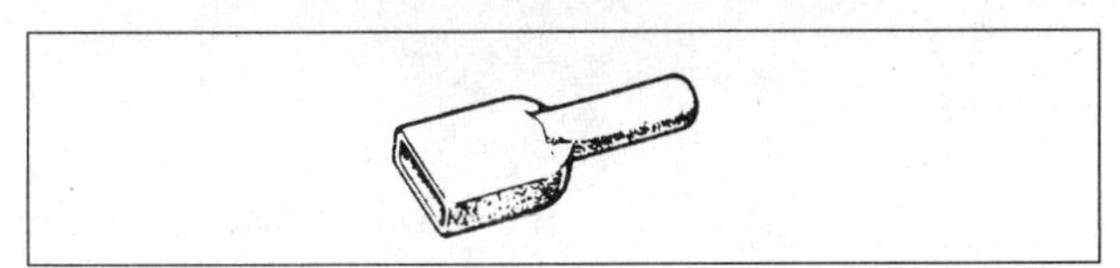

Insulation cover.

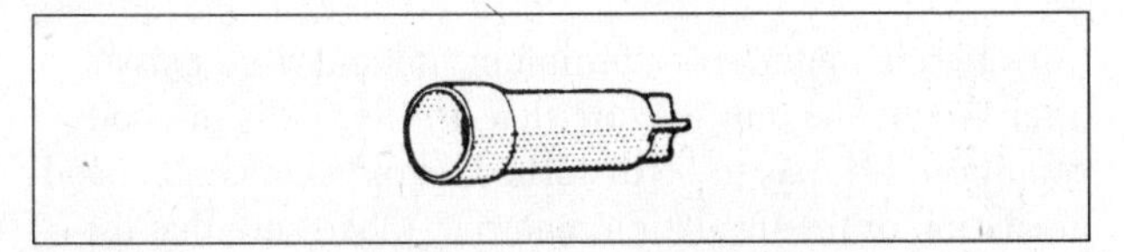

Butt connector for connecting several wires together.

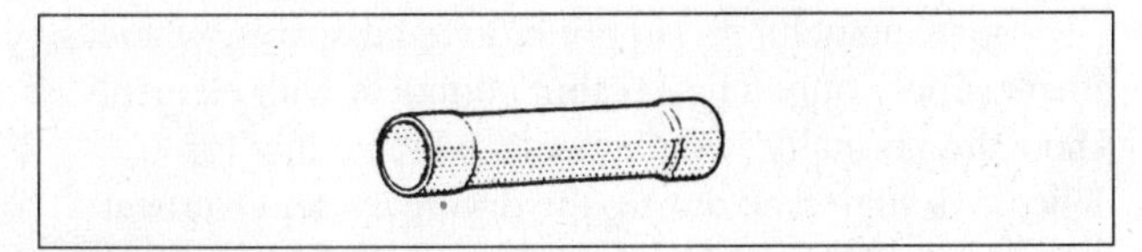

Large in-line connector used for high-amperage wires.

All of the above are 'crimp' fitted to inner and outer of the wires. Make sure that they fit securely and will not easily part.

(see Electrical circuit testing, pages 23-25). Remember that such faults may be intermittent. One reading may be correct and the same test later may prove incorrect. This is because of the movement of the outer insulation of the wire first making, then breaking the electrical connection.

When testing for such intermittent faults, pull or stretch each wire tested. An unbroken wire will not stretch, but one that is broken internally will stretch at the break point. Rectification is a simple matter of renewing the connection with a suitable connector. Do not make the connection by twisting the wires together and covering them with insulation tape. Use only the correct rating of connector and ensure a secure and insulated joint. If a joint is required in a position of cable movement, such as wiring from shell to tub unit components, it is advisable to renew the whole length of wiring, or the joint made in a fixed section of cable. Do not use rigid connections in movable wiring. Take time to do a few simple checks. It saves time and money. Ensure that the harness is secured adequately to the shell of the machine, at the same time allowing for free movement of the motor, heater, etc.

Take care that metal fastening clips do not chafe the plastic insulation around the wires. Also make sure that wires are not in contact with sharp metal edges, such as self-tapping screws.

Before attempting to remove or repair the wiring harness, or any other component in the appliance, isolate the machine from the main electrical supply by removing the plug from the wall socket.

Chapter 5

Drying

Drying components

Both condenser and vented combined washerdriers are largely based on their wash only predecessors. This chapter looks at the components that turn the ordinary automatic washing machine into a washerdrier. The additional parts are often referred to as the 'drying' or 'ventilation group'. Reference to other chapters within the book will be necessary – Temperature control *(pages 85-92)*, Heaters *(pages 79-84)* and Motors *(pages 110-122)*, for example.

Although there are many variations in design and positioning of the drying components in combined washerdriers, they are all based on two basic types. Vented types draw in air from the atmosphere, heat and circulate it through the spun wash load and then vent the resulting moisture-laden air from the machine. Condenser washerdriers use a sealed warm air circulation system that is combined with a condenser unit to remove the moisture from the airflow.

Both systems have many parts in common: a similar configuration of fan assembly, and heater ducting mounted on the upper portion of the outer tub. However, although this configuration is found in a wide variety of makes and models, there are many different variations of component positions possible. A breakdown of the various components of both types of washerdrier system follows.

Combined fan motor, housing and heater detached from the large Ariston drying group of components.

Heater unit

This metal, usually die-cast aluminium unit houses the heating elements (sheathed type) used to heat the air before it enters the inner drum. Often the unit forms the inlet ducting to the drum via the door seal, but some variations have direct access to the drum via the outer tub. Some machines incorporate a flap within the ducting to prevent steam produced during the normal wash cycle condensing within the heater unit. Airflow during the dry cycle pushes the flap valve open. It is possible for the flap to stick in the closed position and cause overheating within the duct. If your machine has such a flap/valve, check that it moves freely.

Insulation material – aluminium foil and fibre glass – often covers the unit. Ensure that, if fitted, it is in good condition. If it has to be disturbed, avoid skin contact and inhalation of the insulation material. Make sure that the aluminium foil and the metal tape used to secure it are clear of all components and electrical connections.

Some manufacturers supply individual spares, whereas others supply only a heater unit complete with elements. There are normally two elements, which allow for delicate or high heat during the drying cycle. Overheat thermostats, TOCs or thermistors are also mounted on or within the unit. Removing fixing screws or bolts allows most heater units to split into two . If the unit has to be

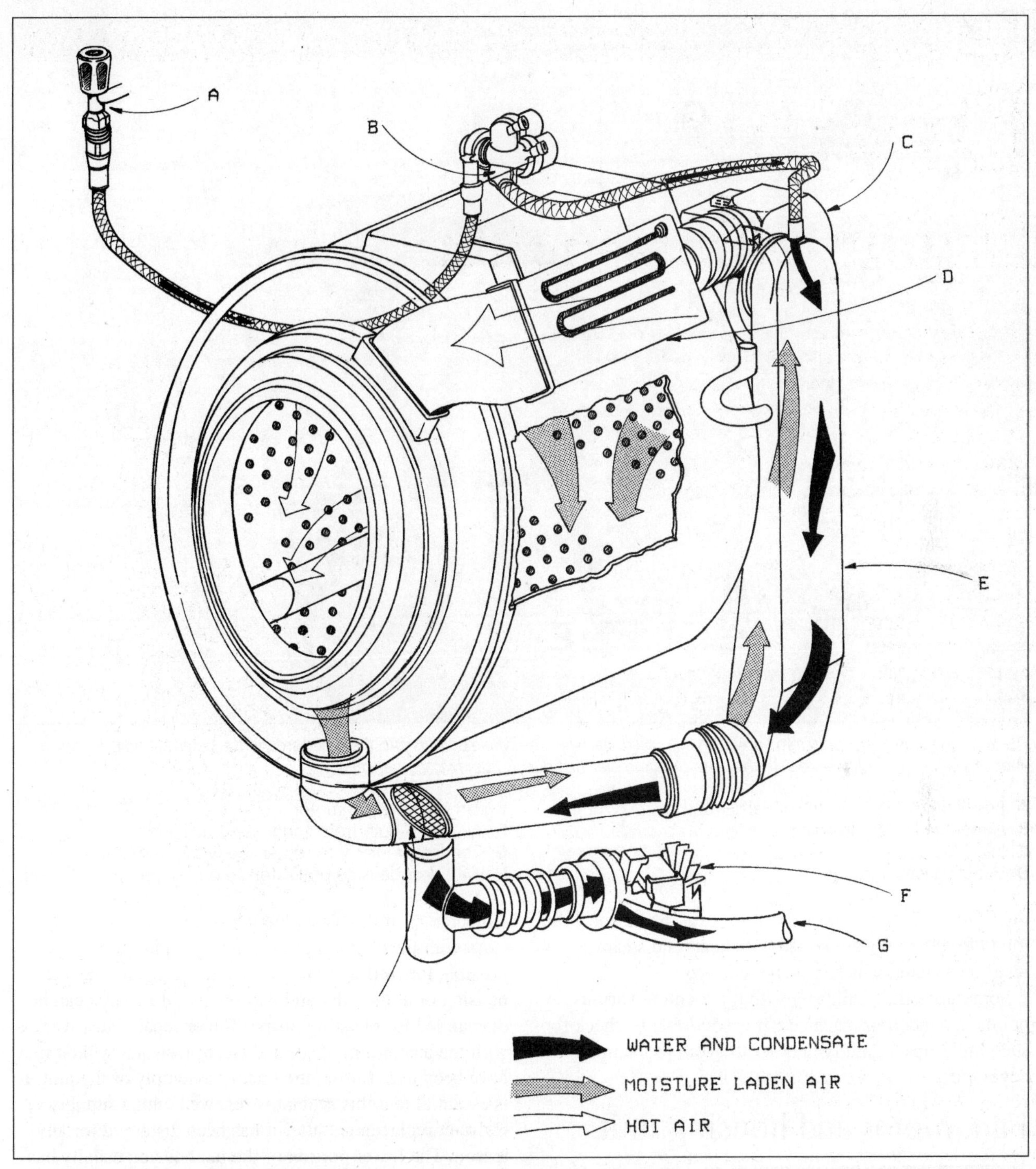

This illustration shows one of a number of variations of condenser systems. This system utilises the existing sump hose in place of an extra tub access point. A self-cleaning filter is also included.

A Cold water inlet supply
B Triple cold valve with one outlet for reduced condenser supply
C Fan motor and housing unit
D Heater duct containing heater elements
E External condenser unit with drain hose to fan housing (in case of condensate build-up)
F Normal outlet pump
G Normal outlet hose
H Self-cleaning filter in sump hose ducting (if fitted)

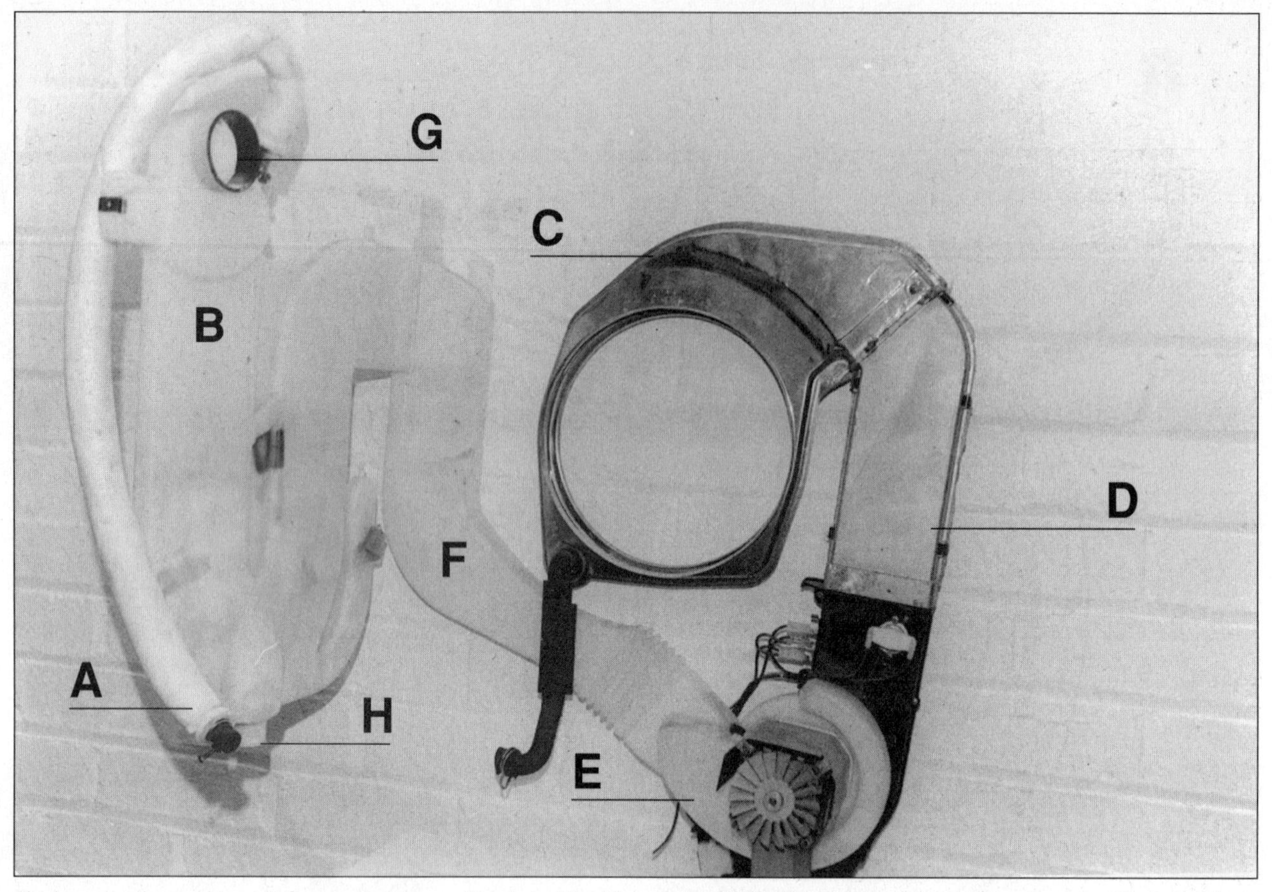

The large condenser drier unit from an Ariston combined machine. The whole unit fits into the space around the outer tub unit. A double door seal is used in this instance with the large circular metal inlet duct (C) sandwiched between.

A Restricted cold supply to condenser unit
B Large condenser moulding
C Inlet duct
D Heater duct
E Fan housing and motor
F Air return duct from condenser unit to fan
G Condenser inlet from outer tub (warm moist air)
H Water/condensate outlet to sump hose/pump

dismantled, for element renewal, for example, inspect the joint closely; many are sealed with heat- and steam-resistant sealant, which must be renewed.

Sometimes fluff builds up within the unit at various points in condenser machines. It is advisable to check for such build-ups regularly in order to avoid problems developing.

Fan, motor and housing unit

The fan, motor and housing may be an integral part of the heater unit casing or condenser unit, or a separate motor and fan housing fixed to the heater unit. In condenser machines, the unit may form the link between the condenser unit and the heater unit. The fan housing may be made of cast aluminium or injection moulded plastic, depending on the make and model of machine. Fan chambers on condenser machines often have mounting positions for thermostats or thermistors to monitor the temperature of the air within the sealed system.

As with the heater unit on condenser machines, it is possible for fluff and lint to accumulate within the fan housing or at the inlet and exit points. Most units can be dismantled for cleaning, inspection or repair. However, as with the heater unit, heat- and steam-resistant sealant may have been used during the factory assembly of the unit. It is essential that this sealant is renewed with a suitably resistant replacement after it has been disturbed for any reason. The lower portion of the fan housing usually has a small drain hole to allow any condensation entering the chamber to drain away. Ensure that this is clear of any kind of blockage.

A centrifugal (barrel style) fan is fixed directly to the fan drive motor shaft within the fan chamber. Various types of fan to shaft fixings are used: threaded shaft and nut, slide fit and grub screw, and taper and collet. The blades of the fan should be kept clean, and, if the fan has to be removed for any reason – for example, to remove

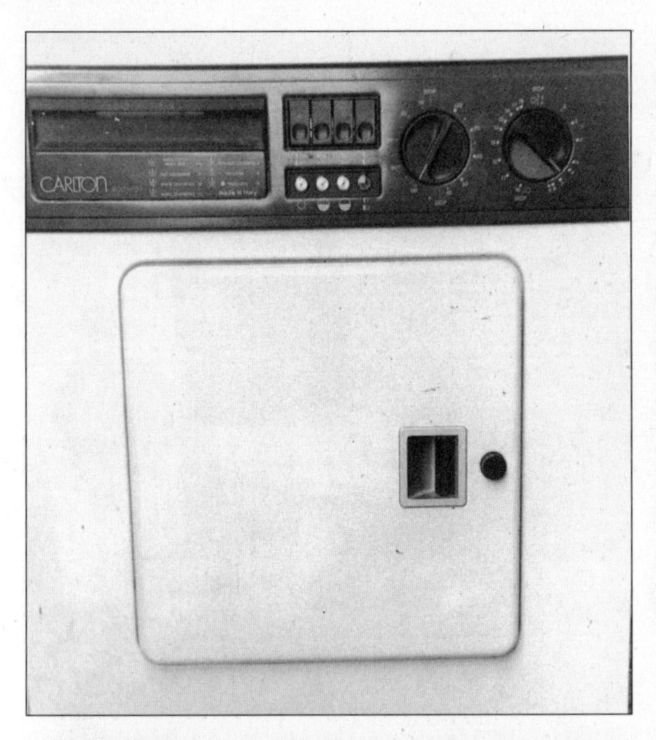

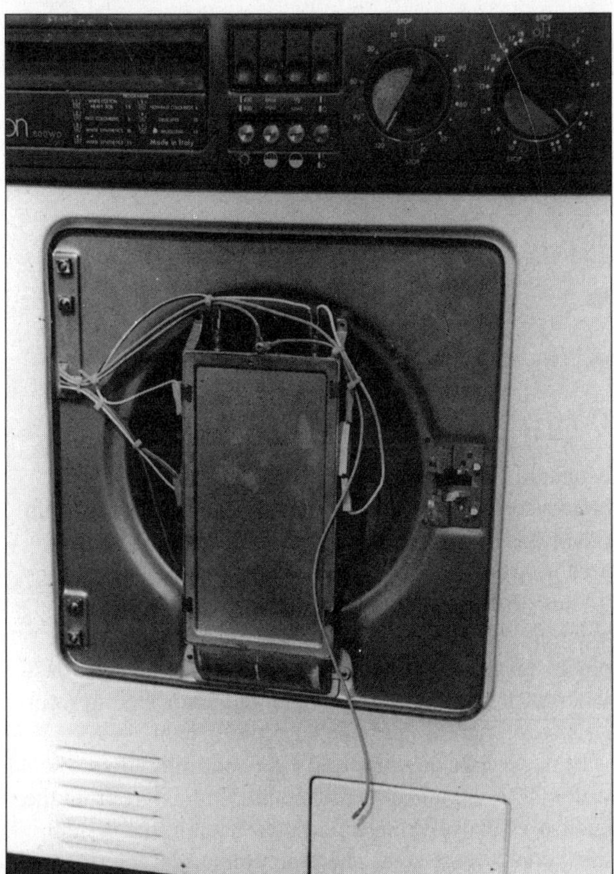

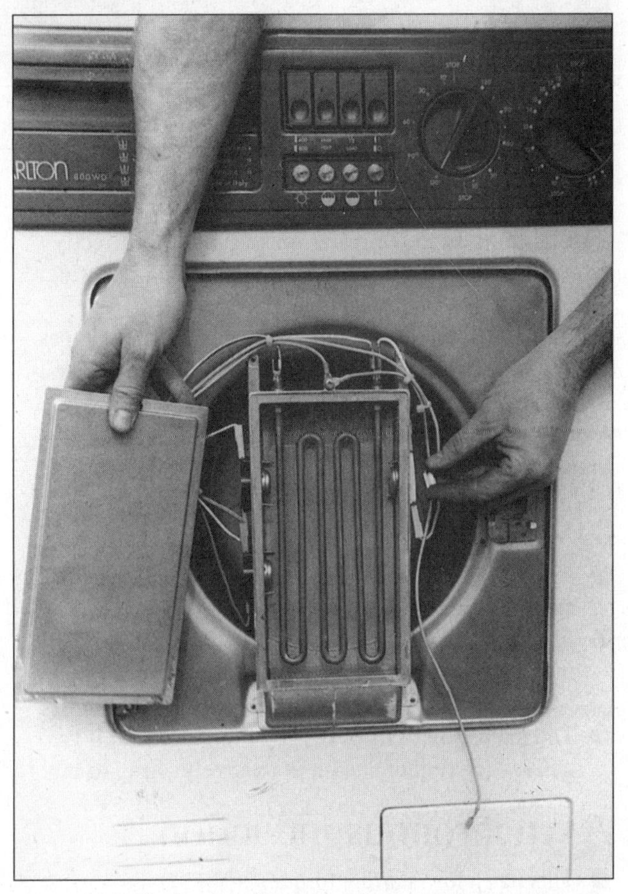

This vented combined washerdrier has an unusual but functional drying arrangement. The heating elements are housed within the all metal door. The fan and motor are located on the front underside of the machine (see air intake vents at front left-hand side). Air is drawn in by the fan unit via the front vents and ducted into the door cavity at an entry point below the door seal. The heated air then enters the drum from the centre of the door protrusion, passes through the wash load and exits via two rear vent hoses.

The two rear vent hoses are connected to the outer tub unit and outer shell of the machine, where two large filters are located. Inlet filters are also included to eliminate particles entering the heater duct, and a flap valve is located on the exit of the heater ducting to eliminate water/steam ingress during normal wash and spin actions of the machine.

fluff build-up from the rear – make sure it is correctly re-positioned on the shaft to avoid jamming or chafing when the fan assembly is rebuilt afterwards.

Motors for rotating the fan may be one of three types: shaded pole induction, asynchronous induction motor (capacitor start) and series wound brush motor.

Some common problems that may occur with all of these motors are listed here.

Shaded-pole induction

- Failure of one or both motor bearing(s) allows the rotor to foul the stator.
- Fluff build-up or an out-of-position fan causes the motor to stall and overheat or burn out if not fitted with a TOC.
- The main benefit of this type of motor is that it is quiet when running and it is relatively cheap to buy.

Asynchronous induction

- This has a low starting torque.
- Failure of the capacitor results in overheating of the motor windings.
- This type of motor is generally reliable and quiet.

Brush motor

- This type of motor is prone to brush wear or sticking brushes.
- This type of motor is noisy in operation.

Condenser unit

Moisture is removed from the airflow within the condenser machine by the creation of a cold area within the sealed system on which it can condense. This cold area may be internal – within the gap between the outer tub and drum – or, more commonly, external. An external, independent unit, usually of semi-transparent plastic material, is fixed to the outer of the tub unit or inner of the shell of the machine, but connected to form an integral part of the sealed airflow system.

Both systems are kept cold by a controlled flow of cold water. To remove the large amount of moisture from the airflow effectively, the water introduced to the system must cool a large area. Therefore, the water entering the system may pass through a trickle bar, or the entry point may be designed to create a capillary action or spray. Each system creates a broad but thin film of cold water within the condenser unit or tub of the machine.

The design and construction of each system allows both the cold water supplied to the unit and the condensate (steam now condensed into liquid) to collect in either the sump of the outer tub or a separate sump of the condenser unit. Periodical operation of the normal outlet pump is then all that is required to discharge the water via the normal outlet hose. The now much cooler, drier air is re-circulated via the fan over the heaters. The time taken to dry the recommended load sizes depends on the efficiency of this sealed system, correct loading and setting of the drying time.

Unlike vented machines, sealed condenser systems do not have accessible fluff/lint filters within the airflow. Some rely on the normal sump/filter to do the job, whereas others have efficient outlet pumps to discharge any fluff down the normal drain/outlet. Some machines have a means of flushing the condenser unit during the water intake of the normal wash cycles. This is normally done by diverting the cold inlet so that both soap dispenser and condenser unit receive unrestricted water supply from the cold inlet valve during filling for wash and rinse cycles. The aim is to flush any fluff/lint that has built up within the unit down into the filter or sump ready to be removed or discharged during normal operation of the machine. In reality, only a portion of the condenser unit is capable of being flushed in this way, so build-ups and blockages can still occur.

Some machines, mainly in the Servis/Hitachi range, have a filter housed within the sump hose ducting which is designed to be self-cleaning by the water passing back and forth within the hose during the wash action rotations. However, blockages can occur if the machine is used to dry several loads without wash cycles in between. If this sort of machine is to be used for continuous dry only cycles, it is wise to intersperse the dry cycles with a short wash cycle (with or without load) to keep the fixed filter clear. It is advisable to see the manufacturer's instructions for details.

As with all components forming the condenser system, sealant may have been used at certain points to prevent both water and air leaks. It is essential that any sealant is renewed during assembly with a sealant identical to the original.

Poor drying performance – condenser machines

- Check that the machine is not overloaded; most can only dry half of what they wash. Read the manufacturer's instructions.
- Ensure the cold water supply is turned on. Condenser

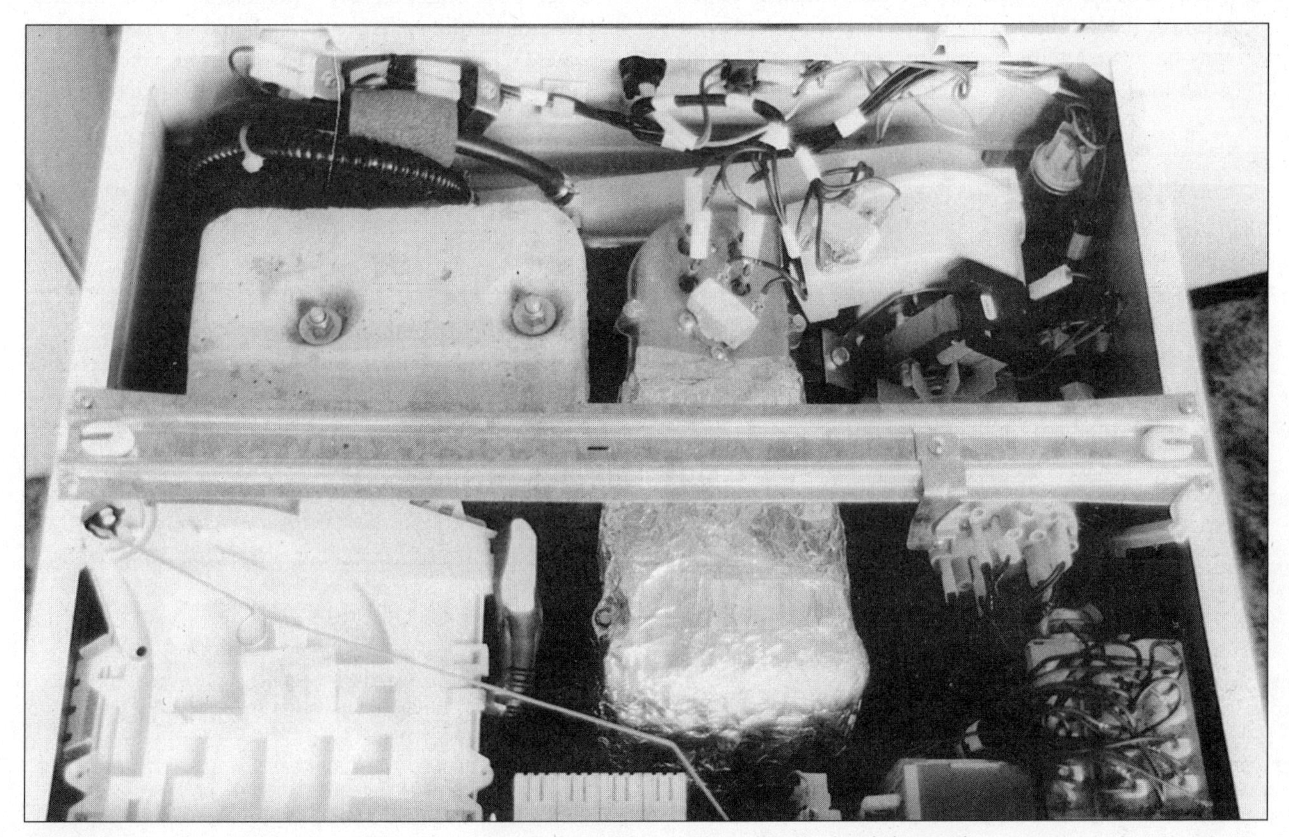

Most heater ducts are insulated and foil covered. Ensure that yours is in good condition and sealed correctly.

machines require a cold water supply during the dry cycle.

- Check that the fan and heaters are alright. Functional testing should prove this. If in doubt, check the continuity of heater and free movement of the circulation fan.
- Check that a trickle of water is passing through the condenser unit. A water flow of less than 350ml per minute (1 pint per 100 seconds) makes most machines ineffective. This usually occurs only with tank-fed cold supplies. Remember the installation requirements of a minimum head of water. Also blockages or restrictions to the valve filter, reducer or spray/trickle bar may cause the flow to be too slow.

If the first four checks prove to be alright, follow the next steps in order.

- Check removable pump filter (if fitted), sump hose catch pot or self-cleaning filter for restrictions. Clean as necessary.
- Check that the pump operates correctly and discharges a full load of water well within the allotted time, that is 1 minute. If in doubt, check the pump and hoses for blockages. Remember, pumps with TOCs may trip part way through the dry cycle through overheating on the long combined cycle *(see Pumps, pages 69-74)*.
- Check that the flow rate through the condenser is not too fast. If too much water is entering the chamber, it may be picked up by the airflow and deposited on the wash load.
- Check the water flow through the condenser unit is evenly spread (by capillary action) giving as large a cool surface area as possible. This is possible only on machines with transparent condenser units *(see page 152)*.
- Try to see if water droplets are being blown through the inlet duct, for example, spots on glass or entry point. If droplets are visible and the water flow through the condenser unit appears correct, a fluff blockage maybe the cause. Strip down and thoroughly clean the unit. Some machines have an extra hose to the unit to flush any fluff away. Problems may also arise from long hair, animal hair and wool. Check all connections and hoses. Ensure that parts fitted with sealant are resealed with the correct sealant. If this is not done, a vapour leak could occur and result in failure of other items or low insulation.
- If all the above tests prove to be correct, the fault may lie in a small weep from one of the inlet valves, that is, it is failing to close correctly. This may wet the wash load via the dispenser as the drying cycle works normally. Check for weeping valves by a process of elimination. With the machine electrically isolated, remove each hose in turn from its corresponding valve and observe the valve outlet. Mop up the slight spill of water from the removal of the hose and observe the outlet closely – perhaps for some minutes. After the initial spillage, there should be no more water. When a constant drip is observed, you have found the offending valve. Although the machine is

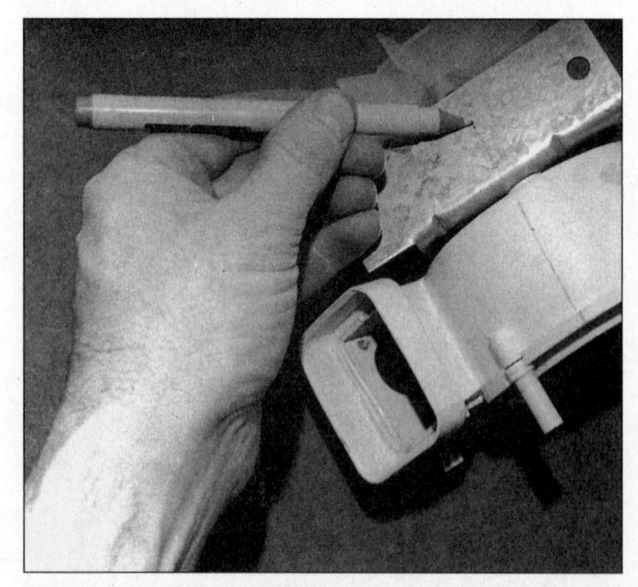

Remove the fan unit and make positioning marks to ensure correct re-assembly.

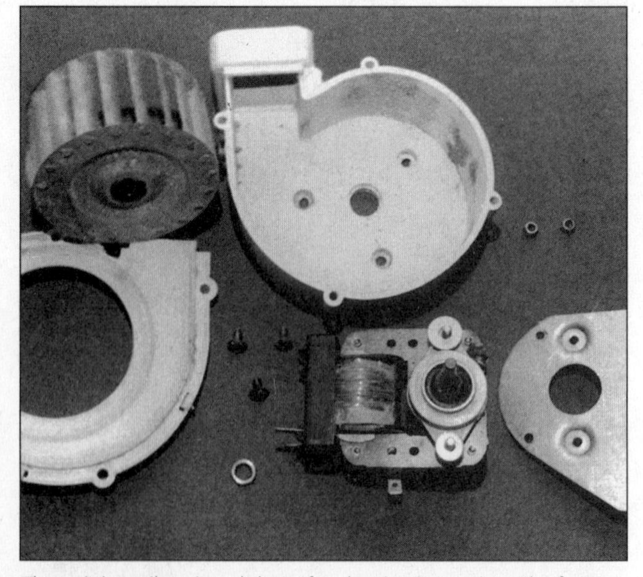

The unit is easily stripped down for cleaning/inspection. The fan, in this instance, is secured to the motor shaft by a brass collet. Once the nut has been slackened, a sharp tap frees it from the shaft.

Thoroughly check the motor bearings for wear or excessive play and for free rotation. The whole unit must be cleaned and correctly sealed before refitting to the machine.

electrically isolated, the water supply to the valves needs to be on for this type of test.

If you are checking a transparent condenser, do so at a safe distance with an RCD protected supply and with only the minimum number of panels removed. Isolate for all other checks, before refitting the panels or continuing with further testing. Note that this is the only test described anywhere in this manual that requires observation while the machine is in operation.

If condensation is found under the top cover of the machine – which may be accompanied by white scale deposits in hard water areas – the cause might be a warm air leak (vapour leak) from the fan housing, heater ducting, inlet point at the door seal or thermostats. Ensure all are correctly positioned and sealed. Leaks on seals or gaskets may be due to imperfections in mouldings or wraggs left during manufacture. Carefully remove and smooth any such defects and ensure a good seal is obtained. Some manufacturers use a sealant compound at these points. Ensure that only the correct heat-resistant sealant is used and do not refit seals without it if it has been used on the original seal.

This Zanussi combined condenser washerdrier does not have any external drying components other than a timer. There is no fan, fan housing or fan motor. Heating for the dry cycle is by the three sheathed heaters mounted in the top left quarter of the tub (above the normal water level). The hose on the right of the outer tub supplies water to a trickle bar mounted within the outer tub. The load is rotated near the heat source, that is, the three heaters, while the opposite side of the tub is kept cool by the cold water trickle. The process works, but drying times can be very long.

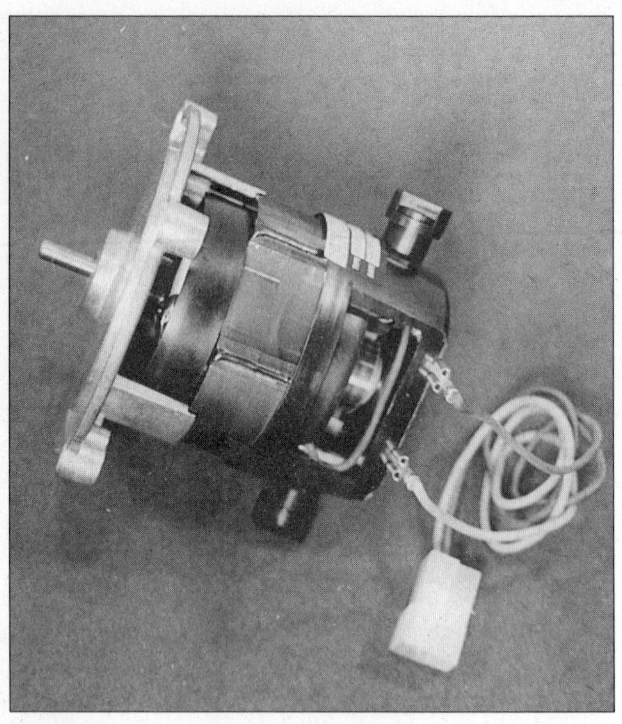

The Hotpoint condenser washerdrier uses a universal brush motor to drive the air circulation fan. Brush wear and sticking are quite common. Brushes are available as spares, but other faults require a complete motor unit.

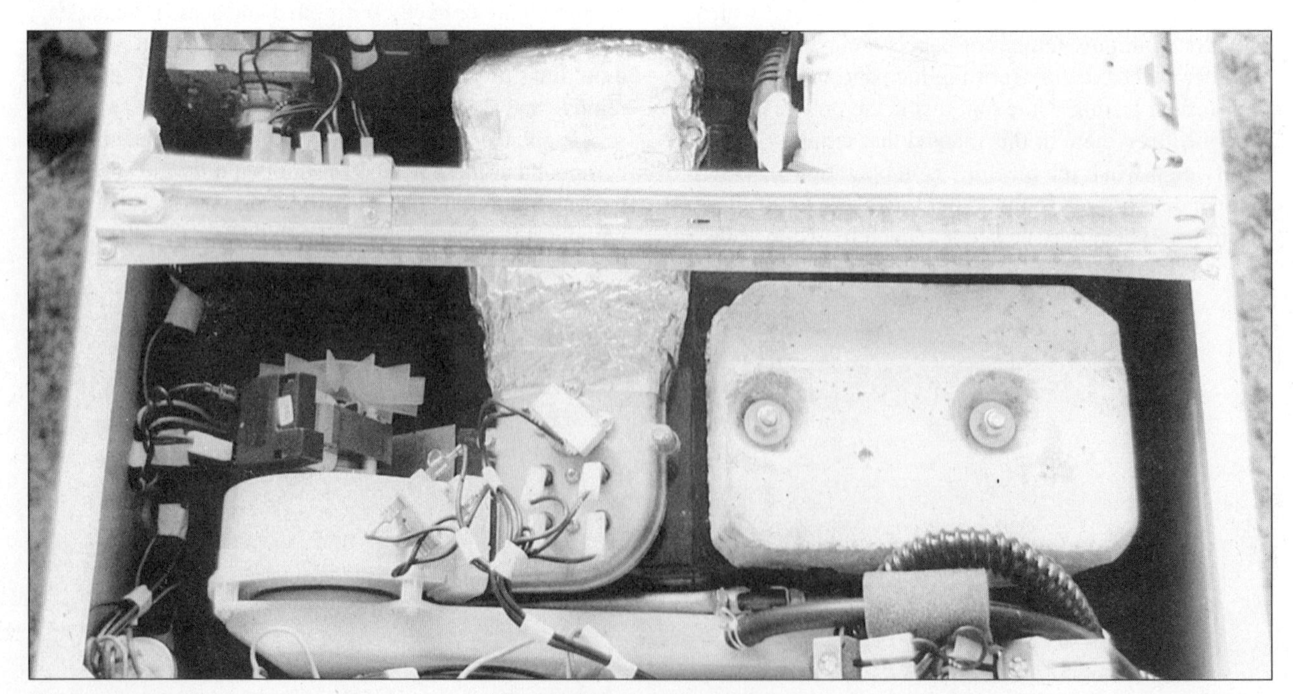

This is a close-up of a rear-mounted condenser unit. The ribbed hose on the right is the restricted cold supply and the hose to the centre is used to flush the condenser unit during normal fill sequences of wash and rinse.

DRYERS WATCH POINTS

1. **Always ensure the mains electrical supply is isolated** when checking; switch off, plug out. If observing a water flow test, do so with the use of an RCD protected circuit and test its operation before use.
2. **Fully isolate the machine** before commencing further investigations, stripdown or repair.
3. **Ensure moving parts, such as the blower motor and fan, have stopped** before continuing with fault finding.
4. **Ensure the drier heating elements have cooled** before checking or stripdown. Also check that the door glass and ducting have cooled sufficiently before you touch them.

Dry only machines

The basic function of the tumbledrier is to circulate warm air through damp clothes for a given period of time selected by the user. This is followed by a cool tumble before switching off at the end of the selected time. This is a relatively simple operation requiring a drum to hold the load, a heater to warm the air and a fan to circulate the warm air through the drum. As with combined washerdrier machines, the drum rotates to allow the whole load to gain maximum benefit from the warm airflow. In most dry only machines the motor that revolves the drum also drives the fan used for air circulation.

As a dedicated machine – dry only – it works much more efficiently than its combined washerdrier counterpart. There are several reasons for this. The drum capacity may be much larger. As it is not necessary to have a watertight outer tub, a larger drum takes up nearly all the available space within the cabinet. Larger machines – with cabinets the same size as a washing machine – can take the same size load, that is, a 5kg (11lb) wash load can be transferred directly into a 5kg (11lb) load tumbledrier. Smaller models with a 2.7kg (6lb) load capacity, convenient if space is restricted, operate in exactly the same way as the larger models.

A larger drum allows for better movement of the clothing through the airflow, ensuring more even and quicker drying, even though the clothing and other items increase in bulk during the drying process. Remember, this increase in the bulk of the load is why combined washerdrier machines can dry only half their wash load.

A larger wattage heater can be used, if necessary, which has a greater surface area for warming the air because more space is available in the machine for such components.

Again, due to the increase in usable space, a larger fan can be used for circulating the air.

Additions to the basic principles have been made over the years. Drum rotation on early and current basic machines is in a single direction. Nowadays, the option is to have reverse drum action similar to wash action – clockwise and anticlockwise. This requires a more complex motor, capable of rotation in both directions, and a more complex timer or motor reversal system. Again, on early and current basic models, temperature control is relatively straightforward with selections of high or low. On more recent and expensive machines, auto sensing is now a feature. This system senses the moisture content of the load and switches off when a pre-selected degree of dryness has been reached. The advantage of auto sensing is that it switches off when the load is dry unlike more basic machines that carry on tumbling and heating for the set time regardless of whether the clothes are dry.

Further refinements and extras include computer-control and intermittent tumble after the drying process is complete, to prevent an unattended load compacting at the end of the cycle. Condenser tumbledriers are also available. These operate in a similar manner to the combined washerdrier system by condensing the moisture rather than venting it. Some require a water supply and outlet, while others use cold air to aid condensing, and a removable container to collect the condensate. The container should be emptied regularly (once every two dry cycles on average) if not permanently drained via a pump and outlet in a similar way to the automatic washer.

Tumbledriers heated by gas are now available. Although commercial machines have had this option for several years, it is relatively new to the domestic user. Because of stringent gas regulations and the obvious need

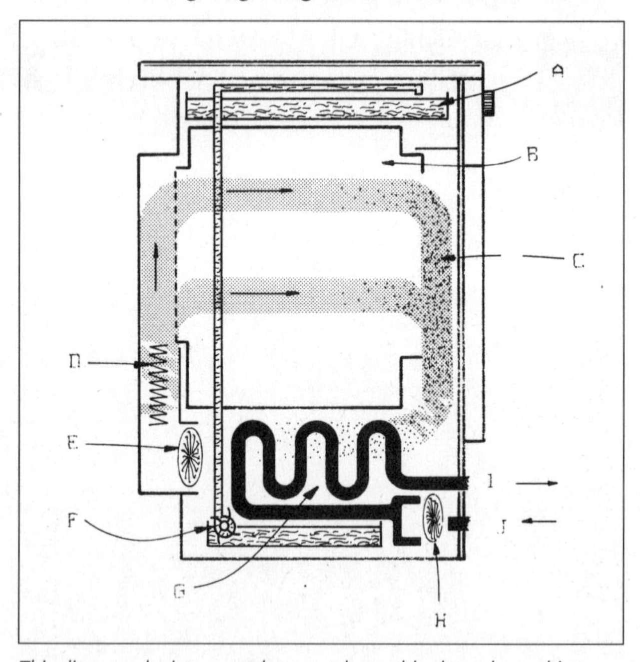

This diagram depicts a condenser style tumble dry only machine which utilises a cold airflow duct to remove the moisture from the sealed warm airflow. Air within the sealed system is circulated by fan (E) over heater unit (D) and into the drum (B). Moisture is picked up by the warm airflow (C) and after passing through a fluff filter, is directed over a cool metal ducting or plate (G). The moisture condenses on the cooler surface and collects in a sump beneath. The condensing plate or duct is kept cool by constant circulation of air created by fan (H). Air is drawn in at (J) and vented at (I). For ease of emptying, the condensate water is transferred by pump (F) to the reservoir tank (A) which requires regular emptying. Alternatively, the machine can be plumbed to a suitable drain/outlet, eliminating the need for manual emptying of the reservoir.

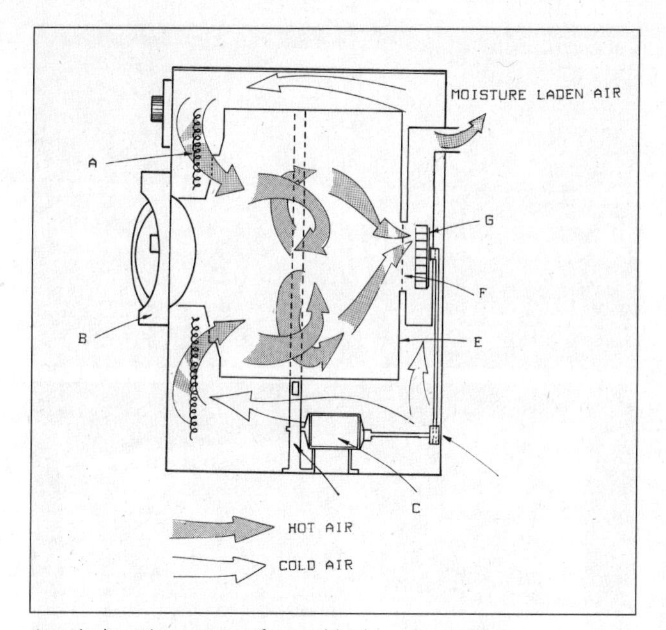

A typical suction system of a tumble dry only machine.

A Spiral heating element
B Door
C Main motor for drum and fan drive
D Fan pulley and elasticated belt
E Drum
F Fluff filter and felt seals
G Suction/vent fan
H Tension pulley and belt

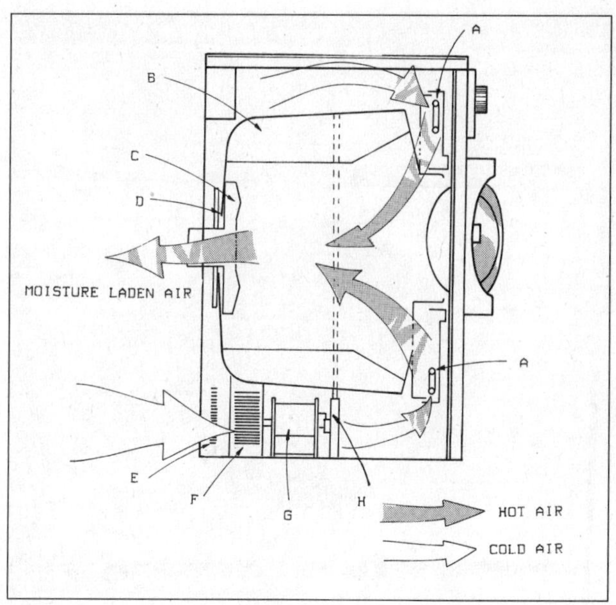

A typical pressurised system of a tumble dry only machine.

A Solid heating elements
B Drum
C Fluff filter
D Felt seal
E Air inlet grille
F Fan (centrifugal type)
G Main motor
H Tension pulley and belt

for safety, do not attempt installation, inspection or repair of these yourself. The installation and repair of gas appliances must be undertaken by those with the proper knowledge, equipment and qualifications.

The main benefit of a larger dry only machine (same size cabinet as a standard washing machine) is that it will dry a full wash load at the same time as another load is being washed, thus saving a lot of time. Do not use both machines from a single socket via an adaptor *(see Electrical basics, pages 10-13)*. If used correctly, a separate drier is also quicker and more efficient than a combined washerdrier.

Tumbledriers operate by tumbling damp clothes in a warm flow of air within the drum. There are two ways in which this airflow may be created – suction and pressurisation. Both systems, however, have the same basic components of drive motor, heater, thermostat, tension pulley and timer. The difference lies in the way the airflow is created.

The suction system

The airflow is created by a large extractor fan, usually belt driven (elastic type) by the same motor as that which rotates the drum by means of a second belt (narrow multi 'V' type). When the drying programme has been selected and switched on, the fan and the drum both rotate. The rotating fan draws air from within the drum and expels it through a vent at the rear of the machine. The cabinet of the machine forms a partially sealed system with an air inlet grille positioned some distance away from the fan vent. Air is drawn in through this inlet grille to replace the air extracted from the drum by the fan. The semi-sealed system directs the cold air drawn in to pass over the heating elements (the position of which may vary depending on make of machine) on its way to the drum, through the clothes, picking up moisture as it does so, and out via the extractor fan vent, thus drying the load in the process. At the onset of the cool tumble (10-15 minutes before the end of the overall selected drying time) the heater is turned off, but drum rotation and fan action continue, creating a cool airflow for the remainder of the pre-set cycle.

Felt seals ensure that air is drawn in only from the drum and not from the shell space, as this would result in poor drying and reduced airflow over the heating element. A removable filter is positioned within the airflow path to trap any fluff or lint and prevent its being blown from the vent.

Popular small front venting tumble dry only machine.

Keep filters clean and renew if damaged.

The pressurised system

The efficiency of this system relies on the machine's shell being sealed to prevent air entering or escaping through any other place than the correct ones. It is essential that panels are fitted correctly and that all sealing felt or foam strips fitted on removable parts, such as back panel, lid or facia parts, are in good order or properly renewed if damaged. Air is drawn in via the inlet grille by a centrifugal fan mounted directly on the main motor shaft. This creates a slight pressure build-up within the sealed shell of the appliance.

The drum inner has an exhaust vent, usually to the rear of the machine, although some models vent to the front via the door or via a duct to a grille at the lower of the front panel. For the pressurised air to escape through the vent, it must first pass through the heating elements, on through the drum and clothing and through a removable fluff/lint trap before escaping through the exhaust vent.

Temperature control

Thermostats are used in machines with both these systems to monitor the exhaust air temperature. They can open circuit the heating element(s) at pre-determined temperature(s). This ensures that an optimum drying temperature is maintained without excessive use of power, that is, heating an already dry load. Normally, as the warm air passes through the damp clothing during the drying process, its temperature drops as it picks up moisture from the wet clothing. On basic machines, the time selected for drying is judged by the individual user, but if the selected time is too long for a particular load, the warm air passing through it when dry does not cool and actuates the exhaust thermostat and open circuits the heater.

The thermostats used are self-setting, so cycling of this action occurs until the remainder of the timed heat cycle has ended and the cool tumble begins. Usually, a double temperature thermostat is used in conjunction with high and low heat settings. It is designed to maintain a pre-determined air temperature throughout the drying cycle, while at the same time providing an overheat safeguard.

Thermostat temperature ratings differ from one make to another and also between models from the same manufacturer, as the temperature ratings are matched to the performance and heater wattage of each machine. Make sure that only the correct replacement thermostat is obtained for your machine. For operation and testing of thermostats, *(see Temperature control pages 85-92)*.

On computer-controlled or auto sensing machines, a combination of fixed thermostats and thermistors or resistance probes is likely to be encountered. The configuration depends on the make and model of the machine.

Functional test

The purpose of the test described here is to check, wherever practical, the main functions of the tumbledrier and its installation.

- Ensure that the door of the machine is closed – turn the programme timer to 20 minutes.
- Check that the drum rotates. If the drier is of the reversing type (drum rotation clockwise and anticlockwise), check for this action. Length of rotation and pause times differ greatly between makes, so be patient and familiarise yourself with the correct action of your machine.
- Open the door and confirm that the drum stops. Confirm that warm air is present in the drum. Note that on some machines closing the door restarts the cycle, whereas other machines need to be restarted by pushing the reset button after the door has been closed.
- Select the other heat settings (if applicable) and repeat the second and third steps to confirm that heating is taking place.
- Allow the timer to advance to the anti-crease cycle and confirm that cold air is now circulating in the drum. Let this action run for 5-10 minutes.
- Monitor the machine noise during all operations. If it is excessive, terminate the test. Isolate the machine before inspection.

If the drier is permanently vented or a vent hose is normally used, check that it is completely clear. Check the outlet of the hose or vent during the second and fifth steps, making sure that there is a good flow of air present.

Some machines may incorporate a second timer of up to 12 hours time delay. It is used to interrupt the main timing control, allowing the timer to be used more economically on reduced rate electrical supply, such as Economy 7. To test this timer, select the shortest time delay possible and ensure that the main timer is also set. Normal action of the machine should commence only after the delay timer has reached the end of the delay selected.

Regular inspection points

Removing fluff is an essential cleaning procedure that relates to the safety and efficiency of tumble dry only machines. Regular inspection of tumbledriers is very important, not only to keep them in good working order, but also to ensure that accumulations of lint and fluff do not build up within the machine, constituting a fire risk.

The following procedure should be carried out on a regular basis. Do not delay longer than 12 months between inspections and inspect heavily used machines

Large rear fan housing of a 4kg (9lb) load early Hoover machine. The fan is driven by an elasticated belt directly from the motor shaft, which also drives the main drum. The motor in this instance is a large shaded-pole motor with multi 'V' belt drive at one end and round pulley at the other.

Ensure thermostats are fitted correctly and free from fluff build-up as this acts as insulation leading to incorrect temperature sensing.

Fault finder

Symptom: Will not work at all

POSSIBLE CAUSES	ACTION
Faulty mains lead, plug or socket.	*(See Plugs and sockets, pages 14-21.)* Check mains lead for continuity.
Door not closed correctly.	Close door again and ensure it is latched correctly.
Door micro-switch failed (O/C).	Check switch action (audible clicks when door is open/closed) and verify continuity, *(see Electrical circuit testing, pages 23-25)*. Inspect-renew if suspect.
Timer fault.	Check switching action of manual timers and continuity, *(see Electrical circuit testing, pages 23-25)*. Check for loose selector knob on shaft of timer. On computer-controlled machines, check wiring edge connections to the circuit board.

Symptom: Machine works but no heat

POSSIBLE CAUSES	ACTION
Insufficient time set.	Remember that the last 15 minutes (approximately) of cycle is a cool tumble on most machines and the heater will not be energized, only drum action.
Open circuit auto reset TOC on heater.	Check heater TOC for continuity, *(see Electrical circuit testing, pages 23-25)*.
Tripped manual reset TOC.	Check thoroughly for blockages or restrictions to air intakes, filter and exhaust vents/hoses before resetting TOC.
Broken/O.C. element.	Check condition of heating element(s) and for continuity. Renew if suspect.
Open circuit (O.C.) exhaust thermostat (if fitted) or thermistors on computer machines.	Check for continuity of stat and of wiring.
Faulty timer.	On manual timers, check action of heater switch and continuity. With computer-controlled machines, check all connections to and from the printed circuit board. If alright, possible fault of control board or relay.

Symptom: Heats and trips TOC but no drum action

POSSIBLE CAUSES	ACTION
Broken drive belt.	Check belt and renew as required.
Door micro-switch O/C (some models only).	Some machines link the door micro-switch into the motor circuit only. Check door is latched correctly and the switch for continuity. Renew if suspect.
Belt slipping.	Check for overload of clothing and condition of belt, *(see Belts, pages 123-126)*. Check drive pulley of motor, position and tension of jockey pulley(s).
Open circuit motor.	Check for free rotation of drum motor. Clean if required. Check for continuity of TOC and windings. If capacitor start, *(see Motors, pages 110-122)*. If centrifugal start, check action/free movement of this system. Some early motors had manual reset TOCs. Check motor thoroughly before resetting.
Faulty timer.	On manual timers, check action of motor switch and continuity; on computer-controlled, check connections to and from printed circuit board.

Symptom: Works temporarily, stops/starts after 10/15 minutes

POSSIBLE CAUSES	ACTION
Motor overheating and tripping its self-setting TOC.	Common fault on several machines for various reasons. Check for the following: fluff blockage in and around the motor or air intake grilles; worn bearings on motor end frames allowing rotor to chafe on stator as motor runs; loose cooling fan on motor; motor overheating due to internal wiring fault; temperamental TOC within motor windings tripping at normal working temperature, often caused by infrequent cleaning of motor; worn drum bearings slowing both drum and motor causing overheat within the motor. The majority of these problems can be avoided by regular cleaning.

Fault finder

Symptom: Machine works but no heat

POSSIBLE CAUSES	ACTION
Drum bearings worn.	Check condition of both front and rear bearings.
Loose fan (pressurised machines).	Check that large centrifugal fan on main motor is secure on shaft and is not rotating independent of it. Ensure that the fan is not damaged or catching on other items. Renew if suspect.
Fan shaft/bearings worn (suction machines).	Check for excessive movement of both pulley and shaft. The shaft should be free to rotate without lateral movement. Renew if found to be worn.
Worn or damaged jockey pulley(s).	This type of fault can give rise to a chattering noise if the pulley is damaged, or a high pitched squeaking noise if dry. Check for lateral movement on pulley shaft. Renew if damaged or suspect in any way.
Loose, worn or broken motor mount.	Inspect motor mounting closely and renew if suspect.
Loose panels or fascia parts.	Ensure all covers and panels are fitted correctly and that all seals (if fitted) are in good order. Make sure all securing screws are fitted and are tight.

Symptom: Machine works but poor drying performance

POSSIBLE CAUSES	ACTION
Too little drying time selected for load.	Reset timer for sufficient time to dry the load. See the manufacturer's instructions for guide to drying times for your machine.
Incorrect heat setting selected.	Check that the low heat setting was not set for fabrics needing extra heat, such as cottons. See manufacturer's instructions for more information.
Heater self-setting TOC tripping.	Check for restrictions in filter and all air vents. If alright, check TOC continuity. If nuisance tripping is suspected (tired TOC), renew with correct replacement.
Open circuit (O.C.) heater.	Check condition and continuity of heater.
Vent thermostat fault.	Check for continuity.
Recycling of exhaust air.	If the machine is not vented correctly, it is possible that the moist air expelled from the exhaust vent could be taken back into the machine via the air intake vent. Check for this action and rectify if found.

Through lack of regular maintenance, fluff has built up on and around the heating elements of this machine. The build-up is charred and has smouldered. This is a potential fire risk. Regular fluff removal is essential.

more frequently. The same procedure should also be carried out if a repair is required between normal routine inspections. A thorough stripdown of the appliance is necessary to ensure that all components are free from lint and fluff and that all wiring connections are secure and in good condition. Do not relegate this procedure to a quick inspection and a cursory cleaning of only those parts that are easily accessible with just the top of the machine removed.

A basic sequence that suits most, if not all, machines is as follows and, as always, should be carried out with the machine fully isolated.

- Check and clean the removable filter. This should be relatively clear as its cleaning should be part of normal usage routine.
- Remove the vent hose (if fitted) and check the outlet duct of the machine for fluff build-up. The full length of the hose or ducting should also be checked, along with external grilles on permanently vented systems. Ensure the whole system is free from restrictions.

When removing fluffing and lint, check all areas of the machine. The motor shown here is in need of thorough cleaning, not only in order to function correctly, but also for safety.

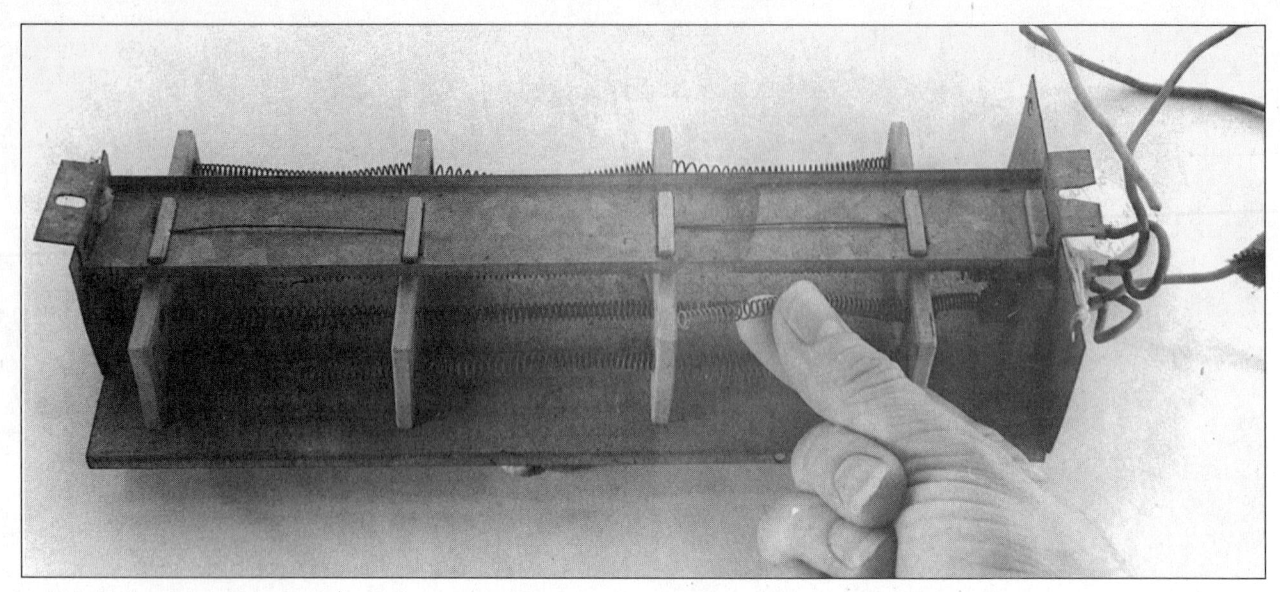

Inspect the heating element closely for fluff contamination, sagging and breakages, as shown here. Complete heater units are not usually expensive.

- Always note components, wiring connections and routing within the machine before stripdown or removal.
- Ensure each component is thoroughly cleaned and remove all traces of lint/fluff. Do this with care, using a small dusting brush and/or cylinder vacuum cleaner.
- Pay particular attention to components such as the timer(s), main motor (cooling grilles, etc.), TOC, thermostats, terminal connections, heating elements and fixing points. Solid elements (sheathed) can be brushed clean but spiral elements must be carefully cleaned to avoid stretching or damage to their ceramic holders. With spiral elements, fluff may build up within the hollow centre. This can best be removed by carefully teasing it free with a needle or similar object.
- Renew any suspect or worn components – overheated/poor connections, worn drive belt(s), tension pulley(s). Ensure correct position and, if required, lubricate drum support bearings and heater support brackets/cord.
- Re-assemble the machine, ensuring that all wiring, protective covers, supports and seals are correctly positioned and fitted.
- Test for earth continuity *(see Electrical basics, pages 10-13 and Electrical circuit testing pages 23-25).*
- Check the condition of both plug and socket, looking for signs of overheating, damage and correct supply *(see Plugs and sockets, pages 14-21).*
- Make a functional test on an RCD protected supply.

It is common for the front lip on the drum to be used as the front support bearing. Ensure that it is smooth and undamaged. It is either supported by a felt strip coated in PTFE or by plastic mounts which act as support bearings. Both types are prone to wear.

Elasticated belts are prone to stretching and fraying. Renew them regularly to maintain peak performance.

Useful hints

Tumble dry only machines are, in essence, simple machines but it is advisable to make yourself aware of the capabilities and limitations of your machine. When cleaning the filter you may be surprised by the quantity of fluff and lint that it has accumulated. However, if used correctly, tumbledriers do not unduly stress the fabrics they dry. It takes around 1,000 tumble dry cycles for a 227g (8 oz) item of clothing to lose 28g (1 oz) in this way.

The length of time your drier takes is proportional to the efficiency and spin speed of your washing machine. Moreover, all fabric contains a degree of moisture when it is dry. Unfortunately, it is possible to over-dry clothing in a tumbledrier and so remove the naturally balanced moisture content of the fabric. This leads to shrinkage, excessive static and wrinkling that will not iron out. Such problems often lead to the machine's being blamed when it is actually the user that is at fault by selecting a dry setting that is too long, too high a heat or both.

Keep the load content and size in mind when setting the timer. If in doubt, refer to the manufacturer's instructions. This not only helps avoid over-drying problems, but also saves money on electricity consumption.

Whenever possible, avoid drying woollen items in a tumbledrier. They are particularly prone to the problems described above.

Articles made from or containing foam, sponge, plastic, wax coatings or printed surfaces should not be dried in a tumbledrier. They can easily be heat damaged and could constitute a fire hazard.

Never use a tumbledrier to dry clothing or materials that have previously been cleaned with flammable spirits or dry cleaning fluids. Ensure such articles are left in a well-ventilated area to air fully and do not subject them to a dry cycle until they have been washed.

Fault codes

Computer-controlled (electronically controlled) machines are often capable of a degree of self-diagnosis. When a fault or problem occurs, a fault code is displayed on the LED or LCD display panel. Occasionally the display describes the fault, but more often it shows alpha numerical codes, such as F1 or E4. The code indicated relates to a particular problem detected by the microprocessor. The manufacturer's instructions give the meaning of each code *(see Timers, pages 98-109, for more detail)*. The machine will not continue with the remainder of the programme setting until the fault has been rectified. Although the fault display codings may differ from one manufacturer to another, many of the faults they indicate are covered in the list of symptoms on pages 158-160.

Faults that are uncommon or difficult to isolate may be caused by poor connections to or faults within the electronic control boards, or penetration by moisture as a result of faulty or incorrectly positioned seals. Ensure that all connections, covers and seals are sound and correctly fitted, and eliminate all other possible faults before changing a control panel. All circuit testing on microprocessor controlled machines must be carried out using a low voltage tester, that is, 1.5 volts, to avoid damage to the electronic components. Do not handle the components of the circuit boards as damage can also be done by static discharged from the fingers.

This type of damage is not always immediately apparent. However, static discharges can corrupt the processor chip's internal programme memory and result in unusual and often intermittent fault symptoms some time after the initial damage was done. Handle the boards only by the edges and only if it is absolutely necessary.

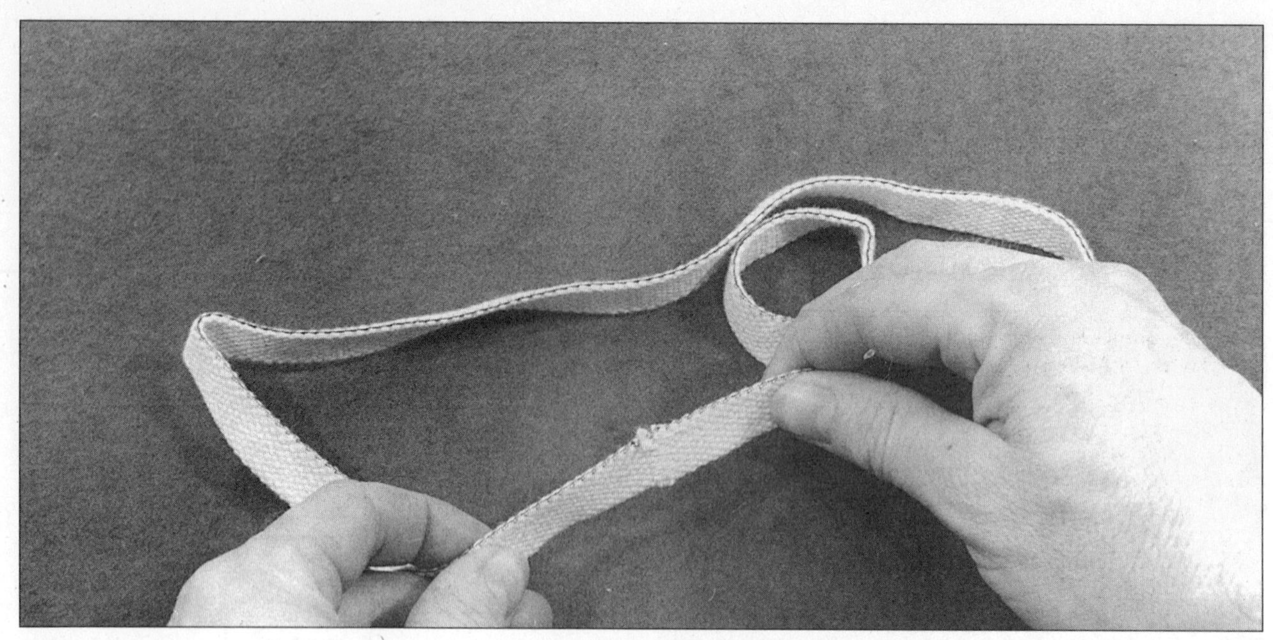

Many Creda machines use a braided webbing as a front bearing. It is glued to the front drum lip and rests on the metal edge of the machine. When the webbing is worn, the machine vibrates badly and becomes very noisy. Renewing the webbing is straightforward, but it is wise to remove all traces of paint from the worn shell lip and make sure that it is smooth. Lubricate both parts with PTFE before re-assembly.

Another feature of electronically controlled machines is that a self-diagnostic functional test routine can be initiated. How the sequence is started differs with makes and models and is usually restricted to service personnel. Most machines need to be turned off (at the socket) for at least 45 seconds to clear any previously set programme. Starting the functional test then requires the pressing of two of the selector buttons simultaneously while switching the machine on. This sequence is used only to highlight the way in which a test sequence may be initiated and does not indicate the way in which any particular machine operates. The number of different ways in which this system can be set makes it impossible to give specific details.

Worn tension pulleys can create a lot of noise or allow the belt to slip or jump out of position constantly. Renew if worn or damaged; also check the supporting shaft for wear. Whenever possible renew both parts as a set .

PTFE lubricant as required by webbing type bearing.

Routine maintenance

When undertaking any checks or repairs, it is well worthwhile ensuring that all components within the machine are free of fluff and lint: motor, motor, relay, micro-switch, pulley, etc.

TOOLS AND MATERIALS

- ☐ Screwdriver(s)
- ☐ Spanner
- ☐ Socket
- ☐ Ceramic insulator
- ☐ TOC

1 This (2.7kg (6lb) load machine has a wrap-around outer shell with no obvious means of access.

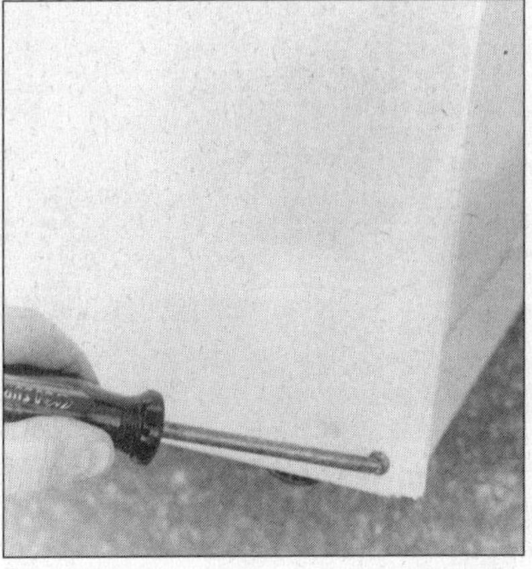

2 First remove the two screws found on the lower edge of each side of the machine.

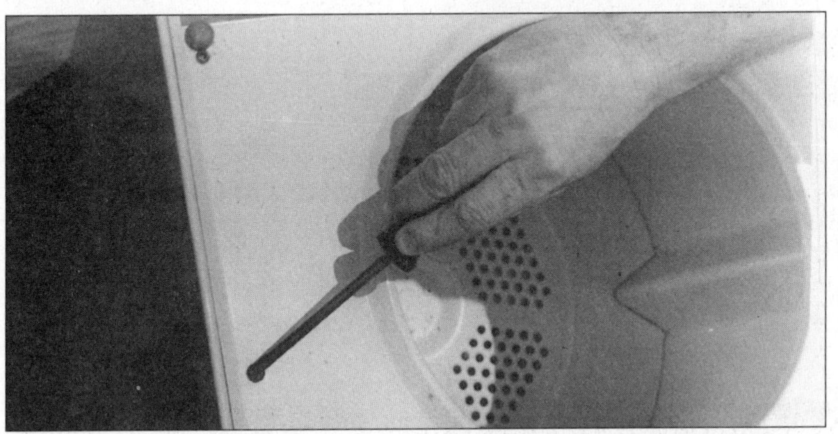

3 Lower the door to expose a further four screws. Make a note of all screws, as they are different sizes.

4 With all eight screws removed, the one piece outer shell of the machine slides out of position vertically.

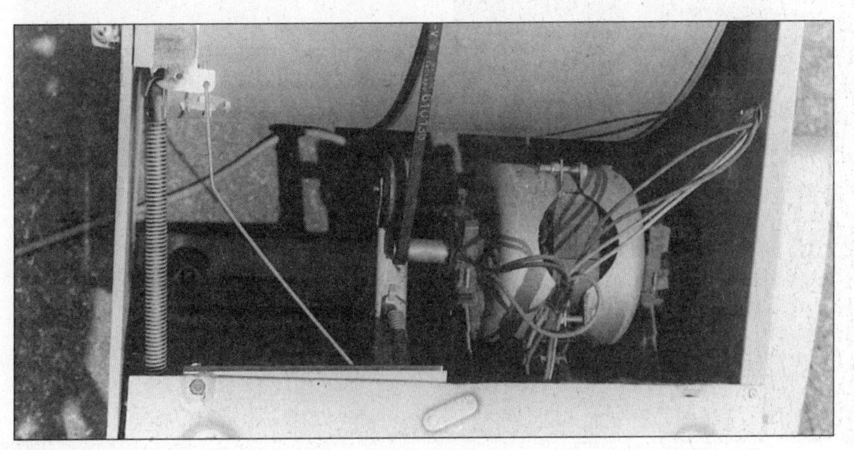

5 Removing the outer shell exposes the induction motor, drum belt, tension pulley system and door micro-switch with actuating arm.

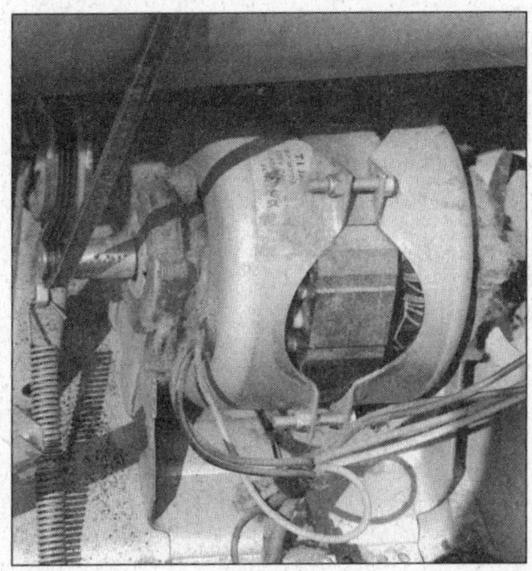

6 Remove fluff and lint from the motor, motor relay, micro-switch and tension pulley system. Inspect the drum belt; this one is alright.

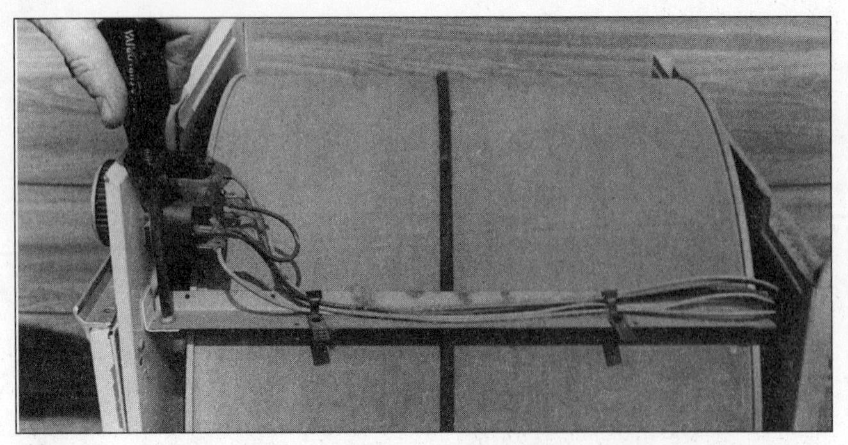

7 To check the front bearing, remove the bolts from the front of the side stays and the wiring to the timer. Make a note of all connections before removal.

8 With the wiring and fixings removed, ease the front of the machine forward to expose the front bearing and drum lip.

9 Closely inspect the plastic support pads, drum lip and sealing felt. In this instance they are sound and require only cleaning.

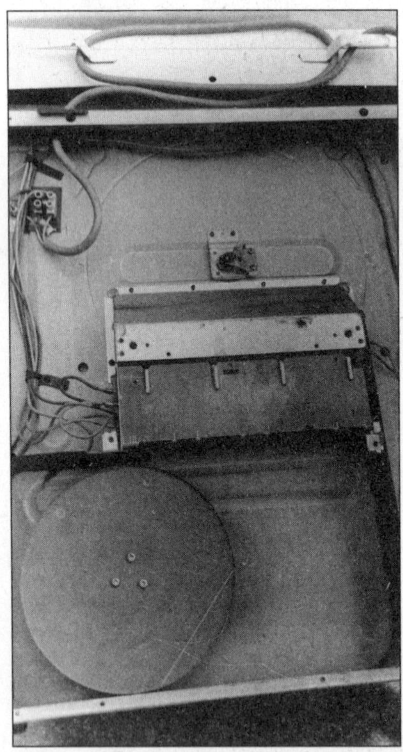

10 Removing the back panel (held in position by self-tapping screws) allows access to the fan, heater unit and TOC.

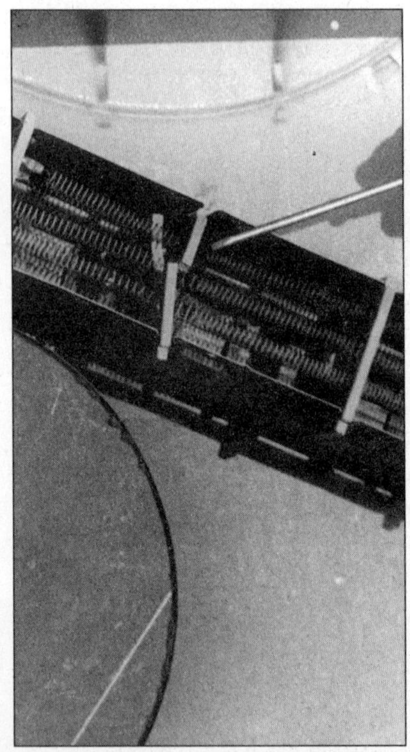

11 Although clean and free from lint, the element has a broken ceramic insulator. The TOC also shows signs of contact damage and both parts should be renewed before re-assembly.

Creda Compact 3

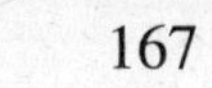

Chapter 6

Further information

Useful tips

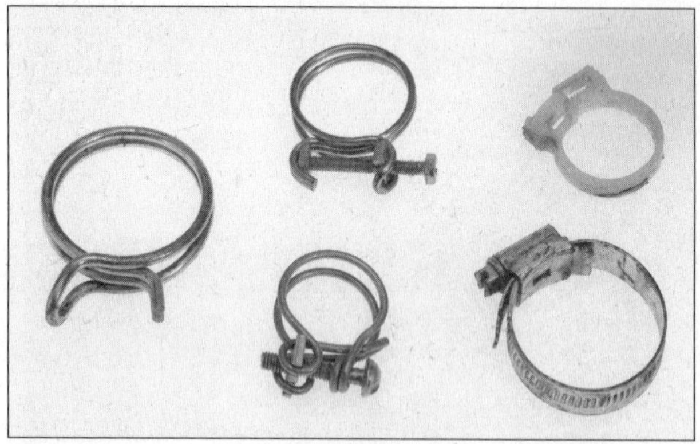

Various types of clips in common use in today's washing machines. Centre: screw type wire clips. Top right: the new type toothed clip. This new clip is much easier and quicker to fit as grips or pliers are used to tighten jaws. Left: a corbin spring clip. Take care when removing this type of clip, as it has a tendency to 'spring' – even under tension. Corbin pliers are best for removal, but, with care, ordinary grips may be used. Lower right: a worm drive or jubilee clip. This is a simple but effective clip.

The door interlock on some machines jams the door shut when it fails. As the fixing screws are behind the locked door, it may be difficult to open the door. It is possible to move the door latch with a screwdriver, as shown. Be careful not to scratch the paint.

Check all hoses thoroughly for perishing and/or cracking. With corrugated hoses (as shown), stretch the hose to ensure a thorough check. It is wise to check any new hose before fitting.

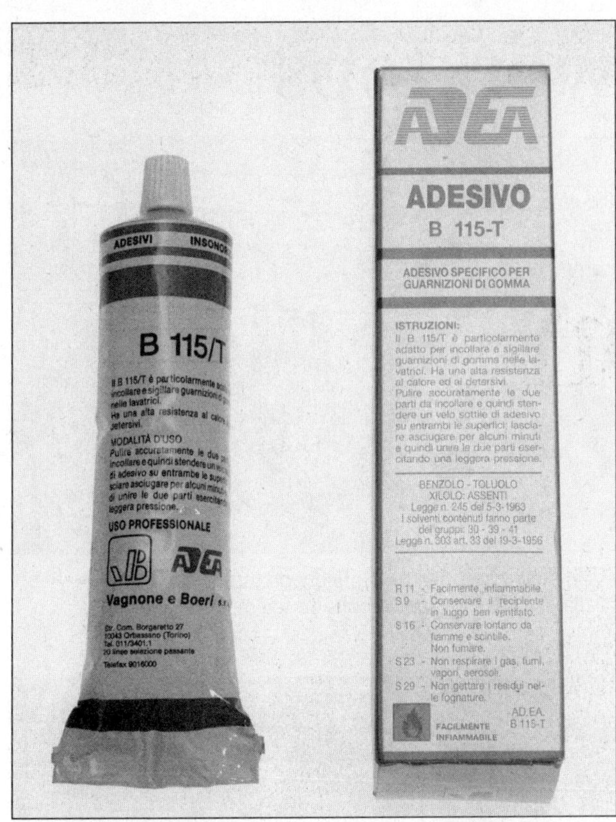

Sealants can be used for pressure system hoses and for aiding the fitting and sealing of new hoses, grommets and so on.

A little washing-up liquid or fabric conditioner can be smeared on grommets and rubber mouldings.

Some machines have a wire surrounding the door seal. This retaining ring can be removed using a flat-bladed screwdriver. The machine shown is a Fagor.

Removing fitted worktops

TOOLS AND MATERIALS

- ☐ Spanner
- ☐ Washing-up liquid

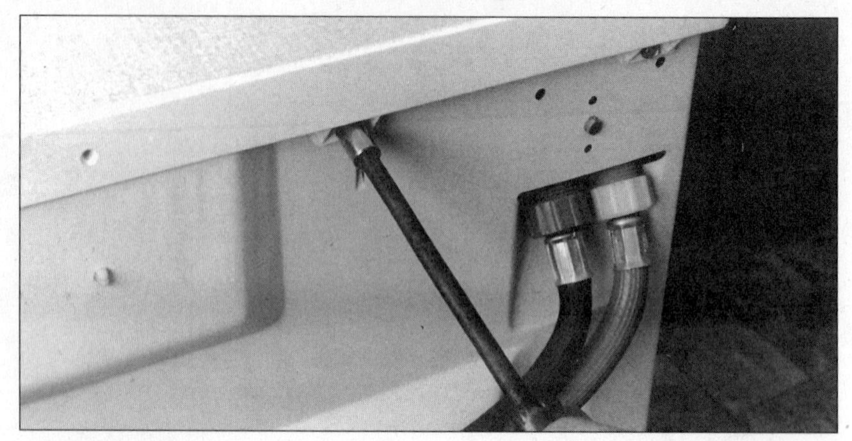

1 Removing the top of some machines that have fitted worktops is not always very straightforward. First, remove the self-tapping bolts on the rear plastic panel section.

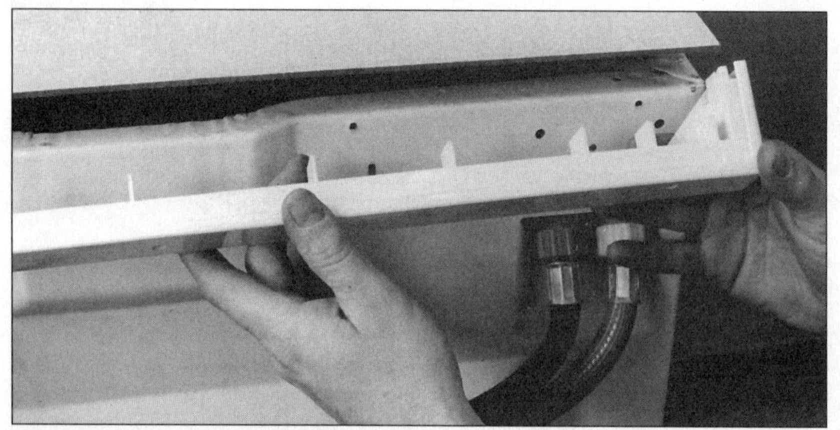

2 Remove the rear plastic surround completely. As always, ensure that the machine is isolated. (Many machines reveal open terminals behind the plastic moulding.)

3 The top can now be moved from its position. Push or pull the top towards the back of the machine and remove. This may require some force, especially if it has not been removed for some time.

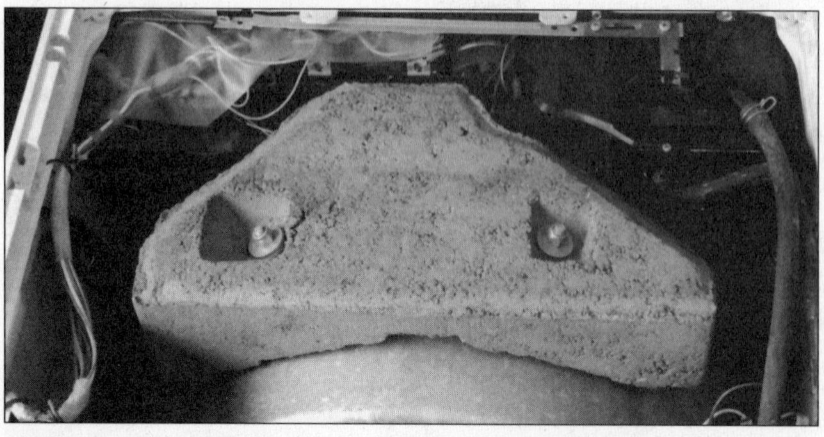

4 Access to the top half of the machine is now possible. To aid refitting the top, a little washing-up liquid may be smeared on the plastic slides of the machine.

Rear access to this type of machine is limited to a much smaller rear opening. Any repairs to this type of machine require the inner and outer tub assembly to be removed via the top of the machine.

Domestic appliance white paint is available for treating damage or rust to the shell of your machine. Spray and brush on types are shown here.

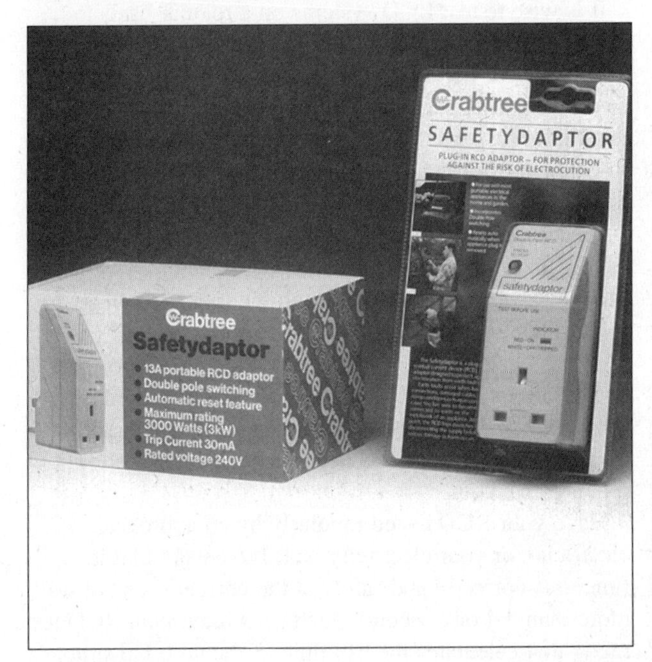

An RCD is essential.

When removing the wire door seal securing band on some machines it is advisable to wear eye protection. This is also essential when drifting out old bearings. Wear sturdy gloves to protect your hands from the rough casting or other sharp edges when handling cast aluminium pulleys.

Low insulation

Low insulation is best described as a slight leak to earth of electricity from the wiring of one or more of the components or wiring in an earthed appliance. If very slight, it will not harm the appliance but it is an indication of faults to come and should be corrected immediately for safety reasons.

How is it caused?

Low insulation may be caused by normal wear and tear over a long period, resulting in a breakdown of the insulating coating on wiring, motor windings, heater elements and so on. Such a breakdown of insulation may not result in a failure of the part and the appliance may still function apparently normally. However, this is no excuse to ignore it, as failure to trace and rectify low insulation is both foolhardy and, in the long run, can be costly in terms of both money and safety.

Faults such as leaking or weeping shaft seals can give rise to water penetrating the motor windings, resulting in low insulation. If not corrected, this could lead to a complete failure of the motor or worse. A simple renewal of the shaft seal and careful cleaning and drying of the TOC and windings may be all that is needed to save money and improve safety for all concerned. It is very important never to compromise on safety by ignoring such symptoms.

Detecting low insulation

An engineer tests for low insulation with an instrument called a metrohm or low insulation tester. The law requires repair engineers to test for low insulation and there is a minimum allowable level. The law requires that the following tests are made by commercial repair engineers.

- Between the earth pin and the plug and all earth connection points within the appliance, the maximum resistance should be 1 ohm, that is, very low resistance – a perfect connection.
- With the appliance turned on but unplugged, between the live pin on the plug and the earth pin on the plug. The minimum resistance should be 1 megohm, ideally no detectable reading, that is, very high resistance – no connection at all.
- Repeat this test between the neutral and the earth pin of the plug repeating both tests at various programme settings.

Testing of individual components can be carried out easily by removing connections to the suspect item and connecting one lead to one of the free terminals and one to the earth terminal. Minimum resistance should be 1 megohm. Then repeat the test using the other connection.

These tests are carried out with a meter designed to test insulation by applying a high voltage (500v), at a very low amperage for safety, to test the insulation quality of the part to which it is connected. Unfortunately, some engineers do not possess such an instrument and so do not check for low insulation. This does not mean that you should not!

A meter for testing low insulation is more expensive than most D-I-Y enthusiasts can afford. An alternative is to utilise an in-line circuit breaker. Plug the machine into the circuit breaker and then plug that into the socket unless an RCD already protects the circuit or socket. Its purpose is to detect low insulation or leakage to earth and turn off the power to the appliance (see Electrical basics, pages 10-13). Although this is not the ideal way of testing for low insulation, it helps to locate it and provide safety for the appliance and its user.

It is wise to test RCD systems on a regular basis to ensure they function correctly and are fully operational when needed. Follow the instructions shown on the unit or on the leaflet accompanying the adaptor. If a fault with the unit is suspected, it should be tested and possibly re-calibrated for maximum performance by a skilled electrician using a special RCD test meter. If a fault is suspected in an RCD unit, have it checked professionally, as it is there for your safety.

If your washing machine or tumbledrier trips an RCD (or similar) system, do not use the appliance until the fault has been rectified. If tripping occurs with no appliances or load on the system, then a fault on the house wiring is indicated and the trip switch should not be reset until the fault has been identified and a correct repair carried out.

Have your RCD tested regularly by an approved electrician or your electricity board to ensure that it functions correctly and safely at the correct speed of no more than 0.4 of a second. Such tests require an RCD test meter that calculates the trip time of the unit. On simple tests the unit may trip but take too long for it to be classified as safe.

LOW INSULATION WATCH POINTS

1. **Ensure that any disconnection or removal of wires is safe** and not earthing via another wire or the metal case of the appliance.

2. **When disconnecting any wires while testing for low insulation, remember that the machine must be isolated from the mains** at all times and the panels or covers must be replaced before the appliance is re-tested. Do not test with exposed wiring.

3. **Before testing for low insulation using a circuit breaker, all earth paths of the appliance should be tested.** Do this by connecting a meter between the earth pin of the plug, and all other metal parts of the appliance in turn. Maximum resistance should be 1 ohm (see Electrical circuit testing pages 23-25).

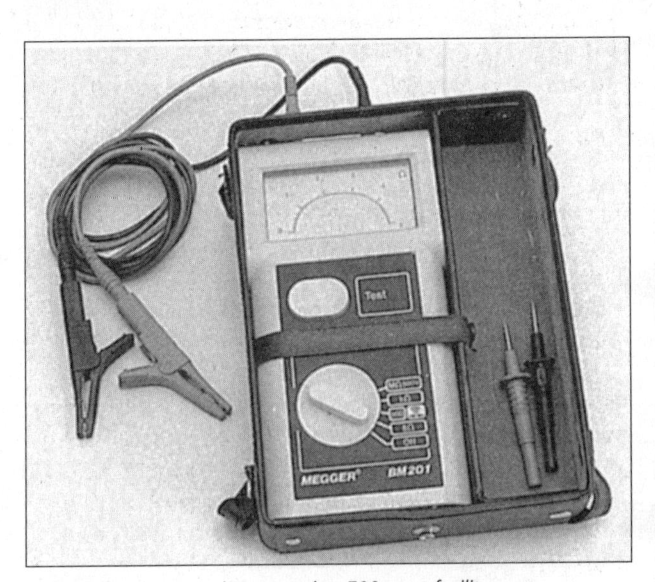

A versatile test meter incorporating 500v test facility.

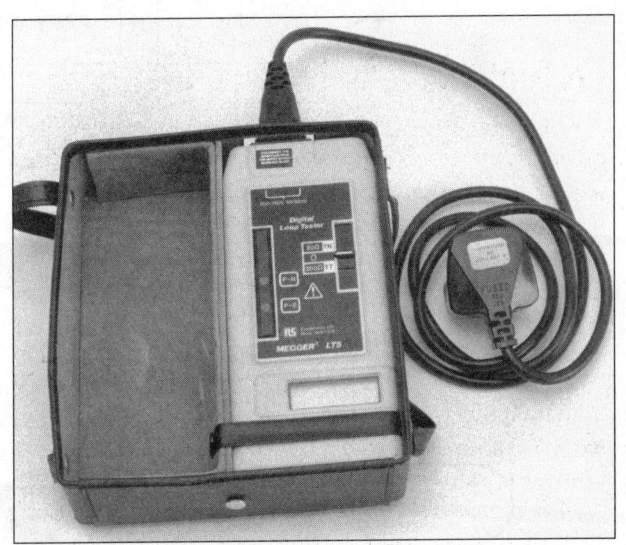

Modern consumer unit with RCD main switch and MCBs on all circuits.

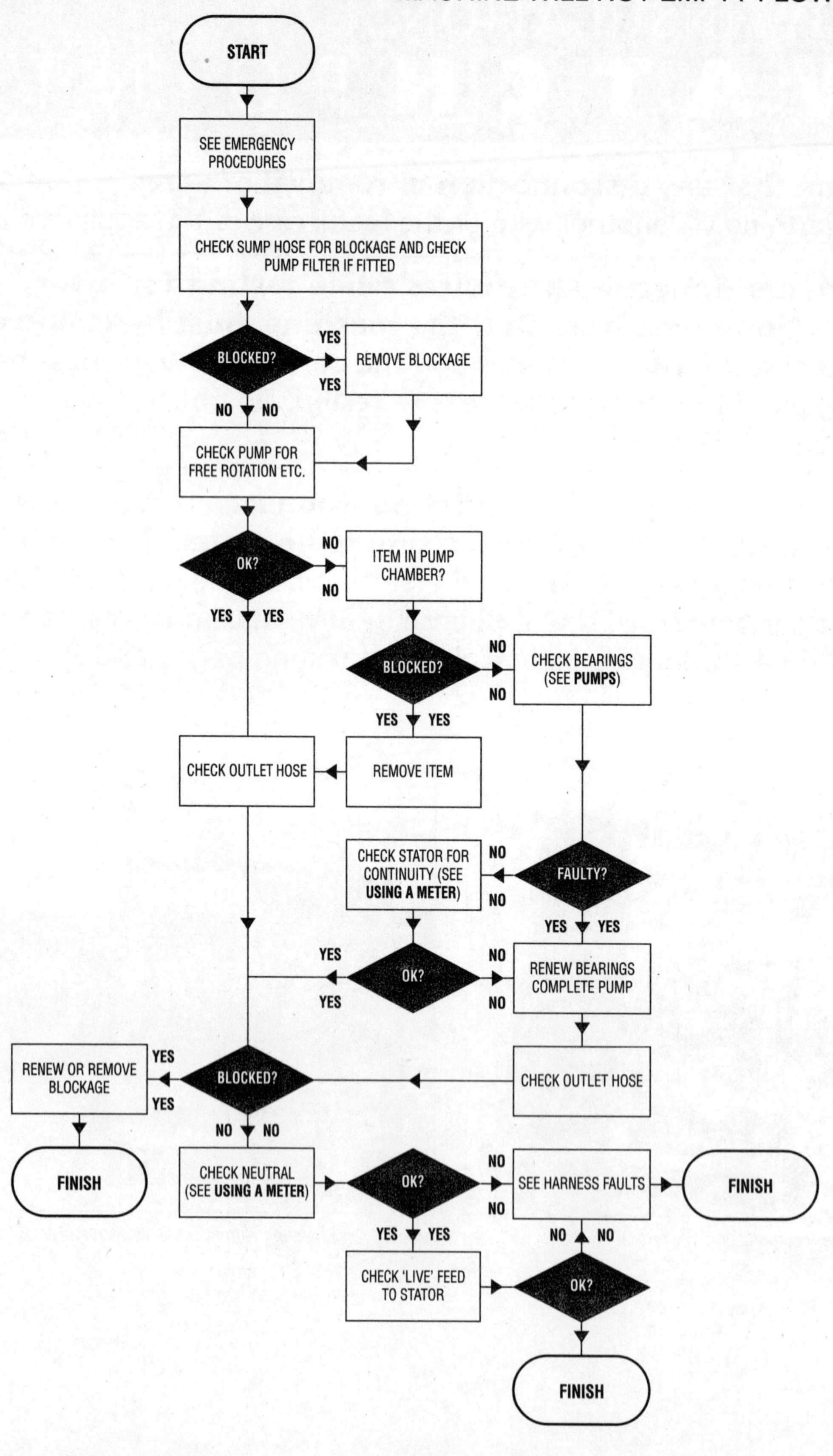
MACHINE WILL NOT EMPTY FLOWCHART
START
SEE EMERGENCY PROCEDURES
CHECK SUMP HOSE FOR BLOCKAGE AND CHECK PUMP FILTER IF FITTED
BLOCKED?
YES
YES
REMOVE BLOCKAGE
NO
NO
CHECK PUMP FOR FREE ROTATION ETC.
OK?
NO
NO
ITEM IN PUMP CHAMBER?
YES
YES
BLOCKED?
NO
NO
CHECK BEARINGS (SEE PUMPS)
YES
YES
CHECK OUTLET HOSE
REMOVE ITEM
CHECK STATOR FOR CONTINUITY (SEE USING A METER)
NO
NO
FAULTY?
YES
YES
OK?
YES
YES
NO
NO
RENEW BEARINGS COMPLETE PUMP
CHECK OUTLET HOSE
RENEW OR REMOVE BLOCKAGE
YES
YES
BLOCKED?
NO
NO
FINISH
CHECK NEUTRAL (SEE USING A METER)
OK?
NO
NO
SEE HARNESS FAULTS
FINISH
YES
YES
NO
NO
CHECK 'LIVE' FEED TO STATOR
OK?
FINISH

LEAK FAULT FINDING FLOWCHART

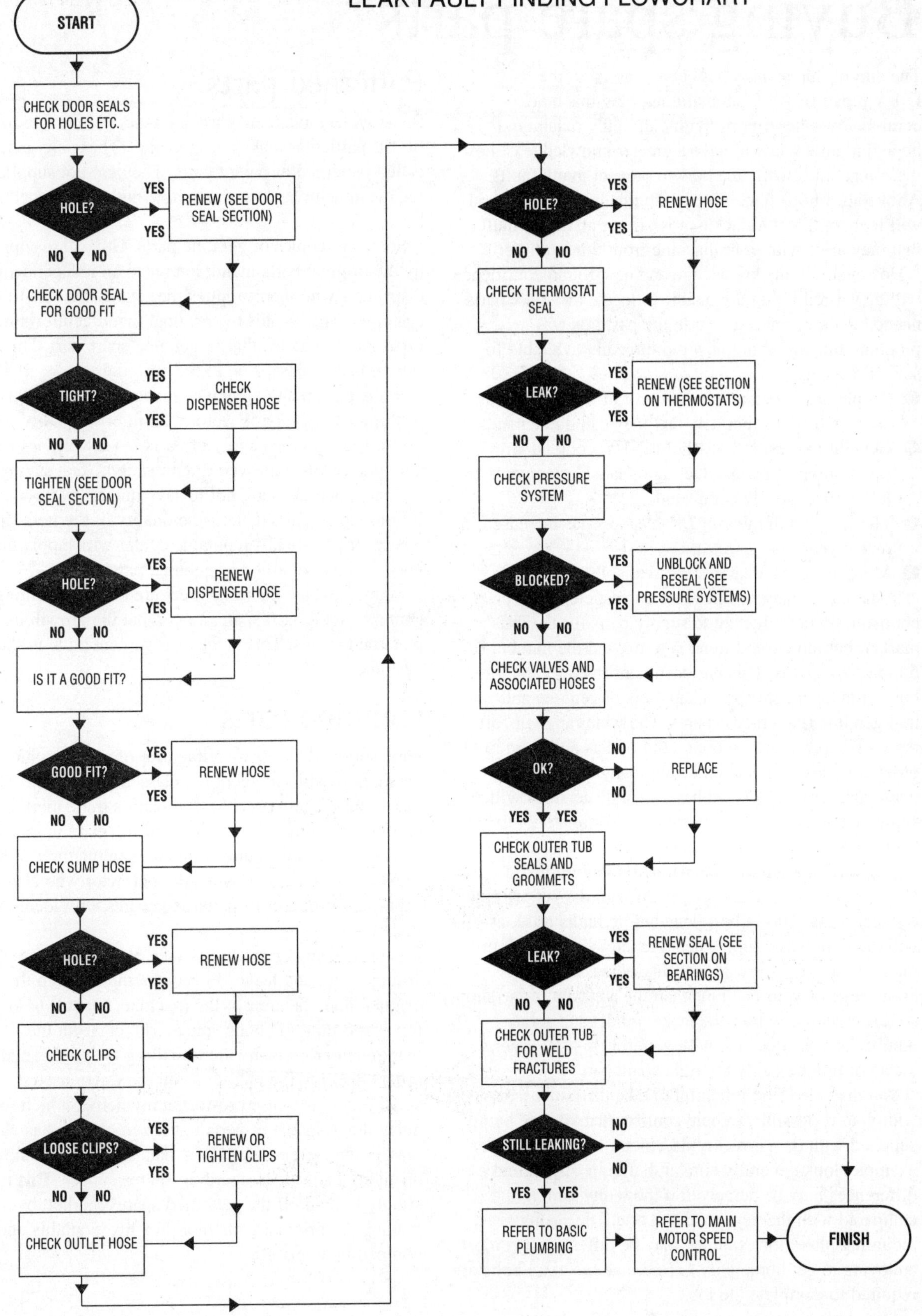

Buying spare parts

The aim of this manual has been to assist in the D-I-Y repair of your automatic washing machine, combined washerdrier or tumble dry only machine. I hope that now you will have a greater knowledge of how these machines work and how to prevent many faults. Above all, I hope that, armed with this information, you will feel confident to tackle most, if not all of the faults that may arise with your machine from time to time.

However, all this knowledge and new-found confidence will be wasted if you are unable to locate the spare parts needed to carry out a repair. In the past this was a problem, but spares are now more readily available for several reasons.

- People are reluctant to pay high call-out and labour charges for jobs that they feel they can do themselves.
- General interest in household D-I-Y, coupled with saving money, gives a feeling of satisfaction when the job is successfully completed.
- The size and number of D-I-Y stores has increased in recent years.
- More pre-packed spares are available.

In the past, many independent domestic appliance companies were reluctant to supply parts for the D-I-Y market, but the current trend is to expand the number of pre-packed spares. This has been confirmed by the three biggest independent spares suppliers of genuine and non-genuine (patterned) spares. The wide range of 'off the shelf' spare parts in both retail outlets and from mail order companies is welcome, and many machine manufacturers who do not have local dealerships will supply parts by post if requested. (Unfortunately this can sometimes be a lengthy process).

One way of obtaining the parts you require is to find a local 'spares and repairs' dealer through the yellow pages or local press. This is best done before faults arise, as you will then not waste time when a repair is necessary. In many instances, you will probably possess more knowledge of your machine than the assistant in the shop, so it is essential to take the make, model and serial number of your machine with you to help him or her locate or order exactly the right spare part .

You may also find it helpful to take the faulty part(s) with you, if possible, to help confirm that you are being supplied with the correct replacement. Most pumps, for example, look generally similar, but quite substantial differences may be perceived if the faulty item is compared with the newly offered one. The casing or mounting plate, for example, may be different. It is most annoying to get home only to find that two extra bolts are required to complete the job.

Patterned parts

Some widely available parts are marked 'suitable for' or 'to fit' particular makes of machine . These are generally called patterned or patent parts. They are not supplied by the manufacturer of your machine but are designed to fit it.

Some are copies of genuine parts. Others are supplied by the original parts manufacturer to an independent distributor who then supplies the retailer for sale to the customer. This avoids the original manufacturer's mark-up as it is not an official or genuine spare part. This saving is then passed on to the customer.

In the past, many appliance manufacturers disliked this procedure, as the parts were of an inferior quality. This is not so today as the supply of parts is very big business and quality has improved dramatically. Great savings can be made, but take care not to save money by buying inferior spare parts. Check the quality of the item first wherever possible. A reputable dealer will supply only good quality patterned or genuine parts.

Many machine manufacturers are now discounting their genuine, authorised spares to combat the growth in patterned spares. This is, obviously, beneficial to the customer.

Genuine parts

Parts supplied by the manufacturer of your machine or by the authorised local agent are classed as 'genuine'. In many cases, they carry the company's trade mark or logo on the packaging. In fact, many of the parts in today's machines are not produced by the manufacturer of the finished machine, but by a sub-contractor who also may supply a distributor of patterned spares with identical items.

Patterned spares producers take on only items that have volume sales and leave slower-moving items to the original manufacturer of the machine. It tends to be a long procedure to obtain spares 'direct' from the manufacturer, as many are unwilling to supply small orders direct to the public. Another system used to deter small orders is to use pro-forma invoicing, which will delay the despatch of parts until your cheque has cleared.

With the increase in D-I-Y, manufacturers are slowly changing their view regarding spares supply. This is simply to fend off the patterned spares market, by making the original parts more readily available and competitively priced.

Parts by post

A free parts list of both patterned and genuine spares for most leading makes of machine can be obtained by writing to:

Dixon Repair Services
(Postal Spares Department)
Cranswick
Driffield
North Humberside
YO25 9QJ.

Enclose a large stamped addressed envelope and details of the make and model number of your machine(s). Parts can be obtained by post at competitive prices and payment can be made by cheque or credit card.

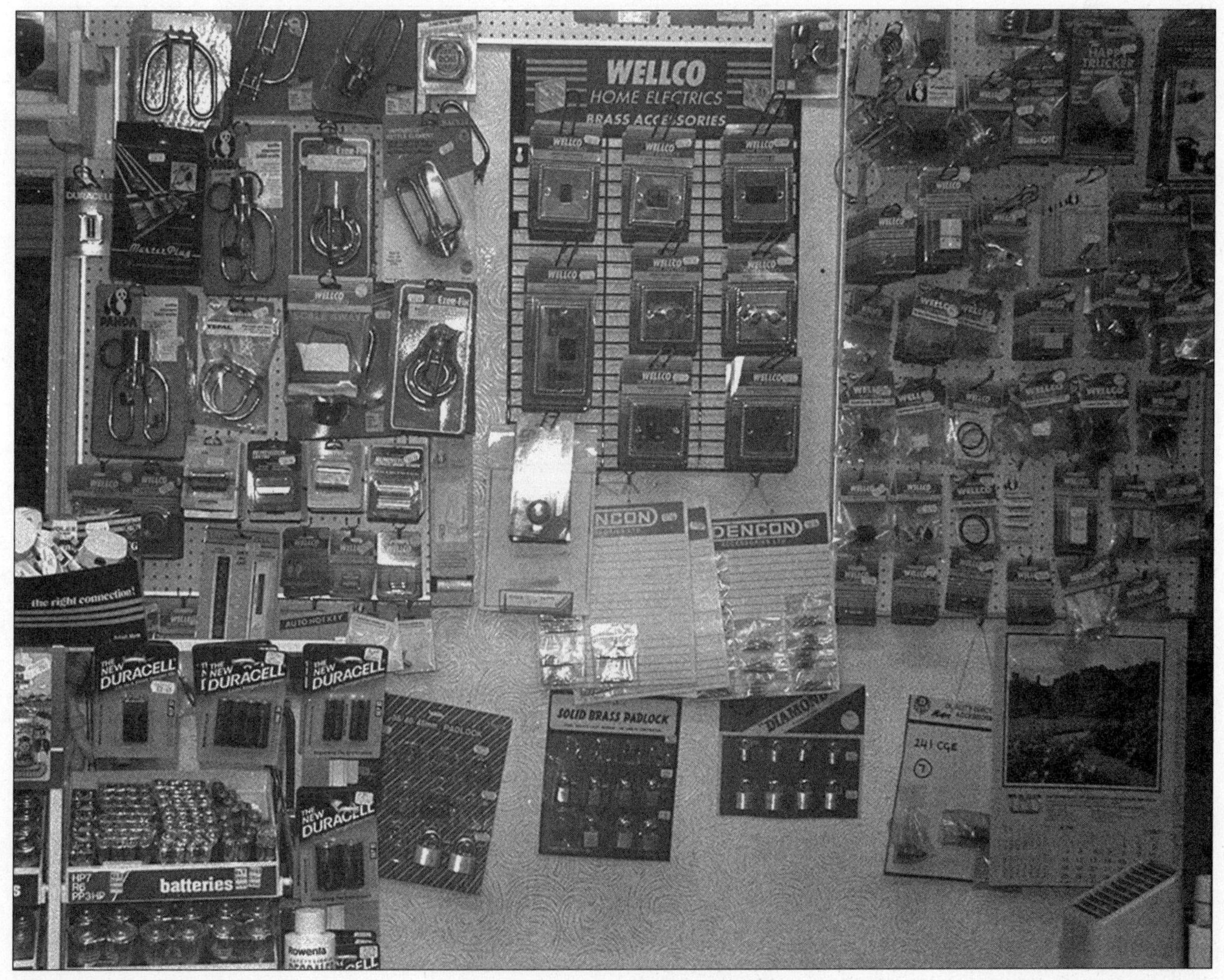

This small but well-stocked shop offers an extremely wide variety of items. Try to locate a similar shop in your own area before faults arise. This will save time and frustration later on.

In conclusion

Finally, the decision between genuine and patterned spares is yours. Cost and availability may have to be taken into consideration, but do not forsake quality for a small financial saving

As a guide, here is a list of manufacturers' names and addresses where parts may be obtained. Addresses other than these can be found in your local Yellow Pages.

MAKE	TELEPHONE NO.
AEG	01753 872 325
Ariston	01322 526 933 (or see Yellow Pages under Merloni Domestic Appliances)
Asko (was Asea)	0181 568 4666
Bauknecht	01345 898 989
Bendix	Yellow Pages
Bosch	0181 573 6789
Candy	0151 334 2781
Colston	see Ariston (above)
Creda	Yellow Pages
Electra	see Local Electricity Board
Electrolux	01325 300 660
Fagor	01707 377 877
Frigidaire	0151 355 0588
Hotpoint	Local Directory
Hoover	Yellow Pages
Indesit	01322 526933 (or see Yellow Pages under Merloni Domestic Appliances)
Kelvinator	0151 334 2781
Miele	Yellow Pages
Philco	0181 902 9626
Philips (now Whirlpool)	01345 898 989
Servis	0121 526 3199
Whirlpool (was Philips)	01345 898 989
Zanussi	Yellow Pages/Local Directory
Zerowatt	0151 334 2781

This information is correct at the time of publication, but may change.

Essential information

Glossary

Amp – Short for ampere. Used to measure the flow of electricity through a circuit or appliance.

Armature – Wire wound centre of brush motor.

Bi-metal – Two different metals which have been joined together. When heated a bi-metal strip bends in a known direction.

Boss – Protection around entry point.

Burn-out – Overheated part of item.

Bus bar – An electrical conductor.

Cable – Conductors covered with a protective, semi-rigid insulating sheath, used to connect the individual components within a wiring system.

Carbon face (seal) – Watertight flat surface seal.

Centrifugal – Force that increases with rotation, causing movement away from its centre.

Circuit – Any complete path for an electric current, allowing it to pass along a 'live' conductor to where it is needed and then to return to its source along a 'neutral' conductor.

Clamp band – Large adjustable clip used for holding door boot.

Closed circuit – A normal circuit that allows power to pass through.

Collet – Tapered sleeve of two or more parts designed to grip a shaft passed through its centre.

Commutator – Copper segment on motor armature.

Component – Individual parts of the machine; for example, pump, valves and motors are all components.

Conductor – The metallic current-carrying 'cores' within cable or flex.

Consumer unit – Unit governing the supply of electricity to all circuits and containing a main on-off switch and fuses or circuit breakers protecting the circuits emanating from it.

Contact – Point at which a switch makes contact.

Continuity – Electrical path with no break.

Corbin – Type of spring hose clip.

Damped – To have reduced movement of suspension, that is, reduce oscillation of tub unit during distribute and spin cycles.

Dispenser – Compartment that holds detergent ready for use.

Dispenser hose – Hose that supplies the tub with detergent and water from the dispenser compartment.

Distribute – To balance load by centrifugal force, that is, a set speed calculated to even out wash load before faster spin.

Door boot – Flexible seal between door and tub.

Door gasket – Flexible seal between door and tub.

Door seal – Flexible seal between door and tub.

Drift – Soft metal rod used for bearing removal.

ELCB – Earth leakage circuit breaker.

E.M.F. – Source of energy that can cause a current to flow in a circuit or device.

Early – Machine not currently on the market.

Energise – To supply power to.

Energized – Having power supplied to.

Flowchart – Method of following complicated steps in a logical fashion.

Functional test – To test machine on a set programme.

Garter ring – Large elastic band or spring used to secure door boot.

Grommet fitting – Method of fitting hoses, etc., requiring no clips.

Harness – Electrical wiring within a machine.

Hertz – Periodic cycle of one second, that is, cycles per second.

'Hunting' – Oscillating.

Impeller – The blades of the water pump.

Insulation – Material used to insulate a device or a region.

Isolate – To disconnect from the electricity supply and water supply.

Laminations – Joined metal parts of stator.

Late – Current machine on market.

Lint – Fluff from clothing that may cause a small blockage.

Live – Supply current carrying conductor.

Make – (1) Manufacturer's name. (2) When a switch makes contact it is said to 'make'.

Micro-processor – Miniature integrated circuit containing programme information.

Miniature circuit breaker (MCB) – A device used instead of fuses to isolate a circuit if fault/overload occurs.

Neutral – Return current carrying conductor.

O.O.B. – Out of balance detection system.

Open-circuit – Circuit that is broken, that is, will not let any power through.

Outer casing – Cabinet, 'shell' of the machine.

PSI – Measurement of water pressure, pounds per square inch, for example, 38psi.

Pawl – Pivoted lever designed to engage ratchet gearing.

Porous – Item that allows water to pass through.

Potentiometer – Variable resistance device.

Processor – Main central processing component of computer control circuitry.

Programmer – see Timer.

PTFE – Special lubricant (mainly for tumbledriers); chemical name Polytetrafluorethylene.

RCCB – Residual current circuit breaker (also known as RCD) .

RCD – Residual current device. See also ELCB and RCCB.

Reciprocating – Mechanical action of backwards and forwards movement.

Reverse polarity – A situation where live and neutral feeds have reversed, for example, connected incorrectly at supply, socket or plug.

Ribbon cable – Flat cable used for low-voltage printed circuit board connections.

Rotor – Central part of an induction motor.

Schematic diagram – Theoretical diagram.

Seal – Piece of pre-shaped rubber that usually fits into a purpose-built groove or between two surfaces, thus creating a watertight seal when subjected to pressure.

Sealant – Rubber substance used for ensuring watertight joints.

Shell – Outer of machine.

Spades – Connections on wires or components; remove gently.

Stat – Abbreviation of thermostat.

Stator – Electrical winding on motor.

Terminal block – A method of connecting wires together safely.

TOC – Thermal overload cut-out. At a pre-set temperature, the TOC will break electrical current to whatever it is in circuit with; it prevents motors, etc. overheating.

Thermistor – A semi-conductor device the resistance of which is affected by temperature.

Thermostat – Device used to monitor temperature.

Thyristor – Electronic switching device.

Triac – Electronic switching device.

Volt – Unit of electrical pressure (potential) difference.

Volatile – When relating to computer control applications, unable to store information when power is turned off. Non-volatile: retains information for a set period even with power turned off.

Watt – Unit of power consumed by an appliance or circuit, the product of the mains voltage and the current drawn (in amps). 1000W = 1 kilowatt (kW).

Essential information

Stain Removal

Stains on washable fabrics fall into two groups:

Group one - stains that will wash out in soap or detergent suds

Type of stain

Beetroot, Blood, Blackcurrant and other fruit juices, Chocolate, Cream, Cocoa, Coffee, Egg, Gravy, Ice lollies, Jam, Meat juice, Mud Milk, Nappy stains, Pickles, Soft drinks, Sauces, Soup, Stews, Syrup, Tea, Tomato ketchup, Wines and Spirits and Washable ink.

Method

Fresh stains - Soak in cold suds to keep the stain from becoming set in the fabric. Then wash in the normal way according to the fabric.
Old dried-in stains - Lubricate with glycerine. Apply a mixture of one part glycerine to two parts water to the stain and leave for 10 minutes. Then treat as fresh stains.
Residual marks - White fabrics only. Bleach out with Hydrogen Peroxide solution. (One part 20 volume hydrogen peroxide to nine parts water). Leave soaking in this solution for 1 hour. Then wash in the normal way. Blood stains may leave residual iron mould marks, which should be treated as for iron mould.
Special Note: For 'built' stains such as egg (cooked), chocolate and mud, scrape off surplus staining matter first before putting to soak. Blood and meat juice stains whether fresh or old should be soaked in cold water first.

Group two - 'treat-'n-wash' stains

What you will need
Glycerine (for lubrication)
Methylated Spirit (handle carefully: *Inflammable - poisonous).*
Turpentine *(Inflammable).*
Amyl Acetate (handle carefully - *Highly inflammable).*
Hydrogen Peroxide.
Proprietary grease solvent - 'Thawpit', 'Beaucare', 'Dab-it-off', etc., *(Do not breathe the vapour: Use in a well-ventilated room.*
Photographic Hypo.
White vinegar (Acetic Acid).
Household ammonia (keep away from eyes).
Cotton wool, Paper tissues, etc.

Handy Hints

1. Act quickly to remove a stain and prevent it 'setting'. The faster you act, the milder the remedy needed.
2. Never run a stain, as this pushes it further into the fabric. 'Pinch out' as much as you can, using a clean cloth or a paper tissue.
3. Never neglect a stain. The more drastic remedies for 'set' stains may harm delicate fabrics. If some stains are left on man-made and drip-dry fabrics in particular, they can be absorbed permanently into the fabric itself. Stains such as iron mould (rust) can weaken cellulosic fabrics and may eventually cause holes.
4. When applying solvents, always work from outside the stain towards the centre to avoid making a ring.
5. Always try a solvent on a hidden part first (e.g. under hem or seam allowance) to make sure it does not harm colours or fabric.
6. Stains on garments to be 'dry-cleaned' should be indicated on the garment (e.g. with a coloured tacking thread). Tell the cleaners what has caused the stain. This facilitates the task of removal and lessens the risk of the stain becoming permanently set by incorrect treatment.

Absorbent pad method

Using two absorbent pads of cotton wool, one soaked with the solvent and other held against the stain. Dab the underside of the stain with solvent and the staining matter will be transferred from the material to the top pad. Change this pad around to a clean part and continue working in this way until no more staining matter comes through. To remove last traces of the stain, wash in usual way.

Type of stain	Solvent	Method
Ballpoint ink	Methylated spirit (INF) (Benzine for acetate and 'Tricel')	Absorbent pad method
Bicycle oil	Proprietary grease solvent	Absorbent pad method or follow manufacturer's instructions
Black lead	Proprietary grease solvent	Absorbent pad method or follow manufacturer's instructions

Type of Stain	Solvent	Method
Chalks and Crayons (washable)		Brush off as much as possible while dry. Then brush stained area with suds (one dessertspoonful to a pint of water). Wash in the usual way
Chalks and Crayons (Indelible)	Methylated spirit. (INF) (Benzine for acetate and 'Tricel'). (INF)	Absorbent pad method
Chewing gum	Methylated spirit. (INF) (Benzine for acetate and 'Tricel'). (INF)	Absorbent pad method. Alternatively rub the gum with an ice cube to harden it. It may then be picked off by hand. Wash as usual to remove final traces
Cod Liver Oil, Cooking Fat, Heavy grease stains	Proprietary grease solvent	Absorbent pad method or manufacturer's instructions
Contact adhesives (e.g. Balsa cement, Evostick')	Amyl Acetate. (INF))	Absorbent pad method
Felt pen inks	Methylated spirit. (INF) (Benzine for acetate and 'Tricel'). (INF)	First, lubricate the stain by rubbing with hard soap and then wash in the usual way. For obstinate stains, use Methylated Spirit and absorbent pad method. Wash again to remove final traces
Grass	Methylated spirit. (INF) (Benzine for acetate and 'Tricel'). (INF)	Absorbent pad method
Greasepaint	Proprietary grease solvent	Absorbent pad method or follow manufacturer's instructions
Hair lacquer	Amyl Acetate. (INF)	Absorbent pad method
Iodine	Photographic Hypo	Dissolve one tablespoon hypo crystals in one pint warm water. Soak the stain for about 5 minutes, watching closely. As soon as the stain disappears, rinse thoroughly, then wash in the usual way
Iron mould (rust marks)	a) Lemon juice (for wool, man-made fibres and all fine fabrics)	Apply lemon juice to the stain and leave it for 10-15 minutes. Place a damp cloth over the stain and iron. Repeat several times, as necessary. Rinse and wash as usual
	b) Oxalic acid solution (for white cotton and linen only) use with care	Dissolve 1/2 teaspoonful oxalic acid crystals in 1/2 pint hot water. Tie a piece of cotton tightly round the stained area (to prevent the solution spreading) and immerse the stained part only. Leave for 2 or 3 minutes. Rinse thoroughly and wash in rich suds
Lipstick and Rouge i) light stains ii) heavy stains	Proprietary grease solvent	Soak then wash in usual way. Absorbent pad method or follow manufacturer's instructions
Marking ink	Marking ink eradicator (from stationers).	Follow instructions on the bottle label carefully
Metal polish	Proprietary grease solvent	Absorbent pad method or follow manufacturer's instructions

Type of Stain	Solvent	Method
Mildew (mould on articles stored damp) a) coloured article		The only treatment is regular soaking, followed by washing in rich suds - this will gradually reduce the marks
b) white cottons and linens without special finishes	Household bleach and vinegar	Soak in one part bleach to 100 parts water with one tablespoonful vinegar. Rinse thoroughly, then wash
c) white, drip-dry fabrics	Hydrogen peroxide solution	Soak in one part hydrogen peroxide (20 volume) and nine parts water until staining has cleared. Rinse thoroughly then wash in the usual way
Nail varnish	Amyl Acetate for all fabrics (INF)	Absorbent pad method
Nicotine (Tobacco juice)	Methylated spirit. (INF) (Benzine for acetate or ''Tricel'. (INF)	Absorbent pad method
Non-washable ink	Oxalic acid solution. (For white cottons and linens only).	See method (b) under iron mould
Paint: Emulsion	Water	Emulsion paint splashes sponged immediately with cold water will quickly be removed. Dried stains are permanent
Paint: Oil	Turpentine or Amyl Acetate (INF)	Absorbent pad method
Perspiration: Fresh stains	Ammonia. **Do not inhale the fumes**.	Damp with water, then hold over an open bottle of household ammonia
Old stains	White vinegar	Sponge with white vinegar, rinse thoroughly, then wash in usual way
'Plasticine' Modelling clay	Proprietary grease solvent or lighter fuel. (INF)	Scrape or brush off as much as possible. Apply solvent with absorbent pad method - wash to remove final traces
Scorch: a) Light marks		Light stains will sometimes respond to treatment as for Group 1 - washable stains
	Glycerine	If persistent, moisten with water and rub glycerine into the stained area. Wash through. Residual marks may respond to soaking in hydrogen peroxide solution
b) Heavy marks	Heavy scorch marks that have damaged the fibres cannot be removed	
Shoe polish	Glycerine and proprietary grease solvent	Lubricate stain with glycerine, then use solvent with absorbent pad method or follow manufacturer's instructions. Wash to remove final traces
Sun tan oil	Proprietary grease solvent	Absorbent pad method or follow manufacturer's instructions
Tar	Eucalyptus oil, Proprietary grease solvent, Benzine or lighter fuel (NF)	Scrape off surplus, then apply solvent with absorbent pad method. Rinse and wash as soon as possible
Verdigris (green stains from copper pipes etc).		Treat as iron mould

Essential information

Index